Learning Guide

for Tortora and Anagnostakos:
Principles of Anatomy and Physiology Sixth Edition

Learning Guide

for Tortora and Anagnostakos:
Principles of Anatomy and Physiology Sixth Edition

Kathleen Schmidt Prezbindowski
College of Mount St. Joseph and
The University of Cincinnati

Gerard J. Tortora
Bergen Community College

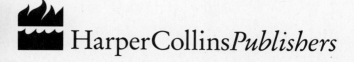
HarperCollins*Publishers*

To Amy and Laurie

Development Editor: Robert Ginsberg
Project Coordination: PC&F, Inc.
Cover Coordinator: Mary Archondes
Cover Design: Circa '86, Inc.
Cover Illustration/Photo: From *The Dance Workshop*. Courtesy Gaia Books Ltd.
Production: Kewal K. Sharma

Learning Guide for Tortora and Anagnostakos: Principles of Anatomy and Physiology, Sixth Edition

Library of Congress Cataloging-in-Publication Data

Prezbindowski, Kathleen Schmidt.
    Learning guide for Tortora and Anagnostakos : principles of
anatomy and physiology / Kathleen Schmidt Prezbindowski, Gerard J.
Tortora. — 6th ed.
        p.   cm.
    ISBN 0-06-045374-5
    1. Human physiology—Examinations, questions, etc.   2. Anatomy,
Human—Examinations, questions, etc.   I. Tortora, Gerard J.
II. Title.
QP34.5.T67,   1990 Suppl.
612'.0076—dc20   89-24471
                                                              CIP

90 91 92 9 8 7 6 5 4 3

Contents

v

Preface

This *Learning Guide* in intended to be used in conjunction with Tortora and Anagnostakos' *Principles of Anatomy and Physiology,* Sixth Edition. The emphasis is on active learning, not passive reading, and the student examines each concept through a variety of activities and exercises. By approaching each concept several times from different points of view, the student sees how the ideas of the text apply to real, clinical situations. In this way the student may be helped not simply to study but to *learn*.

The 29 chapters of the *Learning Guide* parallel those of the Tortora and Anagnostakos text. Each chapter begins with an **Overview** that introduces content succinctly and a **Framework** that visually organizes the interrelationships among key concepts and terms. **Objectives** from the text are included for convenient previewing and reviewing. Objectives are arranged according to major chapter divisions in the **Topic Outline. Wordbytes** present word roots, prefixes, and suffixes that facilitate understanding and recollection of key terms. **Checkpoints** offer a diversity of learning activities that follow the exact sequence of topics in the chapter, with cross-references to pages in the text. This format makes the *Learning Guide* "user friendly" or "student friendly" and enhances the effectiveness of the excellent text. **Answers to Selected Checkpoints** provide feedback to students. A 25-question **Mastery Test** includes both objective and subjective questions as a final tool for student self-assessment of learning. **Answers to the Mastery Test** immediately follow the Mastery Test.

This edition of the *Learning Guide* is especially designed to fit the needs of people with different learning styles. Visual learners will identify the new *Frameworks* as assets in organizing concepts and key terms. Visual and tactile learners will welcome the increased number of coloring and labeling exercises and the addition of related color-code ovals. Further variety in *Checkpoints* is offered by short definitions, fill-ins, tables for completion, and multiple-choice, matching, and arrange-in-correct-sequence activities. Learning on multiple levels is made possible by *For Extra Review* sections that provide enrichment or assistance with particularly difficult topics. *Clinical Challenges* introduce applications especially relevant for students in allied health. These are based on my own experiences as an R.N. and on discussions with colleagues. These challenging exercises enable students to apply abstract concepts to actual clinical settings.

The sixth edition of the *Guide* also provides readily available feedback for self-assessment. Answers to most *Checkpoints* are included at the end of the chapter. The placement of *Answers to the Mastery Test* immediately following the *Mastery Test* offers readily accessible feedback.

A number of persons have contributed to the revision of this book. I would particularly like to thank the following instructors for their helpful comments and suggestions while the manuscript was being revised:

Barry R. Anderson/University of Scranton
Francis Horne/Southwest Texas University
James D. Houston/Westark Community College
Thomas J. King/St. Petersburg Junior College
Ed Krol/Henry Ford Community College
Mark V. Lomolino/University of Arizona
Margaret L. Till/Bloomsburg University of Pennsylvania

I offer particular thanks to the students and faculty at the College of Mount St. Joseph and the University of Cincinnati for their ongoing sharing and support of teaching and learning. I thank my daughters, Amy and Laurie Prezbindowski, my parents and sister, Norine, Harry, and Maureen Schmidt, and close friends for their constant encouragement and caring. I am thankful for the expertise, enthusiasm, and patience of the production staff at Harper & Row, especially Steve Eisen, supplements editor, and Bob Ginsberg, development editor. Special thanks to Jeannine Ciliotta for her skillful editing and creative approach to excellence.

Kathleen Schmidt Prezbindowski

To the Student

This *Learning Guide* is designed to help you do exactly what its title indicates—*learn.* It will serve as a step-by-step aid to help you bridge the gap between goals (objectives) and accomplishment (learning). The 29 chapters of the *Learning Guide* parallel the 29 chapters of Tortora and Anagnostakos' *Principles of Anatomy and Physiology,* Sixth Edition. Each chapter consists of the following parts: *Overview, Framework, Topic Outline with Objectives, Wordbytes, Checkpoints, Answers to Selected Checkpoints,* and *Mastery Test with Answers.* Take a moment to turn to Chapter 1 and look at the major headings in this *Learning Guide.* Now consider each one.

Overview. As you begin your study of each chapter, read this brief introduction, which presents the topic of the chapter. The Overview is a look at the "forest" (or the "big picture") before examination of the individual "trees" (or details).

Framework. The Framework arranges key concepts and terms in an organizational chart that demonstrates interrelationships and lists key terms. Refer back to the Framework frequently to keep sight of interrelationships.

Topic Outline and Objectives. Objectives that are identical to those in the text organize the chapter into "bite-sized" pieces that will help you identify the principal subtopics that you must understand. Preview objectives as you start the chapter, and refer back to them to review. Objectives are arranged according to major chapter divisions in the **Topic Outline.**

Wordbytes present word roots, prefixes, and suffixes that will help you master terminology of the chapter and facilitate understanding and remembering of key terms. Study these, then check your knowledge. (Cover meanings and write them in the margins.) Try to think of additional terms in which these important bits of words are used.

Checkpoints offer a variety of learning activities arranged according to the sequence of topics in the chapter. At the start of each new Checkpoint section, note the related pages in the Tortora and Anagnostakos text. Read these pages carefully and then complete this group of Checkpoints. Checkpoints are designed to help you to "handle" each new concept, almost as if you were working with clay. Through a variety of exercises and activities, you can achieve a depth of understanding that comes only through active participation, not passive reading alone. You can challenge and verify your knowledge of key concepts by completing the checkpoints along your path of learning.

A diversity of learning activities are included in Checkpoints. You will focus on specific facts in the chapter by doing exercises that require you to provide definitions and comparisons, and you will fill in blanks in paragraphs with key terms. You will also color and label figures. Fine color felt-tipped pens or pencils will be helpful since you can easily fill in related color-code ovals. Your understanding will also be tested with matching exercises, multiple-choice questions, and arrange-in-correct-sequence activities.

Two special types of Checkpoints are *For Extra Review* and *Clinical Challenges. For Extra Review* provides additional help for difficult topics or refers you by page number to specific activities earlier in the *Learning Guide* so that you may better integrate related information. Applications for enrichment and interest in areas particularly relevant to nursing or other health sciences students are found in selected *Clinical Challenge* exercises.

When you have difficulty with an exercise, refer to the related pages (listed in the margin of the *Learning Guide*) of *Principles of Anatomy and Physiology* before proceeding further. You will find specific references by page number for all text figures and exhibits.

Answers to Selected Checkpoints. As you glance through the activities, you will notice that a number of questions are marked with solid boxes (■). Answers to these checkpoints are given at the end of each chapter, in order to provide you with some immediate feedback about your progress. Incorrect answers alert you to review related objectives in the *Learning Guide* and corresponding text pages. Each *Learning Guide* (LG) figure included in *Answers to Selected Checkpoints* may be identified with an "A," such as Figure LG 13-1A.

Activities without answers are also purposely included in this book. The intention is to encourage you to verify some answers independently by consulting your textbook and also to stimulate discussion with students or instructors.

Mastery Test. This 25-question self-test provides an opportunity for a final review of the chapter as a whole. Its format will assist you in preparing for standardized tests or course exams that are objective in nature, since the first 20 questions are multiple-choice, true-false, and arrangement questions. The final five questions of each test help you to evaluate your learning with fill-in and short answer questions. Answers for each mastery test are placed at the end of each chapter.

I wish you success and enjoyment in *learning* concepts and relevant applications of anatomy and physiology.

Kathleen Schmidt Prezbindowski

About the Author

Kathleen Schmidt Prezbindowski, Ph.D. (biology), R.N., is professor of biology at the College of Mount St. Joseph, and visiting professor at the University of Cincinnati. She has taught anatomy and physiology, pathophysiology, and biology of aging at both institutions for the past 13 years. She was recently named Distinguished Teacher of the Year at the College of Mount St. Joseph. In 1985, Dr. Prezbindowski received a B.S.N. She has since been codirector of a health promotion program known as PATHS (Positive Adults Taking Health Seriously) in senior centers throughout Ohio. She has written two other Learning Guides and a series of health education videotapes for older persons. Dr. Prezbindowski has also presented and published extensively on mitochondrial membrane structure and innovative college teaching methodology.

Organization of the Human Body

An Introductory Tour of the Human Body

GETTING STARTED

ANATOMY AND
PHYSIOLOGY DEFINED [A]
- Branches of anatomy

WHERE TO GO, HOW TO GET THERE

LEVELS OF
ORGANIZATION [B]
- Chemicals
- Cells
- Tissues (4)
- Organs
- Systems (11)
- Organisms

DIRECTIONS [D]
- Structural plan
- Anatomical position
- Regional names
- Directional terms
- Planes, sections

BODY
CAVITIES [E]
- Dorsal
- Ventral
  - Thoracic
  - Abdomino-
    pelvic

BODY
MEASUREMENT [H]

HOW IT WORKS: FUNCTIONS

LIFE
PROCESSES [C]
- Metabolism
- Excitability
- Conductivity
- Contractility
- Growth
- Differentiation
- Reproduction

HOMEOSTASIS [G]
- ECF
- ICF
- Stressor
- Feedback
  systems

WHEN THE BODY DOES NOT WORK

MEDICAL
IMAGING [F]
- Roentgenogram
- CT
- DSR
- MRI
- US
- PET
- DSA

An Introduction to the Human Body

You are about to embark on a tour, one that you will enjoy for at least several months, and hopefully for many years. Your destination: the far-reaching corners and fascinating treasures of the human body. Imagine the human body as an enormous, vibrant world, abundant in architectural wonders, and bustling with activity. You are the traveler, the visitor, the learner on this tour.

Just as the world is composed of individual buildings, cities, countries and continents, so the human body is organized into cells, tissues, organs, and systems. In fact, a close-up look at a cell (comparable to a detailed study of a building) will reveal structural intricacies—the array of chemicals that comprise cells. Any tour requires a sense of direction and knowledge of the language: directional and anatomical terms that guide the traveler through the body. A tour is sure to enlighten the visitor about the culture in each region—in the human body, the functions of each body part or system. By the time you leave each area (body part or system), you will have gained a sense of how that area contributes to overall stability and well-being (homeostasis).

In planning this trip, an overview of sites to visit will be helpful. So please turn to the Framework for Chapter 1, study it carefully, and refer back to it as often as you wish. It contains the key terms of the chapter and serves as your map and your itinerary. Bon voyage!

TOPIC OUTLINE AND OBJECTIVES

A. Anatomy and physiology defined

1. Define anatomy with its subdivisions, and physiology.

B. Levels of structural organization

2. Define each of the following levels of structural organization that make up the human body: chemical, cellular, tissue, organ, system, and organismic.
3. Identify the principal systems of the human body, list representative organs of each system, and describe the function of each system.

C. Life processes

4. List and define several important life processes of humans.

D. Anatomical and directional terms

5. Define the anatomical position and compare common and anatomical terms used to describe various regions of the human body.
6. Define several directional terms and anatomical planes used in association with the human body.

E. Body cavities and regions

7. List by name and location the principal body cavities and the organs contained within them.

F. Medical imaging

8. Describe the principle and importance of selected medical imaging techniques in the diagnosis of disease.

G. Homeostasis

9. Define homeostasis and explain the effects of stress on homeostasis.

10. Define a feedback system and explain its role in homeostasis.

H. Measuring the human body

WORDBYTES

Now study the following parts of words that may help you better understand terminology in this chapter.

Wordbyte	Meaning	Example	Wordbyte	Meaning	Example
ana-	up	*ana*tomy	physi(o)-	nature	*physio*logy
ante-	before	*ante*rior	post-	after	*post*erior
homeo-	same	*homeo*stasis	sagitta-	arrow	*sagitta*l
inter-	between	*inter*cellular	stasis, stat-	stand, stay	homeo*stasis*
intra-	within	*intra*cellular	-tomo	to cut	ana*tomy*
logo-	study	physio*logy*	viscero-	body organs, function	*viscer*al
meter	measure	*centi*meter			
pariet-	wall	*pariet*al			

CHECKPOINTS

A. Anatomy and physiology defined   (page 6)

A1. Contrast each of the following pairs of terms.

a. Anatomy/physiology

b. Gross anatomy/histology

c. Systemic anatomy/regional anatomy

d. Developmental anatomy/embryology

A2. Look at the two items in each of the following pairs. Tell how their structural differences explain their functional differences.

a. Spoon/fork

b. Hand/foot

c. Incisors (front teeth)/molars

B. Levels of structural organization (pages 6–8)

■ **B1.** Rearrange these terms in order from highest to lowest level of organization, using lines provided.

Term	Level of Organization
Organism	(highest) _____
Cell	_____
Tissue	_____ **Organ** _____
Organ	_____
Chemical	_____
System	(lowest) _____

B2. Complete the following table describing systems of the body. Name two or more organs in each system. Then list one or more functions of each system.

System	Organs	Functions
a.	Skin, hair, nails	
b. Skeletal		
c. Muscular		
d.		Regulates body by nerve impulses
e.	Glands that produce hormones	

System	Organs	Functions
f.	Blood, heart, blood vessels	
g. Lymphatic		
h.		Supplies oxygen, removes carbon dioxide, regulates acid-base balance
i.		Breaks down food and eliminates solid wastes
j.	Kidney, ureters, urinary bladder, urethra	
k. Reproductive		

C. Life processes (pages 8–9)

C1. Complete the list and briefly define each of seven characteristics which distinguish you (or other living organisms) from nonliving things. One is done for you.

a. _____

b. _____

c. _____

d. **Contractility: ability of cells (especially muscle cells) to shorten or contract.**

e. _____

f. _____

g. _____

■ **C2.** Refer to the list of terms in the box, all related to the life process of metabolism. Demonstrate your understanding of these terms by selecting the term that best fits each description provided below.

A. Anabolism	I. Ingestion
B. Catabolism	M. Metabolism
D. Digestion	S. Secretion
E. Excretion	

_____ a. Taking in of foods

_____ b. Breakdown of foods into forms that can be absorbed and used by cells

_____ c. The sum of all chemical processes in the body

_____ d. Synthesis of body's structural and functional components; uses energy

_____ e. Production and release of useful substances by cells

D. Anatomical and directional terms (pages 9–11)

D1. Complete this exercise about the structural plan of the human body.

a. Explain the meaning of the anatomical arrangement of a *tube within a tube*.

b. State one advantage of the following two characteristics of the human body:
vertebral column (backbone) of 26 separate bones

bilateral symmetry

D2. Be sure that you understand the meaning of anatomical position by assuming that position yourself.

D3. Complete the table relating common terms to anatomical terms. (See Figure 1-2, page 10, in your text to check your answers.) For extra practice use common and anatomical terms to identify each region of your own body and that of a study partner.

Common Term	Anatomical Term
a.	Axillary
b. Fingers	
c. Arm	
d.	Popliteal
e.	Cephalic
f. Mouth	
g.	Inguinal
h. Chest	
i.	Cervical
j.	Antebrachial
k. Buttock	
l.	Calcaneal

■ **D4.** Using your own body, a skeleton, a torso, or Figure 1-3 (page 12 in your text), determine relationships among body parts. Write the correct directional term(s) to complete each of these statements.

a. The liver is _____ to the diaphragm.

b. Fingers (phalanges) are located _____ to wrist bones (carpals).

c. The skin on the dorsal surface of your body can also be said to be located on your

_____ surface.

d. The great (big) toe is _____ to the little toe.

e. The little toe is _____ to the great toe.

f. The skin on your leg is _____ to muscle tissue in your leg.

g. Muscles of your arm are _____ to skin on your arm.

h. When you float face down in a pool, you are lying on your

_____ surface.

i. The lungs and heart are located _____ to the abdominal organs.

j. Since the stomach and the spleen are both located on the left side of the abdomen,

they could be described as _____ -lateral.

k. The _____ pleura covers the external surface of the lungs.

■ **D5.** Match each of the following planes with the phrase telling how the body would be divided by such a plane.

> F. Frontal (coronal)   M. Midsagittal (median)
> H. Horizontal (transverse)   S. Sagittal

_____ a. Into superior and inferior portions   _____ c. Into anterior and posterior portions

_____ b. Into equal right and left portions   _____ d. Into right and left portions

E. Body cavities, regions (pages 12–19)

■ **E1.** After you have studied Figures 1-6, 1-7, and 1-8 (pages 14–19 in your text), complete this exercise about body cavities. Circle the correct answer in each statement.

a. The (*dorsal? ventral?*) cavity consists of the cranial cavity and the vertebral canal.

b. The viscera, including such structures as the heart, lungs, and intestines, are all located in the (*dorsal? ventral?*) cavity.

c. Of the two body cavities, the (*dorsal? ventral?*) appears to be better protected by bone.

d. Pleural, mediastinal, and pericardial are terms that refer to regions of the (*thorax? abdominopelvis?*).

e. The (*heart? lungs? esophagus and trachea?*) are located in the pleural cavities.

f. The division between the abdomen and the pelvis is marked by (*the diaphragm? an imaginary line from the symphysis pubis to the superior border of the sacrum?*).

g. The stomach, pancreas, small intestine, and most of the large intestine are located in the (*abdomen? pelvis?*).

h. The urinary bladder, rectum, and internal reproductive organs are located in the (*abdominal? pelvic?*) cavity.

■ **E2.** Complete Figure LG 1-1 according to directions below.

a. Color each of the organs listed on the figure. Select different colors for each organ, and be sure to use the same color for the related color code oval (O).

b. Next, label each organ on the figure.

c. *For extra review.* Draw lines dividing the abdomen into the nine regions, and note which organs are in each region.

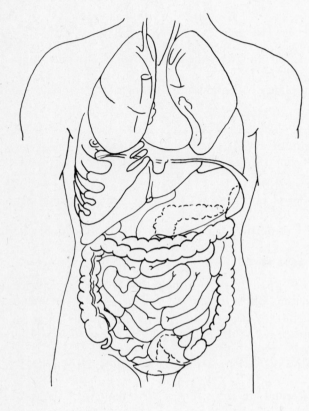

O Appendix   O Lungs
O Diaphragm   O Pancreas
O Gallbladder   O Small intestine
O Heart   O Spleen
O Large intestine   O Stomach
O Liver

Figure LG 1-1 Regions of the ventral body cavity. Complete as directed in Checkpoint E2.

■ **E3.** Using the names of the nine abdominal regions and quadrants, complete these statements. Refer to the figure you just completed and to Figure 1-8 (pages 17–19) in your text.

a. From superior to inferior, the three abdominal regions on the right side are

_____, _____, and

_____.

b. The stomach is located primarily in the two regions named

_____ and _____.

c. The navel is located in the _____ region.

d. The region immediately superior to the urinary bladder is named the

_____ region.

e. The liver and gallbladder are located in the _____ quadrant.

E4. A *clinical challenge*. Complete this exercise about the procedure known as an autopsy.

a. List three reasons why autopsies may be performed.

b. Briefly describe the three phases of an autopsy.

F. Medical imaging (pages 19–22)

■ **F1.** Match the radiographic anatomy techniques listed in the box with the correct descriptions.

CR. Conventional radiograph	MRI. Magnetic resonance imaging
CT. Computed tomography	PET. Positron emission tomography
DSA. Digital subtraction angiography	US. Ultrasound
DSR. Dynamic spatial reconstruction	

_____ a. A recent development in radiographic anatomy, this technique produces moving, three-dimensional images of body organs.

_____ b. This method provides a cross-sectional picture of an area of the body by means of an x-ray source moving in an arc. Results are processed by a computer and displayed on a console.

_____ c. This technique produces a two-dimensional image (roentgenogram) via a single barrage of x-rays. Images of organs overlap, making diagnosis difficult.

_____ d. A method used to study blood vessels by comparing a region before and after injection of a contrast substance (iodine).

_____ e. Utilizes injected radioisotopes that assess function, rather than structure.

_____ f. Two techniques that are noninvasive and do not utilize ionizing radiation. (2 answers)

G. Homeostasis (pages 22–25)

■ **G1.** Describe the types of fluid in the body and their relationships to homeostasis by completing these statements and Figure LG 1-2.

a. Fluid inside of cells is known as _____ fluid (ICF). Color areas containing this type of fluid yellow.

b. Fluid in spaces between cells is called _____. It surrounds and bathes cells and is one form of (*intracellular? extracellular?*) fluid. Color the spaces containing this fluid light green.

c. Another form of extracellular fluid is that located in _____

_____ vessels and _____ vessels. Color these areas dark green.

d. The body's "internal environment" (that is, surrounding cells) is

_____-cellular fluid (ECF) (all green areas in your figure). The condition of maintaining ECF in relative constancy is known as

_____.

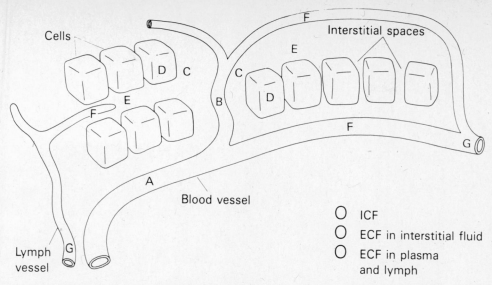

Cells

Interstitial spaces

Blood vessel

Lymph vessel

○ ICF
○ ECF in interstitial fluid
○ ECF in plasma and lymph

Figure LG 1-2 Internal environment of the body: types of fluid. Complete as directed. Areas labeled *A* to *G* refer to Checkpoint G2.

■ **G2.** Refer again to Figure LG 1-2. Show the pattern of circulation of body fluids by drawing arrows connecting letters in the figure in alphabetical order *(A → B → C,* etc.). Now fill in the blanks below describing this lettered pathway: A (arteries and arterioles) → B (capillaries)

→ C (_____) → D (_____)

→ E (_____) → F (_____)

→ G (_____).

G3. List six qualities of ECF that are maintained under optimal conditions when the body is in homeostasis.

G4. Consider how your internal environment is affected by your external environment. Which of the six qualities of your ECF listed above would be altered if you were subjected to the following conditions? Compare your answers with those of a study partner.

a. Taking a two-day hike at high altitude without appropriate food or clothing

b. Invasion of your body by an infectious microorganism that causes fever and diarrhea over a prolonged period of time

G5. Suppose you decide to run a quarter mile at your top speed. Tell how your body would respond to this stress created by exercise. List according to system the changes your body would probably make in order to maintain homeostasis.

a. Integumentary

b. Muscular and skeletal

c. Cardiovascular

d. Respiratory

e. Nervous

f. Endocrine

■ **G6.** Consider an example of a feedback system. Ten workers on an assembly line are producing hand-made shoes. As shoes come off the assembly line, they pile up around the last worker, Worker Ten. When this happens, Worker Ten calls out, "We have hundreds of shoes piled up here." This information (input) is heard by (fed back to) Worker One who determines when more shoes should be begun. Worker One is therefore the ultimate controller of output. This controller could respond in either of two ways. In a *negative feedback system,* Worker One says, "We have an excess of unsold shoes. Let us slow down or stop production until the excess (stress) is relieved—until these shoes are sold." What might Worker One's response be if this were a *positive feedback system*?

G7. Most feedback systems of the body are (*positive? negative?*). Why do you think this might be advantageous and directed toward maintenance of homeostasis?

■ **G8.** Refer to Figure 1-15 (page 24) in your text and match the answers in the box with the statements describing steps in a homeostatic mechanism for controlling elevated blood pressure (BP).

A. Arterioles H. Heart HBP. High blood pressure LBP. Lowered blood pressure	P. Pressure-sensitive nerve cell receptors located in major arteries which send messages to the brain

_____ a. What serves as the *input*?

_____ b. What structures receive this input?

_____ c. What structure then receives instructions from the brain to decrease contractile strength and so eject less blood into major arteries?

_____ d. What structures receive brain messages causing them to dilate so that blood runs out of major arteries and into smaller vessels? (Refer to Figure 21-11, page 622 in your text.)

_____ e. What is the final *output* as a result of the presence of less blood in the major arteries?

■ **G9.** Circle the answers that are classified as signs, rather than as symptoms.

pain   fever   headache   diarrhea   swollen ankles

H. Measuring the human body (page 25)

If you have difficulty answering questions in this section, be sure to read Appendix A in the text.

■ **H1.** Circle ALL correct answers to each question.

a. When you drink all of the milk in a quart container, you are drinking ____ of milk.

A. Almost a liter
B. Almost half a liter
C. 2 liters
D. About 32 oz

b. One milliliter (ml) is equal to:

A. About 1 oz
B. $\frac{1}{100}$ liter
C. 0.1 liter
D. $\frac{1}{1000}$ liter
E. 1 cubic centimeter (cc, or cm^3)

c. If a red blood cell (RBC) is about 7 micrometers (μm) in diameter, how many RBC's could fit side by side across a distance of 1 millimeter (mm)?

A. 0.15
B. 1.5
C. 15
D. 150
E. 1,500

d. One centimeter (cm) is equal to:

A. 0.1 mm
B. 1 mm
C. 10 mm
D. 100 mm
E. 0.1 meter (m)
F. 2 inches
G. 0.4 inch

e. If 20 ml of oxygen can be transported in 100 ml of your blood, how much oxygen can be transported in 1 liter of your blood?

A. 0.2
B. 2
C. 20
D. 200
E. 2,000

f. Since there are 2.2 lb in 1 kilogram (kg), if you weigh 132 lb, you weigh about ____ kg.

A. 60
B. 290
C. 2,112
D. 13.2
E. 132

g. A human liver weighs about 4 lb. This is about ____ in metric units.

   A. 8.8 kg   D. 88 g
   B. 1.8 kg   E. 1,800 g
   C. 180 grams (g)

h. A micrometer is equivalent to:

   A. 0.1 mm
   B. 1000 m
   C. 0.001 mm
   D. 0.001 cm
   E. 1 millionth of a meter

i. The human body contains about 6 liters of blood. This is about ____ pints of blood.

   A. 0.2   D. 12
   B. 2   E. 36
   C. 6

■ **H2.** A *clinical challenge,* Determine Elizabeth's weight and height in metric measurements. Elizabeth weighs 22 lb and is 30 in. Tall. Weight:

_____ kg; height: _____ cm.

■ **H3.** Another *clinical challenge.* Mr. Frederick is to receive 1 liter of 5 percent glucose solution intravenously (IV) over each 8-hour (hr) period. Answer the following questions.

a. How many millimeters (cc)/hr of the IV solution should Mr. Frederick receive?

b. If Mr. Frederick receives no other food or fluid intake, what will be his total IV fluid intake each 24 hr?

c. Each 100 cc of a 5 percent glucose solution contains 5 g of glucose. Each gram of glucose provides about 4 Cal of energy to Mr. Frederick. How many Calories is this patient receiving each day?

■ **H4.** A final metric *clinical challenge.* Tony eats a serving of food that contains 1 oz of sugar, 1 oz protein, and 2 oz of fat. Protein and sugar each provide 4 Cal/g, while fat provides 9 Cal/g. How many Calories are supplied by the serving described above? (*Hint:* How many grams = 1 oz?)

ANSWERS TO SELECTED CHECKPOINTS

B1. From highest to lowest: organism, system, (organ), tissue, cell, chemical.

C2. (a) I. (b) D. (c) M. (d) A. (e) S.

D4. (a) Inferior (caudad). (b) Distal. (c) Posterior. (d) Medial. (e) Lateral. (f) Superficial. (g) Deep. (h) Anterior (ventral). (i) Superior (cephalad). (j) Ipsi-. (k) Visceral.

D5. (a) H. (b) M. (c) F. (d) S.

E1. (a) Dorsal. (b) Ventral. (c) Dorsal. (d) Thorax. (e) Lungs. (f) An imaginary line from the symphysis pubis to the superior border of the sacrum. (g) Abdomen. (h) Pelvic.

E2.

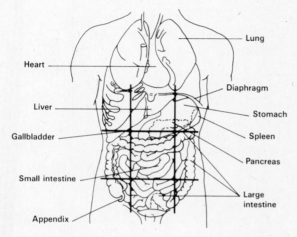

Heart — Lung — Diaphragm — Liver — Stomach — Gallbladder — Spleen — Pancreas — Small intestine — Large intestine — Appendix

Figure LG 1-1A   Regions of the ventral body cavity.

E3. (a) Right hypochondriac, right lumbar, right iliac (inguinal). (b) Left hypochondriac, epigastric. (c) Umbilical. (d) Hypogastric (pubic). (e) Right upper.

F1. (a) DSR. (b) CT. (c) CR. (d) DSA. (e) PET. (f) MRI, US.

G1. (a) Intracellular. (b) Interstitial (or intercellular or tissue) fluid, extracellular. (c) Blood, lymph. (d) Extra-, homeostasis.

G2. C, interstitial (intercellular) fluid; D, intracellular fluid; E, interstitial (intercellular) fluid again; F, blood or lymph capillaries; G, venules and veins or lymph vessels.

G6. "Hurray! We have produced hundreds of shoes; let us go for thousands! Do not worry that the shoes are not selling. Step up production even more."

G8. (a) HBP. (b) P. (c) H. (d) A. (e) LBP.

G9. Fever, diarrhea, swollen ankles.

H1. (a) A, D. (b) D, E. (c) D. (d) C, G. (e) D. (f) A. (g) B, E. (h) C, E. (i) D.

H2. Weight: 10 kg = 22 lb × 1 kg/2.2 lb. Height: 75 cm = 30 in × 2.5 cm/in.

H3. (a) 125 cc/hr = 1,000 cc/8 hr. (b) 3,000 cc/24 hr = 1,000 cc/8 hr. (c) 600 Cal = 150 g × 4 Cal/g. (*Note:* If 100 cc contains 5 g, then 3,000 cc contains 150 g.)

H4. 728 Cal = (28 g/oz × 4 Cal/g sugar) + (28 g/oz × 4 Cal/g protein) + (56 g/2 oz × 9 Cal/g fat).

MASTERY TEST: Chapter 1

Questions 1–10: Circle the letter preceding the one best answer to each question.

1. In a negative feedback system, when blood pressure decreases slightly, the body will respond by causing a number of changes which tend to:

   A. Lower blood pressure
   B. Raise blood pressure

2. The following structures are all located in the ventral cavity EXCEPT:

   A. Spinal cord   D. Gallbladder
   B. Urinary bladder   E. Esophagus
   C. Heart

3. Which pair of common and anatomical terms is mismatched?

   A. Eye/ocular   D. Neck/cervical
   B. Skull/cranial   E. Buttock/gluteal
   C. Armpit/brachial

4. Each answer below includes a correct example of a life process of humans EXCEPT:

   A. Absorption—uptake of sugar by a body cell
   B. Assimilation—the process by which a fat that has entered a cell becomes a part of that cell
   C. Excretion—production and elimination of urine or sweat
   D. Catabolism—synthesis of protein in muscle cell

5. Which is most inferiorly located?

   A. Abdomen   D. Diaphragm
   B. Pelvic cavity   E. Pleural cavity
   C. Mediastinum

6. The spleen, tonsils, and thymus are all organs in which system?

    A. Nervous   D. Digestive
    B. Lymphatic   E. Endocrine
    C. Cardiovascular

7. Which of the following structures is located totally outside of the upper right quadrant of the abdomen?

    A. Liver   D. Spleen
    B. Gallbladder   E. Pancreas
    C. Transverse colon

8. Choose the FALSE statement. (It may help you to mark each statement T for true or F for false as you read it.)

    A. Stress disturbs homeostasis.
    B. Homeostasis is a condition in which the body's internal environment remains relatively constant.
    C. The body's internal environment is best described as extracellular fluid.
    D. Extracellular fluid consists of plasma and intracellular fluid.

9. The system responsible for providing support, protection, and leverage, for storage of minerals, and for production of blood cells is:

    A. Urinary   D. Reproductive
    B. Integumentary E. Skeletal
    C. Muscular

10. Which is most proximally located?

    A. Ankle   C. Knee
    B. Hip   D. Toe

Questions 11–20: Circle T (true) or F (false). If the statement is false, change the underlined word or phrase so that the statement is correct.

T F   11. <u>Anterior and ventral</u> are synonyms.

T F   12. In order to assume the anatomical position, you should <u>lie down with arms at your sides and palms turned backwards</u>.

T F   13. The appendix is usually located in the <u>left iliac</u> region of the abdomen.

T F   14. <u>Anatomy</u> is the study of how structures function.

T F   15. A liter (or 1,000 ml) of fluid is equal to <u>1,000 cubic centimeters (cc)</u>.

T F   16. A <u>tissue</u> is a group of cells and their intercellular substances that are similar in both origin and function.

T F   17. The right kidney is located <u>mostly in the right iliac region of the abdomen</u>.

T F   18. The study of the microscopic structure of cells is <u>cytology</u>.

T F   19. A sagittal section is <u>parallel to and lateral to</u> a midsagittal section.

T F   20. The human body is <u>bilaterally symmetrical</u>.

_____ 21. Region of thorax located between the lungs. Contains heart, thymus, esophagus.

_____ 22. Name a fluid located within cells.

_____ 23. Life process in which unspecialized cells as in early embryo change into specialized cells such as nerve or bone.

_____ 24. A painless diagnostic technique in which sound waves are generated and detected by a transducer, resulting in a sonogram.

_____ 25. One general anatomical characteristic of the structural plan of the human body: the right and left sides of the body are essentially mirror images.

ANSWERS TO MASTERY TEST: ■ Chapter 1

Multiple Choice

1. B   6. B
2. A   7. D
3. C   8. D
4. D   9. E
5. B   10. B

True-False

11. T
12. F. Stand erect facing observer, arms at sides, palms forward
13. F. Right iliac
14. F. Physiology
15. T
16. T
17. F. At the intersection of four regions of the abdomen: right hypochondriac, right lumbar, epigastric, and umbilical
18. T
19. T
20. T

Fill-ins

21. Mediastinum
22. Intracellular
23. Differentiation
24. Ultrasound (US)
25. Bilateral symmetry

FRAMEWORK 2

The Chemical Level of Organization

INTRODUCTION [A] **CHEMICAL REACTIONS [B]** **TYPES OF CHEMICALS**

BASIC CHEMISTRY
■ Elements
■ Atoms
– protons
– neutrons
– electrons
– isotypes
– energy levels
– valence
■ Molecules
■ Compounds
■ Laser

CHEMICAL BONDS
■ Ionic
– cations
– anions
■ Covalent
■ Hydrogen

TYPES
■ Synthesis
– anabolism
■ Decomposition
– catabolism
■ Exchange
■ Reversible

COLLISION THEORY

ENERGY
■ Potential
■ Kinetic
■ Chemical
■ Radiant
■ Electrical

INORGANIC [C]

WATER

INORGANIC ACIDS, BASES, SALTS
■ pH
■ Buffer

ORGANIC [D]

Carbohydrates
■ Energy source
■ H:O = 2:1
■ Mono-, di-, polysaccharide

Lipids
■ Water in-soluble
■ Types
– fats
– phospho-lipids
– steroids
– vitamins A, D, E, K
– prosta-glandins

Proteins
■ Amino Acids
■ Peptide bonds

Nucleic Acids
■ DNA
■ RNA
■ Nucleo-tides

ATP
ADP
AMP

The Chemical Level of Organization

Chemicals comprise the ultrastructure of the human body. They are the submicroscopic particles of which cells, tissues, organs, and systems are constructed. Chemicals are the minute entities that participate in all of the reactions that underlie bodily functions. Every complex compound that molds a living organism is composed of relatively simple atoms held together by chemical bonds. Two classes of chemicals contribute to structure and function: inorganic compounds, primarily water, but also inorganic acids, bases and salts. Organic compounds include carbohydrates, lipids, proteins, nucleic acids, and the energy-storing compound ATP.

As you begin your study of the chemistry of life, carefully examine the Chapter 2 Framework and note the key terms associated with each section.

TOPIC OUTLINE AND OBJECTIVES

A. Introduction to basic chemistry: elements, atoms, molecules, chemical bonds

1. Identify by name and symbol the principal chemical elements of the human body.
2. Explain how ionic, covalent, and hydrogen bonds form.

B. Chemical reactions

3. Define a chemical reaction and explain the basic differences between synthesis, decomposition, exchange, and reversible chemical reactions.
4. Describe how chemical reactions occur and explain the relationship of energy to chemical reactions.

C. Inorganic water, acids, bases, salts; pH, buffers

5. List and compare the properties of water and inorganic acids, bases, and salts.
6. Define pH and explain the role of a buffer system as a homeostatic mechanism that maintains the pH of a body fluid.

D. Organic compounds

7. Compare the structure and functions of carbohydrates, lipids, proteins, deoxyribonucleic acid (DNA), ribonucleic acid (RNA), adenosine triphosphate (ATP), and cyclic AMP.

WORDBYTES

Now study the following parts of words that may help you better understand terminology in this chapter.

Wordbyte	Meaning	Example	Wordbyte	Meaning	Example
di-	two	*di*saccharide	pent-	five	*pent*ose
hex-	six	*hex*ose	poly-	many	*poly*unsaturated
mono-	one	*mono*saccharide	-saccharide	sugar	poly*saccharide*

CHECKPOINTS

A. Introduction to basic chemistry: elements, atoms, molecules, chemical bonds (pages 30–35)

A1. Define *matter* and list the three states in which matter may exist.

A2. Contrast terms in each pair:

a. Element–atom

b. Compound–molecule

■ **A3.** Refer to Figure LG 2-1, and do the following exercise.

a. Color yellow the four boxes containing the elements that make up about 96 percent of the weight of the body. These four elements are relatively (*simple? complex?*) in atomic structure among the total 103 elements in the periodic table.

b. Color light green the boxes containing calcium and phosphorus. These two elements together contribute about _____ percent of body weight.

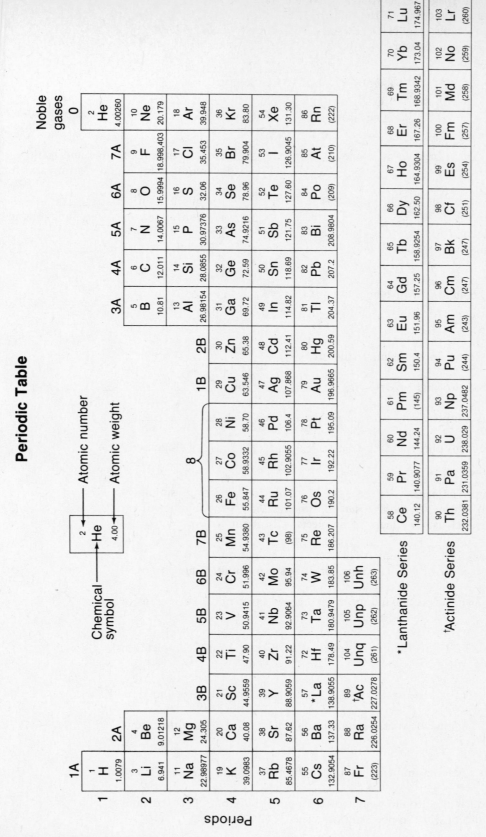

Figure LG 2-1 Periodic table.

23

■ **A4.** Refer to Exhibit 2-2 (page 40) in your text. Match the elements listed in the box with descriptions of their functions in the body. *For extra review.* Locate each of these elements in the periodic table (Figure LG 2-1) and write the atomic number of each element after its description. The first one has been done for you.

> C. Carbon   I. Iodine
> Ca. Calcium   N. Nitrogen
> Fe. iron   O. Oxygen

____C____ a. Found in every organic molecule

____6____

_____ b. Component of all protein molecules

and DNA and RNA _____

_____ c. Vital to normal thyroid gland

function _____

_____ d. Essential component of hemoglobin

_____

_____ e. Constituent of bone and teeth; required for blood clotting and for

muscle contraction _____

_____ f. Constituent of water; functions in

cellular respiration _____

■ **A5.** Complete this exercise about atomic structure.

a. An atom consists of two main parts: _____ and _____ .

b. Within the nucleus are positively charged particles called _____

and uncharged (neutral) particles called _____. Together these

are known as _____ .

c. Electrons bear (*positive? negative?*) charges. The number of electrons spinning around the nucleus is (*greater than? equal to? smaller than?*) the number of protons in the atomic nucleus. The mass of an electron is about

_____ the mass of a proton or a neutron.

d. Electrons are arranged in orbits. The inner orbit has a maximum capacity of

_____ electrons, whereas a maximum of _____ electrons may be present in the next orbit.

■ **A6.** Refer to Figure LG 2-1 and Figure 2-2, page 31 in your text, and complete this exercise about atomic structure.

a. The atomic number of potassium (K) is _____. Locate this number above the K

on Figure LG 2-1. This means that a potassium atom has _____ protons and

_____ electrons.

b. Identify the number located immediately under the K in Figure LG 2-1. This num-

ber is _____, and is the atomic weight of the potassium atom. This indicates the total number of nucleons in the potassium nucleus.

■ **A7.** Complete this exercise about isotopes.

a. If you subtract the atomic number from the atomic weight, you will identify the

number of neutrons in an atom. Most potassium atoms have _____ neutrons. Different isotopes of potassium all have 19 protons, but have varying numbers of (*electrons? neutrons?*).

b. The isotope ^{14}C differs from ^{12}C in that ^{14}C has _____ neutrons.

c. Radioisotopes have a nuclear structure that is unstable and can decay,

emitting _____.

■ **A8.** The type of chemical bonding that occurs between atoms depends on the number of electrons in the outer orbits of the bonding atoms. Do this exercise about electrons and bonding.

a. Write the chemical symbols of several atoms in the periodic table that have only

one electron in the outer orbit. _____

b. Now write the chemical symbols of atoms that have one electron missing from an

otherwise complete outer orbit. _____

c. Potassium has one (*extra? missing?*) electron in its outer orbit. It is therefore more likely to be an electron (*donor? acceptor?*). When it gives up an electron (which is a negative entity), it becomes more (*positive? negative?*); in other words potassium forms K^+, which is a(n) (*anion? cation?*).

d. Chlorine has a valence of (*+1? −1?*), indicating that it has one (*extra? missing?*) electron in its outer orbit. As potassium chloride is formed, chlorine (*gains? gives up?*) an electron and so becomes the (*anion? cation?*) Cl^-.

e. Such bonding of potassium and chlorine is (*covalent? ionic?*) bonding.

■ **A9.** Match the description with the types of bonds listed in the box.

┌───┐
│ C. Covalent   H. Hydrogen   I. Ionic   │
└───┘

_____ a. Atoms lose electrons if they have just one or two electrons in their outer orbits; they gain electrons if they need just one or two electrons to complete the outer orbit.

_____ b. This bond is a bridgelike, weak link between a hydrogen atom and another atom such as oxygen or nitrogen.

_____ c. This type of bond is most easily formed by carbon (C) since its outer orbit is half-filled.

_____ d. These bonds are only about 5 percent as strong as convalent bonds and so are easily formed and broken. They are vital for holding large molecules like protein or DNA in proper configurations.

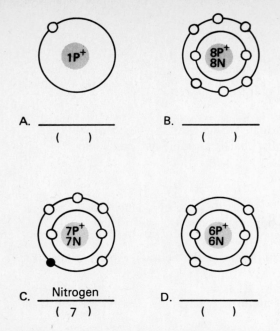

A. _____   B. _____
   (   )   (   )

C. __Nitrogen__   D. _____
   (7)   (   )

Figure LG 2-2 Atomic structure of four common atoms in the human body. Complete as directed in Checkpoint A10.

■ **A10.** Refer to Figure LG 2-2 and complete this exercise.

a. Write atomic numbers in the parentheses ( ) under each atom. (*Hint:* Count the number of electrons or protons = atomic number.) One is done for you.

b. Write the name or symbol for each atom on the line under that atom. (*Hint:* You may identify these by atomic number in Figure LG 2-1.

c. Based on the number of electrons that each of these atoms is missing from a complete outer orbit, draw the number of covalent bonds that each of the atoms can be expected to form. Note that this is the valence, or combining capacity, of these atoms. One is done for you.

d. *For extra review.* Draw the atomic structure of an ammonia molecule (NH_3). Note that this is a covalently bonded compound.

■ **A11.** Circle the molecules that are classified as compounds.

H   Glucose

H_2 N_2

H_2O A molecule that contains two or more different
                     kinds of atoms

B. Chemical reactions (pages 35–36)

■ **B1.** Identify the kind of chemical reaction described in each statement below.

> D. Decomposition reaction   R. Reversible reaction
> E. Exchange reaction   S. Synthesis reaction

_____ a. The end product can revert to the
          original combining molecules.

_____ b. Two reactants combine to form an
          end product; for example, many glu-
          cose molecules bond to form
          glycogen.

_____ c. Such a reaction is partly synthesis
          and partly decomposition (as in
          Figure LG 2-3b, page 30).

_____ d. All chemical reactions involve making
          or breaking of bonds. This type of
          reaction involves only breaking of
          bonds.

_____ e. This type of reaction is catabolic, as
          in digestion of foods, such as starch
          digestion to glucose.

_____ f. This type of reaction is most likely to
          release heat.

B2. Write a paragraph describe the collision theory explanation for chemical reactions.
Include three factors that determine whether a collision will lead to a chemical
reaction.

■ **B3.** Contrast energy forms by completing this exercise.

a. _____ is the capacity to do work. (*Kinetic? Potential?*) energy

   is energy of motion, whereas _____ energy is stored energy.

b. In anabolic (or synthesis) reactions, chemical energy is (*required? released?*) as
   chemical bonds are formed. The opposite type of metabolic process, called

   _____, involves release of energy as when chemical bonds of
   foods are broken.

c. _____ energy is the energy of sunlight that warms your skin

   on a hot day. Another source of radiant energy is _____.

d. Conduction of nerve impulses involves (*mechanical? electrical?*) energy.

C. Inorganic compounds, water, acids, bases, salts; pH, buffers (pages 37–42)

■ **C1.** Write *inorganic* after phrases that describe inorganic compounds and *organic* after those that describe organic compounds.

a. Held together almost entirely by covalent bonds: _____
b. Tend to be very large molecules which serve as good building blocks for body

structures: _____
c. The class that includes water, the most abundant compound in the

body: _____
d. The class that includes carbohydrates, proteins, fats, and nucleic

acids: _____

■ **C2.** Answer these questions about the functions of water.

a. Water normally is a (*solute? solvent?*).
b. Water requires a (*large? small?*) amount of heat in order to change it into a gas. As a result, for each small amount of perspiration (sweat) evaporated during heavy exercise, a (*large? small?*) amount of heat can be released. This is fortunate since dehydration is less likely to occur.
c. Name a location where water's function as a lubricant is essential.

■ **C3.** Match the descriptions below with the answers in the box.

A. Acid	B. Base	S. Salt

_____ a. A substance that dissociates into hydroxyl ions (OH^-) and one or more cations. Example: NaOH
_____ b. A substance that dissociates into hydrogen ions (H^+) and one or more anions. Example: H_2SO_4

_____ c. A substance that dissolves in water forming cations and anions neither of which is H^+ or OH^-. Example: $CaCl_2$

■ **C4.** In the space below draw the outline of a cell. Write inside it symbols for ions or electrolytes concentrated in intracellular fluid. Write outside it symbols for ions concentrated in extracellular fluid.

C5. Draw a diagram showing the pH scale. Label the scale from 0 to 14. Indicate by arrows increasing acidity (H^+ concentration) and increasing alkalinity (OH^- concentration).

■ **C6.** Choose the correct answers regarding pH.

a. Which pH is most acid?

    A. 4   C. 10

    B. 7

b. Which pH has the highest concentration of OH^- ions?

    A. 4   C. 10

    B. 7

c. Which solution has pH closest to neutral?

    A. Gastric juice (digestive juices of the stomach)

    B. Blood

    C. Lemon juice

    D. Milk of magnesia

d. A solution with pH 8 has 10 times (*more? fewer?*) H^+ ions than a solution with pH 7.

e. A solution with pH 5 has (*10? 20? 100?*) times (*more? fewer?*) H^+ ions than a solution with pH 7.

C7. Contrast:

a. Strong acid/weak acid

b. Strong base/weak base

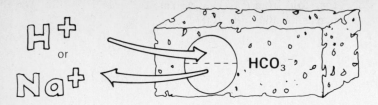

a. H⁺ and Na⁺: only one of you can join me at any one time.

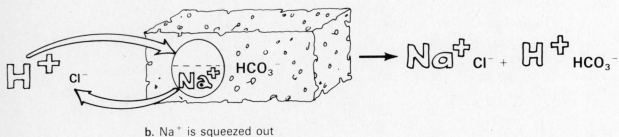

b. Na⁺ is squeezed out as H⁺ is drawn in.

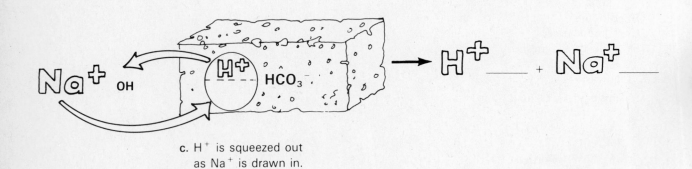

c. H⁺ is squeezed out as Na⁺ is drawn in.

○ H^+ and arrows showing movement of H^+

○ Na^+ and arrows showing movement of Na^+

Figure LG 2-3 Buffering action of carbonic acid-sodium bicarbonate system. Complete as directed in Checkpoint C8.

■ **C8.** Refer to Figure LG 2-3 and complete this exercise about buffers.

a. In Figure LG 2-3a, you can see why buffers are sometimes called "chemical sponges." They consist of an anion (in this case bicarbonate, or HCO_3^-) that is combined with a cation, which in this buffer system may be either H^+ or Na^+. When H^+ is drawn to the HCO_3^- "sponge," the weak acid

_____ is formed. When Na^+ attaches to HCO_3^-, the weak

base _____ is present. Imagine H^+ and Na^+ continually trading places as they are "drawn in or squeezed out of the sponge."

b. Buffers are chemicals that help the body to cope with (*strong? weak?*) acids or bases that are easily ionized and cause harm to the body. When a strong acid, such as hydrochloric acid (HCl), is added to body fluids, buffering occurs, as shown in Figure LG 2-3b. The easily ionized hydrogen ion (H^+) from HCl is "absorbed by the sponge" as Na^+ is released from the buffer/sponge to combine with chloride

(Cl^-). The two products, _____ and

_____, are weak (less easily ionized).

c. Now complete Figure LG 2-3c by writing the names of products resulting from buffering of a strong base. Select different colors for H^+ (and its arrow) and Na^+ (and its arrow) to show this exchange reaction.

d. When the body continues to take in or produce excess strong acid, the concentra-

tion of the _____ member of this buffer pair will decrease as it is used up in the attempt to maintain homeostasis of pH.

e. Though buffer systems provide rapid response to acid-base imbalance, they are limited since one member of the buffer pair can be used up. And they can convert only strong acids or bases to weak ones; they cannot eliminate them. Two systems of the body that can actually eliminate acid or basic substances are

_____ and _____.

D. Organic compounds (pages 42–50)

D1. Organic compounds all contain carbon (C). State two reasons why carbon is an ideal element to serve as the primary structural component for living systems.

D2. Carbohydrates carry out important functions in your body.

a. State one primary function of carbohydrates.

b. State two secondary roles of carbohydrates.

■ **D3.** Complete these statements about carbohydrates.

a. The ratio of hydrogen of oxygen in all carbohydrates is _____ : _____.

The general formula for carbohydrates is _____.

b. The chemical formula for a *hexose* is _____. Two common

hexoses are _____ and _____.

c. One common pentose sugar found in DNA is _____. Its
name and its formula ($C_5H_{10}O_4$) indicate that it lacks one oxygen atom from the

typical pentose formula which is _____.

d. When two glucose molecules ($C_6H_{12}O_6$) combine, a disaccharide is formed. Its for-

mula is _____, indicating that in a synthesis reaction such
as this, a molecule of water is (*added? removed?*). Refer to Figure LG 2-4.
Identify which reaction (*1? 2?*) on that figure demonstrates synthesis.

_____

e. Continued dehydration synthesis leads to enormous carbodydrates called

_____. One such carbohydrate is _____.

f. When a disaccharide such as sucrose is broken, water is introduced to split the
bond linking the two monosaccharides. Such a decomposition reaction is termed

_____, which literally means "splitting using water." (See
Figure LG 2-9, page 43 in your text.) On Figure LG 2-4, hydrolysis is shown by
reaction (*1? 2?*).

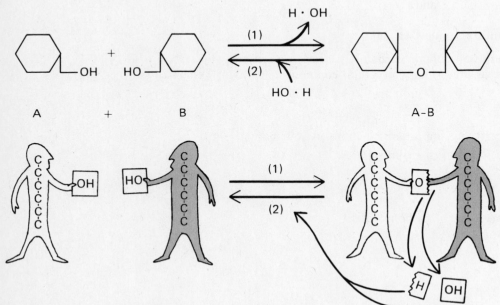

Figure LG 2-4 Dehydration synthesis and hydrolysis reactions. Refer to Checkpoint D3.

■ **D4.** Contrast carbohydrates with lipids in this exercise. Write C next to any statement true of carbohydrates, and write L next to any statement true of lipids.

_____ a. These compounds are insoluble in water. (*Hint:* Water is ineffective for washing such compounds out of clothes.)

_____ b. These compounds are organic.

_____ c. These substances have a hydrogen to oxygen ratio of about 2:1.

_____ d. Very few oxygen atoms (compared to the numbers of carbon and hydrogen atoms) are contained in these compounds. (Look carefully at Figure 2-10, page 45 in your text.

■ **D5.** Circle the answer that correctly completes each statement.

a. A triglyceride consists of:

A. Three glucoses bonded together
B. Three fatty acids bonded to a glycerol

b. A typical fatty acid contains (see Figure 2-10, page 45 in your text):

A. About three carbons   B. About 16 carbons

c. A polyunsaturated fat contains _____ than a saturated fat.

A. More hydrogen atoms   B. Fewer hydrogen atoms

d. Saturated fats are more likely to be derived from:

A. Plants, for example, corn oil or peanut oil
B. Animals, as in beef or cheese

■ **D6.** Match the types of lipids listed in the box with their functions given below..

B. Bile salts	E. Estrogens
Car. Carotene	Pho. Phospholipids
Cho. Cholesterol	Pro. Prostaglandins

_____ a. Present in egg yolk and carrots; leads to formation of vitamin A, a chemical necessary for good vision

_____ b. The major lipid in cell membranes

_____ c. Present in all cells, but infamous for its relationship to "hardening of the arteries"

_____ d. Substances that break up (emulsify) fats before their digestion

_____ e. Steroid sex hormones produced in large quantities by females

_____ f. Important regulatory compounds with many functions, including mimicking of hormones

■ **D7.** Complete this exercise about fat-soluble vitamins.

a. Name four fat-soluble vitamins (consult Exhibit 2-4, page 44 in your text, for

help): _____ _____ _____ _____

b. Since these four vitamins are absorbed into the body along with fats, any problem with fat absorptions is likely to lead to:

c. The fat-soluble vitamin that helps with calcium absorption and so is necessary for

normal bone growth is vitamin _____ .

d. Vitamin _____ may be administered to a patient before surgery in order to prevent excessive bleeding since this vitamin helps in the formation of clotting factors.

e. Normal vision is associated with an adequate amount of vitamin _____ .

f. Vitamin _____ has a variety of possible functions including the promotion of wound healing and prevention of scarring.

■ **D8.** Contrast proteins with the other organic compounds you have studied so far by completing this exercise.

a. Carbohydrates, lipids, and proteins all contain carbon, hydrogen, and oxygen. A

fourth element, _____ , makes up a substantial portion
(16 percent) of proteins.

b. Just as large carbohydrates are composed of repeating units (the

_____), proteins are composed of building blocks called

_____ .

c. As in the synthesis of carbohydrates or fats, when two amino acids bond together, a water molecule must be (*added? removed?*). This is another example of (*hydrolysis? dehydration?*) synthesis.

d. The product of such a reaction is called a _____ -peptide. (See Figure 2-11, page 47 in your text.) When many amino acids are linked together

in this way, a _____ -peptide results. Several polypeptides

form a _____ .

D9. There are at least _____ different kinds of amino acids, much like a 20-letter alphabet that forms protein "words." It is the specific sequence of amino acids that determines the nature of the protein formed. In order to see the significance of this fact, do the following activity. Arrange the five letters listed below in several ways so that you form different five-letter words. Use all five letters in each word.

e   i   l   m   s

___ ___ ___ ___ ___   ___ ___ ___ ___ ___

___ ___ ___ ___ ___   ___ ___ ___ ___ ___

Note that although all of your words contain the same letters, the sequence of letters differs in each case. The resulting words have different meanings, or functions. Such is the case with proteins too, except that protein "words" may be thousands of "letters" (amino acids) long, involving all or most of the 20-amino acid "alphabet." Proteins must by synthesized with accuracy. Think of the drastic consequences of omission or misplacement of even a single amino acid. (How different are *limes, miles, smile, slime!*)

■ **D10.** Match the description below with the four levels of protein organization in the box.

P. Primary structure	S. Secondary structure
Q. Quaternary structure	T. Tertiary structure

_____ a. Specific sequence of amino acids

_____ b. Twisted and folded arrangement (as in spirals or pleated sheets)

_____ c. Irregular three-dimensional shape

_____ d. Two or more tertiary patterns bonded together

D11. List six principal classes of proteins. Consider how you would describe to a friend how each of these six kinds of proteins helped your friend today.

■ **D12.** Complete Figure LG 2-5 as follows:
a. Circle one nucleotide.
b. Label the complementary set of nucleotides on the right side of the figure to

produce the double-stranded structure of DNA. Bases are paired _____ to

_____ and _____ to _____.
c. DNA structure, like that of proteins, depends on sequence. Proteins differ from one

another by their specific sequence of _____, whereas DNA molecules (for example, DNA determining blood type and DNA controlling eye

color) differ from one another by the sequence of _____.
(Note again, as with protein structure, the dire consequences of an omission or substitution in this "four-letter alphabet": A C G T.)
d. How does a strand of RNA differ from DNA?

D13. *For extra review.* Cover Figure LG 2-5 and (without glancing at it) extend the length of the DNA polynucleotide by adding three nucleotides on the left side only. Check your drawing and correct where necessary. Then complete the right side.

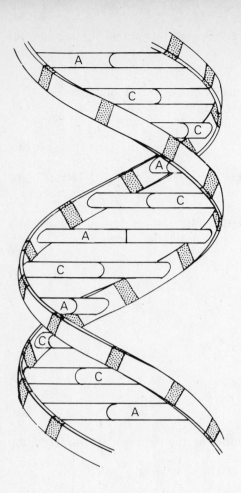

Figure LG 2-5 Structure of DNA. P, phosphate, S, sugar (deoxyribose); A, adenine; T, thymine; G, guanine; C, cytosine. Complete as directed in Checkpoint D12.

■ **D14.** Describe the structure and significance of ATP by completing these statements.

a. ATP stands for adenosine triphosphate. Adenosine consists of a base that is a component of DNA, that is, _____, along with the five-carbon sugar named _____.

b. The "TP" of ATP stands for _____. The final (*one? two? three?*) phosphates are bonded to the molecule by high energy bonds.

c. When the terminal phosphate is broken, a great deal of energy is released as ATP is split into _____ + _____.

d. ATP, the body's primary energy-storing molecule, is constantly reformed by the reverse of this reaction as energy is made available from foods you eat. Write this reversible reaction.

D15. A compound similar to ATP, but with only one phosphate, is named

_____. A cyclic form of this molecule can be made with the help

of the enzyme _____. Cyclic AMP has important regulatory
functions and will be discussed in Chapter 18.

ANSWERS TO SELECTED CHECKPOINTS

A3. (a) O (65 percent), C (18 percent), H (10 percent), N (3 percent); simple. (b) 3.

A4. (b) N, 7. (c) I, 53. (d) Fe, 26. (e) Ca, 20. (f) O, 8.

A5. (a) Nucleus, electrons. (b) Protons (p^+), neutrons (n^0), nucleons. (c) Negative, equal to, 1/2000, (d) 2, 8.

A6. (a) 19, 19, 19. (b) 39.0983 (rounded off to 39).

A7. (a) 20 (39–19), neutrons. *Note:* Varied weights of isotopes account for the uneven number of atomic weight, such as 39.0983 for potassium. (b) 8. (c) Radiation.

A8. (a) Examples: H, Na, K. (b) Examples: F, Cl, Br, I. (c) Extra, donor, positive, cation. (d) −1, missing, gains, anion. (e) Ionic.

A9. (a) I. (b) H. (c) C. (d) H.

A10. (a, b: see Figure LG 2-2A)

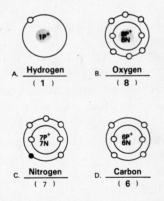

Figure LG 2-2A Atomic structure of four common atoms in the human body.

(c) H̱   —O—   (N),   —C̱—

(d)

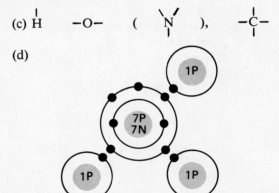

A11. H_2O, glucose, a molecule that contains two or more different kinds of atoms.

B1. (a) R. (b) S. (c) E. (d) D. (e) D. (f) D.

B3. (a) Energy, kinetic, potential. (b) Required, catabolism or decomposition (c) Radiant, light. (d) Electrical.

C1. (a) Organic. (b) Organic. (c) Inorganic. (d) Organic.

C2. (a) Solvent. (b) Large, large. (c) The mucous lining of digestive or respiratory tract or the serous fluid of pleura, pericardium, or peritoneum.

C3. (a) B. (b) A. (c) S.

C4.

Na^+ K^+ Cl^- HPO_4^{2-}

C6. (a) A. (b) C. (c) B. (d) Fewer. (e) 100, more.

C8. (a) Carbonic acid, or H_2CO_3; sodium bicarbonate, or $NaHCO_3$. (b) Strong, the salt sodium chloride (NaCl) and the weak acid H_2CO_3 ($H \cdot HCO_3$). (c) Water (H_2O or $H \cdot OH$) and sodium bicarbonate ($NaHCO_3$). (d) $NaHCO_3$ or weak base. (e) Respiratory (lungs), urinary (kidneys).

D3. (a) 2:1, $(CH_2O)_n$. (b) $C_6H_{12}O_6$; glucose, fructose, galactose. (c) Deoxyribose, $C_5H_{10}O_5$. (d) $C_{12}H_{22}O_{11}$, removed, 1. (e) Polysaccharides; starch, glycogen, cellulose. (f) Hydrolysis, 2.

D4. (a) L. (b) C L. (c) C. (d) L.

D5. (a) B. (b) B. (c) B. (d) B.

D6. (a) Car. (b) Pho. (c) Cho. (d) B. (e) E. (f) Pro.

D7. (a) A, D, E, K. (b) Symptoms of deficiencies of fat-soluble vitamins. (c) D. (d) K. (e) A. (f) E.

D8. (a) Nitrogen. (b) Monosaccharides, amino acids. (c) Removed, dehydration. (d) Di, poly, protein.

D10. (a) P. (b) S. (c) T. (d) Q.

D12.

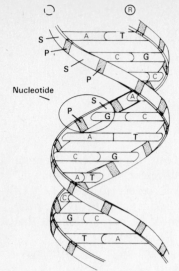

Figure LG 2-5A Structure of DNA. P, phosphate, S, sugar (deoxyribose); A, adenine; T, thymine; G, guanine; C, cytosine.

(a) Phosphate-sugar-base (such as P-S-A or P-S-T in Figure LG 2-5A). (b) A-T, C-G. (c) Amino acids, bases (A, T, C, G). (d) RNA is single-stranded, its sugar is ribose, and it contains the base uracil (U), not thymine (T).

D14. (a) Adenine, ribose. (b) Triphosphate, two. (c) ADP + phosphate (P). (d) ATP ⟶ ADP + P + energy.

MASTERY TEST: Chapter 2

Questions 1–11: Circle the letter preceding the one best answer to each question.

1. All of the following statements are true EXCEPT:

   A. A reaction in which two amino acids join to form a dipeptide is called a dehydration synthesis reaction.
   B. A reaction in which a disaccharide is digested to form two monosaccharides is known as a hydrolysis reaction.
   C. About 65 to 75 percent of living matter consists of organic matter.
   D. Strong acids ionize more easily than weak acids do.

2. Choose the one TRUE statement.

   A. A pH of 7.5 is more acidic than a pH of 6.5.
   B. Anabolism consists of a variety of decomposition reactions.
   C. An atom such as chlorine (Cl), with seven electrons in its outer orbit, is likely to be an electron donor (rather than electron receptor) in ionic bond formation.
   D. Polyunsaturated fats are more likely to reduce cholesterol level than saturated fats are.

3. In the formation of the ionically bonded salt NaCl, Na^+ has:

   A. Gained an electron from Cl^-
   B. Lost an electron to Cl^-
   C. Shared an electron with Cl^-
   D. Formed an isotope of Na^+

4. Over 99 percent of living cells consist of just six elements. Choose the element that is NOT one of these six.

   A. Calcium   E. Nitrogen
   B. Hydrogen   F. Oxygen
   C. Carbon   G. Phosphorus
   D. Iodine

5. Which of the following describes the structure of a nucleotide?

   A. Base-base
   B. Phosphate-sugar-base
   C. Enzyme
   D. Dipeptide
   E. Adenine-ribose

6. Which of the following groups of chemicals includes only polysaccharides?

A. Glycogen, starch
B. Glycogen, glucose, galactose
C. Glucose, fructose
D. RNA, DNA
E. Sucrose, polypeptide

7. $C_6H_{12}O_6$ is most likely the chemical formula for:

A. Amino acid
B. Fatty acid
C. Hexose
D. Polysaccharide
E. Ribose

8. Which of the following is an artificial sweetener used as a replacement for saccharin?

A. Aspartame
B. Fructose
C. Carotene
D. Deoxyribose
E. Olestra

9. Which of the following substances are used mainly for structure and regulatory functions and are not normally used as energy sources?

A. Lipids
B. Proteins
C. Carbohydrates

10. Which is a component of DNA, but not of RNA?

A. Adenine
B. Phosphate
C. Guanine
D. Ribose
E. Thymine

11. All of the following answers consist of pairs of proteins and their correct functions EXCEPT:

A. Contractile—actin and myosin
B. Immunological—collagen of connective tissue
C. Catalytic—enzymes
D. Transport—hemoglobin

Questions 12–20: *Circle T (true) or F (false). If the statement is false, change the underlined word or phrase so that the statement is correct.*

T   F   12. Oxygen can form <u>three bonds since it requires three electrons</u> to fill its outer energy level.

T   F   13. <u>Oxygen, water, NaCl, and glucose</u> are inorganic compounds.

T   F   14. There are about <u>four</u> different kinds of amino acids found in human proteins.

T   F   15. The number of protons always equals the number of <u>neutrons</u> in an atom.

T   F   16. <u>K^+ and Cl^- are both cations.</u>

T   F   17. A strong acid has <u>more</u> of a tendency to contribute H^+ to a solution than a weak acid has.

T   F   18. <u>NaHCO_3 is the weak base</u> of the carbonic acid-bicarbonate buffer system.

T   F   19. ATP contains <u>more</u> energy than ADP.

T   F   20. <u>Prostaglandins, cyclic AMP, steroids, and fats</u> are all classified as lipids.

Questions 21–25: fill-ins. Identify each organic compound below. Write names of compounds on lines provided.

21. _____

22. _____

23. _____

24. _____

25. _____

ANSWERS TO MASTERY TEST: ■ Chapter 2

Multiple Choice

1. C   7. C
2. D   8. A
3. B   9. B
4. D   10. E
5. B   11. B
6. A

True-False

12. F. Two bonds since it requires two electrons
13. F. Water and NaCl (Oxygen is not a compound; glucose is organic.)
14. F. 20
15. F. Electrons
16. F. K^+ is a cation
17. T
18. T
19. T
20. F. Prostaglandins, steroids, and fats

Fill-ins

21. Amino acid
22. Monosaccharide or hexose or glucose
23. Polysaccharide or starch or glycogen
24. Adenosine triphosphate (ATP)
25. Fat or lipid or triglyceride (unsaturated)

FRAMEWORK 3

Cells

STRUCTURE

FUNCTIONS

CHANGES

STRUCTURE

GENERALIZED ANIMAL CELL; PLASMA CELL MEMBRANE [A, B]

Structure
- Fluid mosaic model
- – phospholipids
- – proteins

Functions
- Boundary
- Contacts
- Receptors
- Permea-bility

CYTOPLASM, ORGANELLES [C]
- Nucleus
- Ribosomes
- ER
- Golgi
- Mitochondria
- Lysosomes
- Peroxisomes
- Cytoskeleton
- Centrosome, centrioles
- Flagella, cilia

CELL INCLUSIONS, EXTRACELLULAR MATERIALS [D]
- Fluid
- Amorphous
- Fibrous

FUNCTIONS

MOVEMENT ACROSS MEMBRANES [B]

Passive
- Diffusion
- Osmosis
- Filtration (BP)
- Dialysis

Active
- Na$^+$-K$^+$ pump
- Endocytosis
- Exocytosis

GENE ACTION: PROTEIN SYNTHESIS [E]

Transcription
Translation

Recombinant DNA

CELL DIVISION [F]

Somatic
- Mitosis

Reproductive
- Meiosis

CHANGES

ABNORMAL: CANCER [G]

AGING [H]

MEDICAL TERMS [I]

The Cellular Level of Organization

Cells are the basic structural units of the human body, much as buildings comprise the cities of the world. Every cell possesses certain characteristics similar to those in all other cells of the body. For example, membranes surround and compartmentalize cells, just as walls support virtually all buildings. Yet individual cell types exhibit variations that make them uniquely designed to meet specific body needs. In an architectural tour, the structural components of cathedral or mansion, lighthouse or fort, can be recognized as specific for activities of each type of building. So too the number and type of organelles differ in muscle or nerve, cartilage or bone cells. Cells carry out significant functions such as movement of substances into, out of, and throughout the cell; synthesis of the chemicals that cells need (like proteins, lipids, and carbohydrates); and cell division for growth, maintenance, repair, and formation of offspring. Just as buildings change with time and destructive forces, cells exhibit aging changes. In fact, sometimes their existence is haphazard and harmful, as in cancer.

As you begin your study of cells, first carefully study the Chapter 3 Framework, noting relationships among concepts and key terms.

TOPIC OUTLINE AND OBJECTIVES

A. Generalized animal cell

1. Define a cell and list its principal parts

B. Movement of materials across plasma membrane

2. Explain the structure of the plasma (cell) membrane, and describe how materials move across it.

C. Cytoplasm, organelles

3. Describe the structure and functions of the following cellular structures: nucleus, ribosomes, agranular and granular endoplasmic reticulum (ER), Golgi complex, mitochondria, lysosomes, peroxisomes, cytoskeleton, centrioles, flagella, and cilia.

D. Cell inclusions, extracellular materials

4. Distinguish between cell inclusions and extracellular material.

E. Gene action: protein synthesis

5. Define a gene and explain the sequence of events involved in protein synthesis.

F. Normal cell division: somatic, reproductive

6. Discuss the stages, events, and significance of somatic and reproductive cell division.

G. Abnormal cell division: cancer (CA)

7. Describe cancer (CA) as a homeostatic imbalance of cells.

H. Cells and aging

8. Explain the relationship of aging to cells.

I. Medical terminology

9. Define medical terminology associated with cells.

Now study the following parts of words that may help you better understand terminology in this chapter.

Wordbyte	Meaning	Example	Wordbyte	Meaning	Example
cyto-	cell	*cyto*logist	-oma	tumor	carcin*oma*
-elle	small	organ*elle*	-philic	loving	hydro*philic*
hydro-	water	*hydro*static	-phobic	fearing	hydro*phobic*
hyper-	above	*hyper*tonic	prot-	first	*prot*ein
hypo-	under	*hypo*tonic	stasis, -static	stand, stay	meta*stasis*
iso-	equal	*iso*tonic	-tonic	tension or pressure	hyper*tonic*
meta-	beyond	*meta*stasis			

CHECKPOINTS

A. Generalized animal cell (page 54)

A1. List the four principal parts of a generalized animal cell.

A2. Define extracellular materials.

B. Plasma cell membrane, movement of materials across plasma membranes (pages 55–62)

■ **B1.** Refer to Figure LG 3-1 and do this exercise. Color the following parts of the plasma membrane.

○ a. Phospholipid polar "head" _____

○ b. Phospholipid nonpolar "tail" _____

○ c. Integral protein _____

○ d. Integral protein with carbohydrate attached _____

○ e. Peripheral protein _____

On the lines provided next to each, write one or more functions for each part that you colored.

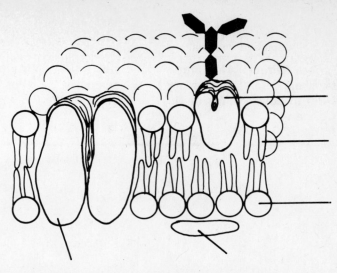

Figure LG 3-1 Plasma membrane. Color as indicated in Checkpoint B1.

■ **B2.** From the complex and varied structure of membranes, it is clear that membranes do more than just "sit there" around cells. State four important functions of plasma membranes.

B3. Explain why substances must move across your plasma membranes in order for you to survive and maintain homeostasis.

■ **B4.** Complete parts a–e of Table LG 3-1.

Table LG 3-1 Summary of mechanisms for movement of materials across membranes

Name of Process	What Moves (Solute or Solvent?)	Where (Direction, e.g., High to Low Concentration)?	Membrane Necessary (Yes or No)?	Carrier Necessary (Yes or No)?	Active or Passive?
a. Diffusion					
b. Facilitated diffusion					
c.	Solvent	From high to low concentration of water across semipermeable membrane			
d.		From high pressure area to low pressure area			
e. Dialysis				No	
f. Active transport		From low to high concentration of solute			
g. Phagocytosis			Yes		
h.	Solvent	From outside cell to inside cell			
i. Exocytosis					Active

46

■ **B5.** Select from the following list of terms to identify passive transport processes described below.

Dia. Dialysis	Fil. Filtration
Dif. Diffusion	O. Osmosis
Fac. Facilitated diffusion	

_____ a. Net movement of any substance (such as cocoa powder in hot milk) from region of higher concentration to region of lower concentration; membrane not required.

_____ b. Same as (a) except movement across a semipermeable membrane with help of a carrier; ATP not required

_____ c. Net movement of water from region of high water concentration (such as 2 percent NaCl) to region of low water concentration (such as 10 percent NaCl) across semipermeable membrane; important in maintenance of normal cell size and shape.

_____ d. Movement of molecules from high pressure zone to low pressure zone, for example, in response to force of blood pressure

_____ e. Movement of small molecules across a semipermeable membrane, leaving large molecules behind; the principle used in artificial kidney machines

B6. A large part of the human diet is starch. When starch is digested, it is broken down to glucose. Answer these questions related to absorption of glucose.

a. Explain why glucose cannot readily pass through plasma membranes by simple diffusion.

b. By what process does glucose cross plasma membranes as it is absorbed into cells lining the digestive tract?

c. List three factors that would speed up this rate of movement of glucose.

■ **B7.** Complete the following exercise about osmosis in blood.

a. Human red blood cells (RBCs) contain intracellular fluid that is osmotically similar

to _____ NaCl.

A. 2.0 percent   C. 0 percent (pure water)
B. 0.90 percent

b. A solution that is hypertonic to RBCs contains (*more? fewer?*) solute particles and (*more? fewer?*) water molecules than blood.

c. Which of these solutions is hypertonic to RBCs?

A. 2.0 percent NaCl   C. pure water
B. 0.90 percent NaCl

d. If RBCs are surrounded by hypertonic solution, water will tend to move (*into? out of?*) them, so they will (*crenate? hemolyze?*).

e. A solution that is _____-tonic to RBCs will maintain the shape and size of the RBC. An example of such a solution is

_____.

f. Which solution will cause RBCs to hemolyze?
   A. 2.0 percent NaCl   C. pure water
   B. 0.90 percent NaCl

g. Which solution has the highest osmotic pressure?
   A. 2.0 percent NaCl   C. pure water
   B. 0.90 percent NaCl

■ **B8.** When blood is dialyzed in an artificial kidney, why must nutrients and certain ions be added to the blood?

B9. Contrast *active* with *passive transport* processes.

■ **B10.** Complete parts f–i of Table LG 3-1.

■ **B11.** Describe active transport in the following exercise.

a. Glucose is a chemical compound that may be actively transported from areas of (*high? low?*) concentration to areas of (*high? low?*) concentration, as from digestive tract cells into the bloodstream.

b. Active transport utilizes an _____ protein within the

membrane, as well as an energy source, _____. In fact, a typi-

cal body may use up to _____ percent of its ATP for active transport.

B12. State one advantage of use of liposomes with antibiotic or cancer drug therapy.

C. Cytoplasm, organelles (pages 62–73)

C1. The cell is compartmentalized by the presence of organelles. Of what advantage is this?

■ **C2.** Color and then label all of the parts of Figure LG 3-2 listed with color code ovals.

■ **C3.** Describe parts of the nucleus by completing this exercise.

a. The gellike fluid in the nucleus is called _____.

b. Nuclear storage sites of RNA used in protein synthesis are known as

_____.

c. During cell reproduction chromatin forms rod-shaped bodies called

_____.

d. Chromosomes consist of subunits known as _____ which are

composed of four types of proteins (called _____) plus DNA.
A fifth type of histone helps to maintain adjacent nucleosomes in the configuration

of a _____.

○ Granular (rough) endoplasmic reticulum
○ Chromatin
○ Nucleolus
○ Plasma (cell) membrane
○ Mitochondrion
○ Cytoplasm

○ Centriole
○ Microtubule
○ Lysosome
○ Nuclear membrane
○ Agranular (smooth) endoplasmic reticulum
○ Golgi complex

Figure LG 3-2 Generalized animal cell. Color and label as directed in Checkpoint C2.

■ **C4.** Describe the sequence of events in synthesis, secretion, and discharge of protein from the cell by numbering these five steps in correct order.

_____ A. Proteins accumulate and are modified, sorted, and packaged in cisternae of Golgi complex.

_____ B. Proteins pass through ER to *cis* cisternae of Golgi complex.

_____ C. Secretory granules move toward cell surface where they are discharged.

_____ D. Proteins are synthesized at ribosomes on rough ER.

_____ E. Vesicles containing protein pinch off of Golgi complex *trans* cisternae to form secretory granules.

■ **C5.** Match the organelles listed in the box with the correct descriptions of their functions.

Cen. Centrioles	G. Golgi complex	Mit. Mitochondria
Cil. Cilia	L. Lysosomes	N. Nucleus
ER. Endoplasmic reticulum	Mf. Microfilaments	P. Peroxisomes
F. Flagella	Mt. Microtubules	R. Ribosomes

_____ a. Site of direction of cellular activities by means of genes located here

_____ b. Cytoplasmic sites of protein synthesis; may occur attached to *ER* or scattered freely in cytoplasm

_____ c. System of membranous channels providing pathways for transport within the cell and surface areas for chemical reactions; may be granular or agranular

_____ d. Stacks of cisternae with vesicular ends; involved in packaging and secretion of proteins and lipids and synthesis of carbohydrates

_____ e. Called "suicide packets" since they may release enzymes which lead to autolysis of the cell

_____ f. Similar to lysosomes, but smaller; contain the enzyme catalase that breaks down hydrogen peroxide

_____ g. Cristae-containing structures, called "powerhouses of the cell" since ATP production occurs here

_____ h. Form part of cytoskeleton; involved with cell movement and contraction

_____ i. Part of cytoskeleton, proves support and gives shape to cell; form flagellae, cilia, centrioles, and spindle fibers; made of protein tubulin

_____ j. Help organize spindle fibers used in cell division

_____ k. Long, hairlike structures that help move entire cell, for example, sperm cell

_____ l. Short, hairlike structures that move particles over cell surfaces

■ **C6.** *For extra review.* Considering the functions of organelles listed in the above exercise, choose the answers (organelles) that fit the following descriptions.

_____ a. Abundant in white blood cells which use these organelles to help in phagocytosis of bacteria

_____ b. Present in large numbers in muscle and liver cells which require much energy

_____ c. Located on surface of cells of the respiratory tract; help to move mucus

_____ d. Absent in mature nerve cells, resulting in inability of these cells to reproduce if destroyed

D. Cell inclusions, extracellular materials (page 73)

D1. Contrast *organelle* with *inclusion.*

D2. List three examples of cell inclusions.

■ **D3.** List three examples of *extracellular materials.*

■ **D4.** Complete the table about two major classes of matrix materials: *amorphous* and *fibrous*. Indicate to which category each of the substances listed below belongs. Then give an example of a location in the body where each type of matrix is found.

Matrix Material	Class	Location
a. Collagen		
b. Hyaluronic acid	Amorphous	
c. Chondroitin sulfate		
d. Elastin		Elastic fibers of skin and blood vessels

E. Gene action: protein synthesis (pages 73–77)

E1. State the significance of protein synthesis in the cell.

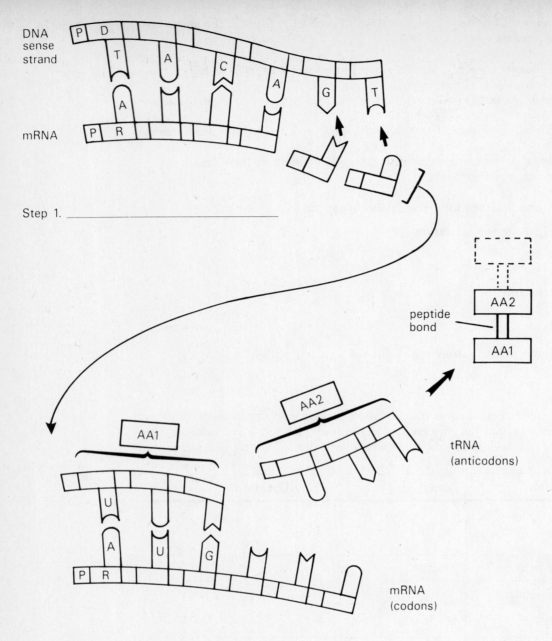

DNA sense strand

mRNA

Step 1. _____

peptide bond

AA2

AA1

AA1

AA2

tRNA (anticodons)

mRNA (codons)

Step 2. _____

○ A. Adenine   ○ D. Deoxyribose
○ C. Cytosine   ○ R. Ribose
○ G. Guanine   ○ AA1. Amino acid 1
○ T. Thymine   ○ AA2. Amino acid 2
○ U. Uracil   ○ Peptide bond
○ P. Phosphate

Figure LG 3-3 Protein synthesis. Label and color as directed in Checkpoint E2.

■ **E2.** Refer to Figure LG 3-3 and describe the process of protein synthesis in this exercise.

   a. Write the names of the two major steps of protein synthesis on the two lines in Figure LG 3-3.

   b. Transcription takes place in the (*nucleus? cytoplasm?*). In this process, a portion of

       one side of a double-stranded DNA serves as a mold or _____ and is called the (*sense? antisense?*) strand. Nucleotides of

       _____ line up in a complementary fashion next to DNA nucleotides (much as occurs between DNA nucleotides in replication of DNA prior to mitosis). Select different colors and color the components of DNA nucleotides (P, D, and bases T, A, C, and G).

   c. Complete the strand of mRNA by writing letters of complementary bases. Then color the parts of mRNA (P, R, and bases U, A, C, and G).

   d. The second step of protein synthesis is known as _____ since it involves translation of one "language" (the nucleotide base code of

       _____ RNA) into another "language" (the correct sequence of

       the 20 _____ to form a specific protein).

   e. Translation occurs in the (*nucleus? cytoplasm?*), specifically at a (*chromosome? lysosome? ribsosome?*) to which mRNA attaches.

   f. Each amino acid is transported to this site by a particular _____RNA characterized by a specific three-base unit or (*codon? anticodon?*). This portion of tRNA, for example, U A C, binds to a complementary portion of mRNA (*codon? anticodon?*). In step 2 of Figure LG 3-3, complete the labeling of bases and color all components of tRNA and mRNA. Be sure you select colors that correspond to those used in step 1.

   g. As the ribosome moves along each mRNA codon, additional amino acids are

       transferred into place by _____RNA. _____ bonds form between adjacent amino acids with the help of enzymes from the

       _____. Now color the amino acids and the peptide bond that joins them.

■ **E3.** Summarize protein synthesis by completing this exercise.

   a. A typical gene contains about _____ pairs of nucleotides. The mRNA strand

       transcribed from this gene would consist of about _____ nucleotides (containing

       _____ bases).

   b. Certain sections of DNA that do not code for proteins are known as (*introns? exons?*). The regions of mRNA that copy this DNA must be deleted from mRNA. Imagine that a 900-base mRNA strand remains after the deletion process. This

       strand would consist of about _____ mRNA codons (3 bases per codon). During translation, how many amino acids could be assembled along this strand of

       mRNA? _____ (*Note:* This summary simplifies numbers, omitting certain codons, such as for termination.)

E4. List six therapeutic substances that have been produced by bacteria containing recombinant DNA.

F. Normal cell division: somatic and reproductive (pages 77–82)

F1. Once you have reached adult size, do your cells continue to divide? (*Yes? No?*) Of what significance is this fact?

■ **F2.** Describe aspects of cell division by completing this exercise.

a. Division of any cell in the body consists of two processes: division of the

_____ and division of the _____.

b. In the formation of mature sperm and egg cells, nuclear division is known as

_____. In the formation of all other body cells, that is,

somatic cells, nuclear division is called _____.

c. Cytoplasmic division in both somatic and reproductive division is known as

_____. In (*meiosis? mitosis?*) the two daughter cells have the same hereditary material and genetic potential as the parent cell.

d. Chromosomes consist of DNA present in the form of units called

_____ bound by protein (histones). Humans have about

_____ functional genes.

e. Prior to cell division DNA replicates. The helix partially uncoils and the double

strand separates at the points where _____ connect so that new complementary nucleotides can attach here. (See Figure LG 2-5, page 36.) In this way the DNA of a single choromosome is doubled to form two identical

_____ during the _____-phase preceding a mitotic division.

■ **F3.** Carefully study the phases of mitosis shown in Figure 3-17, page 79, in your text. Then check your understanding of major events of each phase by doing this matching exercise.

A. Anaphase	P. Prophase
I. Interphase	T. Telophase
M. Metaphase	

_____ a. Centromeres divide and drag chromosomes along microtubules towards opposite poles of the cell.

_____ b. Nuclear membrane disappears; chromatin coils into distinct chromosomes (chromatids). Centrioles move to opposite ends of cell, enabling microtubules to form mitotic spindles. The longest mitotic phase.

_____ c. The series of events is essentially the reverse of prophase; cytokinesis occurs.

_____ d. Chromatids line up with their centromeres along the equatorial plate.

_____ e. Cell is not involved in cell division; it is in a metabolic phase which follows telophase.

■ **F4.** Do the following exercise about reproductive cell division.

a. Meiosis is a special type of nuclear division that occurs only in

_____ and is one process in the production of sex cells

or _____ .

b. Cells that begin the process of meiosis are ovarian or testicular cells that have been undergoing mitosis up to this point. Each cell initially contains _____ chromosomes, which is the (*diploid or 2n? haploid or n?*) number for humans. These chromosomes consist of _____ homologous pairs.

c. By the time that cells complete meiosis, they will contain (*46? 23?*) chromosomes, the (*2n? n?*) number.

d. Meiosis differs from mitosis in that meiosis involves (*one? two? three?*) divisions with replication of DNA only (*once? twice?*). The first division is known as the

_____ division, whereas the second division is the

_____ division.

e. Replication of DNA occurs (*before? after*) reduction division, resulting in doubling of chromosome number in humans from 46 chromosomes to

_____ "chromatids." (Chromatid is a name given to a chromosome while the chromosome is attached to its replicated counterpart, that is, in the doubled state.) Doubled or paired chromatids are held together by means of a (*centromere? centrosome?*).

f. A unique event that occurs in Prophase I is the lining up of chromosomes in

homologous pairs, a process called _____. Note that this process (*does? does not?*) occur during mitosis. The four chromatids of the

homologous pair are known as a _____. The proximity of

chromatids permits the exchange or _____ of DNA between maternal and paternal chromatids. What is the advantage of crossing-over to sexual reproduction?

g. Another factor that increases possibilities for variety among sperm or eggs you can produce is the random (or chance) assortment of chromosomes as they line up on either side of the equator in Metaphase (*I? II?*).

h. Division I reduces the chromosome number to _____ chromatids. Division II results in the separation of doubled chromatids by splitting

of _____ so that resulting cells contain only _____ chromosomes.

i. Meiosis is only one part of the process of forming sperm or eggs. Each original diploid cell in a testis will lead to production of (*1? 2? 4?*) mature sperm, while each diploid ovary cell produces (*1? 2? 4?*) mature eggs (ova) plus three small cells

known as _____.

G. Abnormal cell division: cancer (CA) (pages 83–85)

■ **G1.** Complete this exercise about cancer.

a. Cancer cells can arise only from cells that (*can? cannot?*) undergo mitosis.

b. A term which means "tumor or abnormal growth" is _____.
A cancerous growth is a (*benign? malignant?*) neoplasm, whereas a noncancerous tumor is a (*benign? malignant?*) neoplasm.

c. Which type of growth is more likely to spread and to possibly cause death? (*Benign? malignant?*) A term which means the "spread" of cancer cells is

_____.

d. In the process of metastasis, cancer cells compete with normal cells for

_____. They also migrate into space previously occupied by

other cells since they do not follow rules of _____. By what routes do cancers reach distant part of the body?

e. What may cause the pain associated with cancer?

■ **G2.** Match terms in the box with the definitions below.

```
A. Adenocarcinoma      M. Myeloma
Car. Carcinoma         O. Osteosarcoma
Ch. Chondrosarcoma     S. Sarcoma
```

_____ a. General term for malignant tumor arising from any connective tissue

_____ b. Malignant tumor derived from bone (a connective tissue)

_____ c. General term for malignant tumor of epithelial tissue

_____ d. Malignant tissue derived from epithelial tissue of a gland

_____ e. Malignant tumor arising from cartilage (a connective tissue)

_____ f. A malignancy of tissue that forms blood

■ **G3.** Contrast methods used to categorize cancers in this checkpoint.

a. (*Grading? Staging?*) refers to the degree of microscopic similarity of cancer cells to normal healthy cells. Which cancer suggests a better prognosis or outcome? (*Grade II? Grade IV?*)

b. In the TNM method of *staging* of cancers, *T* refers to the extent or

_____ of the tumor, *N* indicates the progression of lymph

_____ involvement. Which cancer is more likely to have spread extensively to secondary sites? (*M1? M4?*)

G4. Complete this exercise about causative factors associated with cancer.

a. Define *carcinogen*.

Name a carcinogen linked to cancer among cigarette smokers.

b. Identify one or more types of cancer associated with the viruses named below.

HTLV-1 _____

EBV _____

HBV _____

c. Define *oncogenes*.

d. Do oncogenes appear to be involved in all cancers? (*Yes? No?*) Explain the apparent cause of the rare childhood eye cancer known as retinoblastoma.

G5. A *clinical challenge.* Are all cancer cells alike within a single tumor? (*Yes? No?*) Of what significance is this in planning cancer treatment?

H. Cells and aging (pages 85–86)

H1. Describe changes that normally occur in each of the following during the aging process.

a. Number of cells in the body

b. Collagen fibers

c. Elastin

H2. In a sentence or two, summarize the main points of each of these theories of aging.

a. Free radical

b. Immune system

c. Pituitary gland hormone

I. Medical terminology (pages 86)

■ **I1.** Match the terms in the box describing alterations in cells or tissues with descriptions below.

D. Dysplasia	HT. Hypertrophy
HP. Hyperplasia	M. Metaplasia

_____ a. Increase in size of tissue or organ by increase in size (not number) of cells, as growth of your biceps muscle with exercise

_____ b. Increase in size of tissue or organ due to increase in number of cells, such as a callus on your hand or breast tissue during pregnancy

_____ c. Change of one cell type to another normal cell type, such as change from single row of tall (columnar) cells lining airways to multilayers of cells as response to constant irritation of smoking

_____ d. Abnormal change in cells in a tissue as due to irritation or inflammation. May revert to normal if irritant removed, or may progress to neoplasia

■ **I2.** Match the terms in the box with the definitions below.

> A. Atrophy   O. Oncology
> B. Biopsy   S. Senescence
> N. Necrosis

_____ a. Death of a group of cells

_____ b. Decrease in size of cells with decrease in size of tissue or organ

_____ c. Study of tumors

_____ d. Removal and examination of tissue from the living body for diagnosis

_____ e. Process of growing old

ANSWERS TO SELECTED CHECKPOINTS

B1.

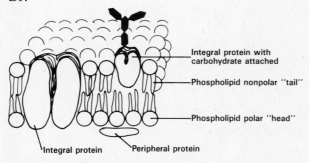

Integral protein with carbohydrate attached

Phospholipid nonpolar "tail"

Phospholipid polar "head"

Integral protein   Peripheral protein

Figure LG 3-1A   Plasma membrane.

(a) Water-soluble and facing outwards, so compatible with watery environment surrounding the membrane. (b) Forms a water-insoluble barrier between cell contents and extracellular fluid; nonpolar molecules such as lipids, oxygen, and carbon dioxide pass readily between "tails". (c) Channels for passage of water or ions, carriers. (d) Receptors. (e) Serve as enzymes or help cell to change shape.

B2.   Flexible boundary, contacts with other cells, receptors sites, selective permeability or movement across membranes.

B4.

Name of Process	What Moves (Solute or Solvent?)	Where (Direction, e.g., High to Low Concentration)?	Membrane Necessary (Yes or No)?	Carrier Necessary (Yes or No)?	Active or Passive?
a. Diffusion	Solute or solvent	From high to low concentration of solute	No	No	Passive
b. Facilitated diffusion	Solute, such as glucose	From high to low concentration of solute	Yes	Yes	Passive
c. **Osmosis**	Solvent	From high to low concentration of water across semipermeable membrane	Yes	No	Passive
d. Filtration	Solute or solvent	From high pressure area to low pressure area	Yes (or other barrier such as filter paper)	No	Passive
e. Dialysis	Small solutes	From high to low concentration of solutes	Yes	No	Passive

B5.   (a) Dif. (b) Fac. (c) 0. (d) Fil. (e) Dia.

B7.   (a) B. (b) More, fewer. (c) A. (d) Out of, crenate. (e) Iso-, 0.90 percent NaCl (physiological saline) or 5.0 percent glucose. (f) C. (g) A.

B8.   Because small ions (even "good ones" needed by the body) are removed by dialysis.

B10.

Name of Process	What Moves (Solute or Solvent?)	Where (Direction, e.g., High to Low Concentration)?	Membrane Necessary (Yes or No)?	Carrier Necessary (Yes or No)?	Active or Passive?
f. Active transport	Solutes, such as glucose or ions (Na⁺, K⁺)	From low to high concentration of solute	**Yes**	**Yes**	**Active**
g. Phago-cytosis	Large particles	From outside of cell to inside of cell	Yes	**No**	**Active**
h. **Pino-cytosis**	Solvent	From outside cell to inside cell	**Yes**	**No**	**Active**
i. Exo-cytosis	Large molecules or particles	From inside of cell to outside of cell	**Yes**	**No**	Active

B11. (a) Low, high (b) Integral, ATP, 40.

C2.

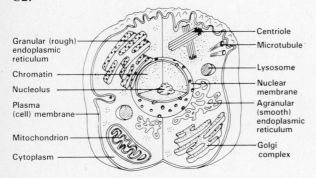

Figure LG 3-2A   Generalized animal cell.

Labels: Granular (rough) endoplasmic reticulum, Chromatin, Nucleolus, Plasma (cell) membrane, Mitochondrion, Cytoplasm, Centriole, Microtubule, Lysosome, Nuclear membrane, Agranular (smooth) endoplasmic reticulum, Golgi complex.

C3.   (a) Karyolymph. (b) Nucleoli. (c) Chromosomes. (d) Nucleosomes, histones, coil.

C4.   3, 2, 5, 1, 4 (or D, B, A, E, C).

C5.   (a) N. (b) R. (c) ER. (d) G. (e) L. (f) P. (g) Mit. (h) Mf. (i) Mt. (j) Cen. (k) F. (l) Cil.

C6.   (a) L. (b) Mit. (c) Cil. (d) Cen.

D3.   Body fluids, such as interstitial fluid and plasma; mucus; amorphous and fibrous matrix.

D4.

Matrix Material	Class	Location
a. Collagen	**Fibrous**	**Collagenous (white) fibers in tendons, bone**
b. Hyaluronic acid	Amorphous	**Joints, eyeballs**
c. Chondroitin sulfate	**Amorphous**	**Cartilage, bone, and blood vessels**
d. Elastin	**Fibrous**	Elastic fibers of skin and blood vessels

E2.   (a) See Figure LG 3-3A. (b) Nucleus, template, sense, messenger RNA (mRNA). (c) See Figure LG 3-3A. (d) Translation, messenger, amino acids. (e) Cytoplasm, ribosome. (f) See Figure LG 3-3A. (g) t, peptide, ribosome.

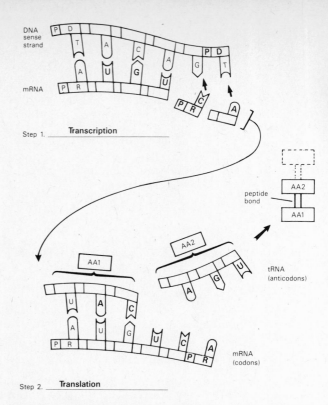

Step 1.   Transcription

Step 2.   Translation

Figure LG 3-3A   Protein synthesis.

E3.   (a) 1,000, 1,000, 1,000 (b) Introns, 300, 300.

F2.   (a) Nucleus, cytoplasm. (b) Meiosis, mitosis, (c) Cytokinesis, mitosis. (d) Genes, 100,000. (e) Nitrogen bases, chromatids, inter-.

F3.   (a) A. (b) P. (c) T. (d) M. (e) I.

F4.   (a) Gonads (ovaries or testes), gametes. (b) 46, diploid or 2n, 23. (c) 23, n. (d) Two, once, reduction, equatorial. (e) Before, "46 × 2" (or 46 doubled), centromere. (f) Synapsis, does not, tetrad, crossing-over, permits exchange of DNA between chromatids and greatly increases possibilities for variety among sperm or eggs and thus offspring. (g) I. (h) "23 × 2" (or 23 doubled chromatids), centromeres, 23. (i) 4, 1, polar bodies.

G1. (a) Can. (b) Neoplasm, malignant, benign. (c) Malignant, metastasis. (d) Nutrients and space, contact inhibition, blood or lymph, or by detaching from an organ, such as liver, and "seeding" into a cavity (as abdominal cavity) to reach other organs which border on that cavity. (e) Pressure on nerve, obstruction of passageway, loss of function of a vital organ.

G2. (a) S. (b) O. (c) Ca. (d) A. (e) Ch. (f) M.
G3. (a) Grading, Grade II. (b) Size, node, M (metastasis) 4.
I1. (a) HT. (b) HP. (c) M. (d) D.
I2. (a) N. (b) A. (c) O. (d) B. (e) S.

MASTERY TEST: Chapter 3

Questions 1–10: Circle the letter preceding the one best answer to each question.

1. The organelle which carries out the process of autolysis, and is therefore called a "suicide packet" is:

   A. Peroxisome
   B. Mitochondrion
   C. Centrosome
   D. Lysosome
   E. Endoplasmic reticulum

2. Which statement about proteins is FALSE?

   A. Proteins are synthesized on ribosomes
   B. Proteins are so large that they normally will not pass out of blood that undergoes filtration in kidneys.
   C. Proteins are so large that they will not pass through the dialysis membrane used in artificial kidneys.
   D. Proteins are so large that they can be expected to create osmotic pressure important in maintaining fluid volume in blood.
   E. Proteins are small enough to pass through pores of typical plasma (cell) membranes.

3. Choose the FALSE statement about prophase of mitosis.

   A. It occurs just after metaphase.
   B. It is the longest phase of mitosis.
   C. Nucleoli become less visible.
   D. Centrioles move apart and become connected by spindle fibers.

4. Choose the FALSE statement about the Golgi complex.

   A. It is usually located near the nucleus.
   B. It consists of stacked membranous sacs known as cisternae.
   C. It is involved with protein and lipid secretion.
   D. It pulls chromosomes toward poles of the cell during mitosis.

5. Choose the FALSE statement about genes.

   A. Genes contain DNA.
   B. Genes contain information which controls heredity.
   C. The human body has a total of 46 genes.
   D. Genes are transcribed by messenger RNA during the first step of protein synthesis.

6. Choose the FALSE statement about protein synthesis.

   A. Translation occurs in the cytoplasm.
   B. Messenger RNA picks up and transports an amino acid during protein synthesis.
   C. Messenger RNA travels to a ribosome in the cytoplasm.
   D. Transfer RNA is attracted to mRNA due to their complementary bases.

7. All of the following structures are part of the nucleus EXCEPT:

A. Nucleolus   D. Chromosomes
B. Karyolymph   E. Centrosome
C. Chromatin

8. The fact that the aroma of cookies baking in the kitchen reaches you in the living room is due to:

A. Active transport   D. Diffusion
B. Dialysis   E. Phagocytosis
C. Osmosis

9. Bone cancer is an example of what type of cancer?

A. Sarcoma   C. Carcinoma
B. Myeloma

10. All of the following are examples of extracellular materials EXCEPT:

A. Melanin
B. Hyaluronic acid
C. Chondroitin
D. Collagenous fibers
E. Elastic fibers

Questions 11–20: Circle T (true) or F (false). If the statement is false, change the underlined word or phrase so that the statement is correct.

T F 11. Plasma membranes consist of a double layer of <u>carbohydrate molecules</u> with proteins embedded in the bilayer.

T F 12. Every cell of the body has <u>the same appearance</u> as the generalized animal cell.

T F 13. <u>Cristae</u> are folds of membrane in mitochondria.

T F 14. Sperm move by means of lashing their long tails named <u>cilia</u>.

T F 15. Hypertrophy refers to increase in size of a tissue related to increase in the <u>number</u> of cells.

T F 16. <u>Diffusion, osmosis, and filtration</u> are all passive transport processes.

T F 17. A 5 percent glucose solution is <u>hypotonic</u> to a 10 percent glucose solution.

T F 18. Most healthy cells <u>lack</u> the property of contract inhibition.

T F 19. The mitotic phase that follows metaphase is <u>anaphase</u>.

T F 20. <u>Cytokinesis</u> is another name for mitosis.

Questions 21–25: fill-ins. Complete each sentence with the word which best fits.

_____ 21. The first meiotic division is known as the ____ division.

_____ 22. ____ proteins are loosely bound to the cell membrane and easily separated from it.

_____ 23. White blood cells engulf large solid particles by the process of ____.

_____ 24. ____ DNA consists of DNA combined from several sources, such as from human cells and from bacteria.

_____ 25. The sequence of bases of mRNA which would be complementary to DNA bases in the sequence A-T-T-C-A-C would be ____.

ANSWERS TO MASTERY TEST: ■ Chapter 3

Multiple Choice

1. D   6. B
2. E   7. E
3. A   8. D
4. D   9. A
5. C   10. A

True-False

11. F. Phospholipid molecules
12. F. Some of the characteristics of
13. T
14. F. Flagella
15. F. Size
16. T
17. T
18. F. Do have or possess
19. T
20. F. Nuclear divsion (Cytokinesis is cytoplasmic division.)

Fill-ins

21. Reduction
22. Peripheral
23. Phagocytosis
24. Recombinant
25. U-A-A-G-U-G

FRAMEWORK 4

Tissues

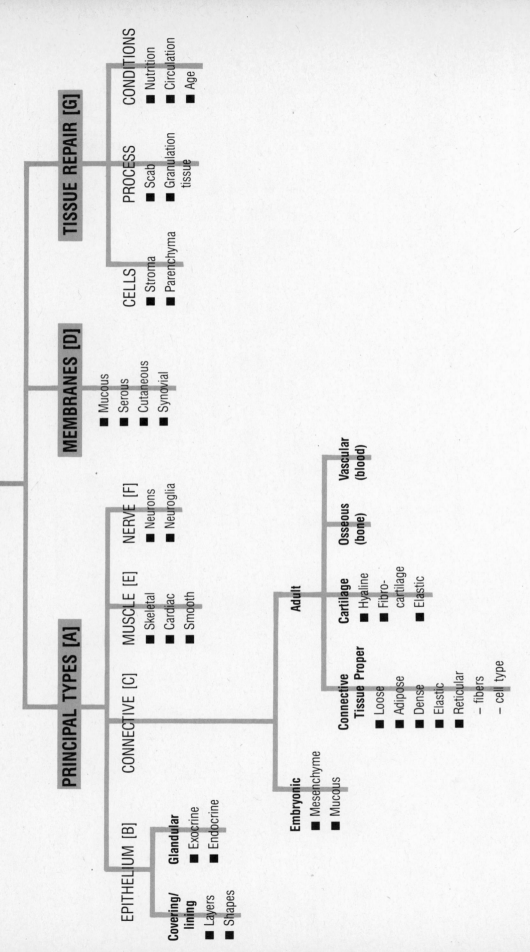

PRINCIPAL TYPES [A]

EPITHELIUM [B]

Covering/ lining
- ■ Layers
- ■ Shapes

Glandular
- ■ Exocrine
- ■ Endocrine

CONNECTIVE [C]

Embryonic
- ■ Mesenchyme
- ■ Mucous

Adult

Connective Tissue Proper
- ■ Loose
- ■ Adipose
- ■ Dense
- ■ Elastic
- ■ Reticular
  - – fibers
  - – cell type

Cartilage
- ■ Hyaline
- ■ Fibro- cartilage
- ■ Elastic

Osseous (bone)

Vascular (blood)

MUSCLE [E]
- ■ Skeletal
- ■ Cardiac
- ■ Smooth

NERVE [F]
- ■ Neurons
- ■ Neuroglia

MEMBRANES [D]
- ■ Mucous
- ■ Serous
- ■ Cutaneous
- ■ Synovial

TISSUE REPAIR [G]

CELLS
- ■ Stroma
- ■ Parenchyma

PROCESS
- ■ Scab
- ■ Granulation tissue

CONDITIONS
- ■ Nutrition
- ■ Circulation
- ■ Age

The Tissue Level of Organization

Tissues consist of cells and the intercellular (between cells) materials secreted by these cells. All of the organs and systems of the body consist of tissues that may be categorized into four classes or types: epithelium, connective, muscle, and nerve. Each tissue exhibits structural characteristics (anatomy) that determine the function (physiology) of that tissue. For example, epithelial tissues consist of tightly packed cells that form the secretory glands and also provide protective barriers, as in the membranes that cover the surface of the body and line passageways or cavities. Tissues are constantly stressed by daily wear and tear and sometimes by trauma or infection. Tissue repair is therefore an ongoing human maintenance project.

As you begin this chapter, first carefully examine the Chapter 4 Framework, and note relationships among concepts and key terms there.

TOPIC OUTLINE AND OBJECTIVES

A. Types of tissue

1. Define a tissue and classify the tissues of the body into four major types.

B. Epithelial tissue

2. Compare the distinguishing characteristics of epithelial and connective tissues, and list the structure, location, and function for the following types of epithelium: simple squamous, simple cuboidal, simple columnar (nonciliated and ciliated), stratified squamous, stratified cuboidal, stratified columnar, transitional, and pseudostratified columnar.
3. Define a gland and distinguish between exocrine and endocrine glands.

C. Connective tissue

4. Identify the distinguishing characteristics of connective tissue.
5. Discuss the intercellular substance, fibers, and cells that constitute connective tissue.

6. List the structure, function, and location of loose (areolar) connective tissue, adipose tissue, dense, elastic, and reticular connective tissues; cartilage; osseous (bone) tissue; and vascular (blood) tissue.

D. Membranes

7. Define an epithelial membrane and list the locations and functions of mucous, serous, cutaneous, and synovial membranes.

E. Muscle tissue

8. Contrast the three types of muscle tissue with regard to structure, location, and modes of control.

F. Nervous tissue

9. Describe the structural features and functions of nervous tissue.

G. Tissue repair

10. Describe the conditions necessary for tissue repair.

Now study the following parts of words that may help you better understand terminology in this chapter.

Wordbyte	Meaning	Example	Wordbyte	Meaning	Example
acin-	grape	*acin*ar gland	macro-	large	*macro*phage
adip(o)-	fat	*adip*ose	multi-	many	*multi*cellular
areol-	small space	*areol*ar	pseudo-	false	*pseudo*stratified
crine-	to secrete	endo*crine*	squam-	thin plate, scale	*squam*ous
endo-	within	*endo*crine			
exo-	outside	*exo*crine	strat-	layer	*strat*ified
inter-	between	*inter*cellular	uni-	one	*uni*cellular

CHECKPOINTS

A. Types of tissue   (page 92)

A1. Define tissue.

Define histology.

A2. Complete the table about the four main classes of tissues.

Tissue	General Functions
a. Epithelium	
b. Connective tissue	
c.	Movement
d.	Initiates and transmits nerve impulses that coordinate body activities

B. Epithelial tissue   (pages 92–101)

B1. Name the two subtypes of epithelial tissue.

B2. List five examples of locations of epithelium.

■ **B3.** Describe the structure of epithelium in this exercise.
   a. Epithelium consists mostly of (*closely packed cells? intercellular material with few cells?*).
   b. Epithelium is penetrated by (*many? no?*) blood vessels.

■ **B4.** Complete these sentences describing *basement membrane.*
   a. The structure that attaches epithelium to underlying connective tissue is called the

   _____. This membrane consists of (*cells? extracellular materials?*).
   b. The portion of this membrane adjacent to epithelium is secreted by (*epithelium? connective tissue?*). It consists of the protein named (*collagen? reticulin?*) along with glycoproteins, and is known as the (*reticular lamina? basal lamina?*).
   c. Many basement membranes consist of a second, underlying layer also. This is secreted by (*epithelium? connective tissue?*), and is called the

   _____ lamina.

B5. Name three adhesion proteins and describe several functions of these proteins.

■ **B6.** Which tissues consist of cells that wear out constantly and are replaced by mitosis throughout life?

■ **B7.** *A clinical challenge.* The fact that neoplastic epithelial cells lack normal adhesion proteins and are sloughed off from tissues such as the cervix of uterus or lining of airways serves as the basis for the diagnostic test known as

_____.

■ **B8.** Study diagrams of epithelial tissue types (A–F) in Figure LG 4-1 and then write the following information on lines provided on the figure.
   a. The name of each tissue
   b. One or more locations of each type of tissue
   c. One or more functions of each type of tissue

Now color the following structures on each diagram:
O Nucleus   O Intercellular material
O Cytoplasm   O Basement membrane

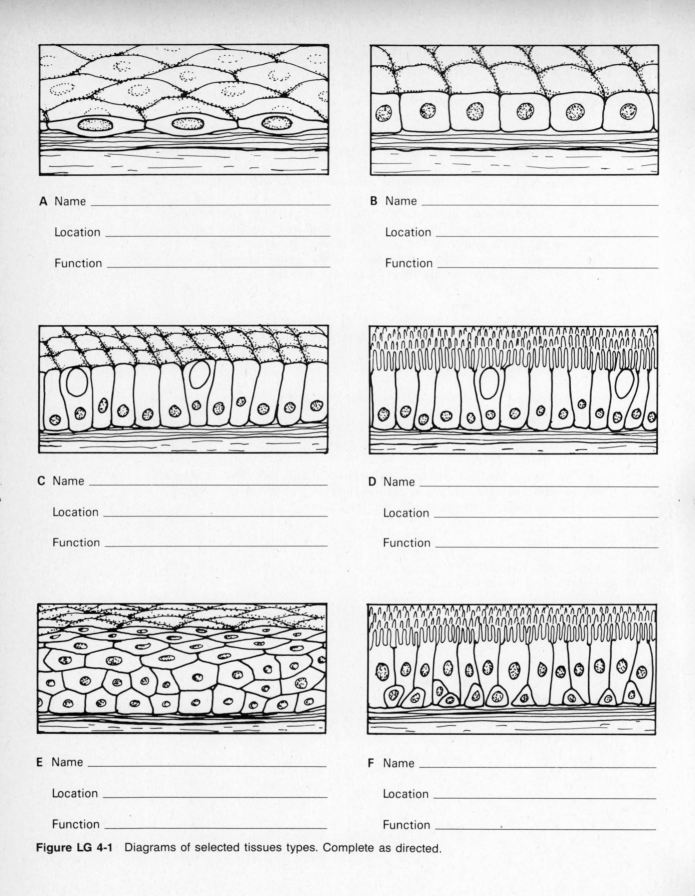

A Name _____

    Location _____

    Function _____

B Name _____

    Location _____

    Function _____

C Name _____

    Location _____

    Function _____

D Name _____

    Location _____

    Function _____

E Name _____

    Location _____

    Function _____

F Name _____

    Location _____

    Function _____

Figure LG 4-1 Diagrams of selected tissues types. Complete as directed.

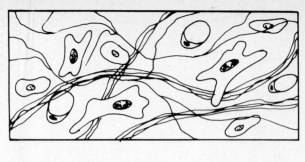

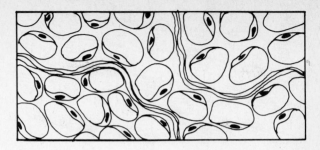

G Name_____

Location _____

Function _____

H Name _____

Location _____

Function _____

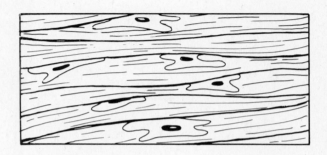

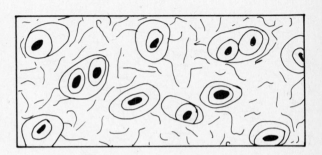

I Name _____

Location _____

Function _____

J Name _____

Location _____

Function _____

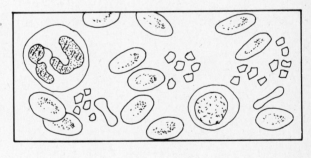

K Name _____

Location _____

Function _____

L Name _____

Location _____

Function _____

Figure LG 4-1   Continued.

M Name _____

Location _____

Function _____

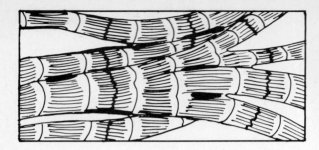

N Name _____

Location _____

Function _____

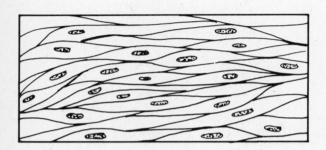

O Name _____

Location _____

Function _____

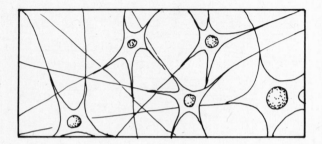

P Name _____

Location _____

Function _____

Figure LG 4-1 Continued.

B9. Check your understanding of the most common types of epithelium listed in the box by writing the name of the type after the phrase which describes it.

Pseudostratified columnar	Simple squamous
Simple columnar, ciliated	Stratified squamous
Simple columnar, unciliated	Stratified transitional
Simple cuboidal	

a. Lines the inner surface of the stomach and

intestine: _____

b. Lines urinary tract, as in bladder, permitting

distension: _____

c. Lines mouth; present on outer surface of

skin: _____

d. Single layer of cube-shaped cells; found in kidney tubules and ducts of some glands:

_____

e. Lines air sacs of lungs where thin cells are required for diffusion of gases into blood:

_____

f. Not a true stratified; all cells on basement membrane, but some do not reach surface of

tissue: _____

g. Endothelium and mesothelium contain

_____

B10. Write a sentence describing each of these modifications of the epithelial lining of the intestine.

a. Microvilli

b. Goblet cells

c. Cilia

B11. Keratin is a (*carbohydrate? protein?*). What is its function?

List three locations of nonkeratinized stratified squamous epithelium.

B12. What are *glands?* Why are glands studied in this section on epithelium?

■ **B13.** Write EXO before descriptions of *exocrine* glands, and ENDO before descriptions of *endocrine* glands. (Endocrine glands will be studied further in Chapter 18.)

_____ a. Their products are secreted into ducts that lead either directly or indirectly to the outside of the body.

_____ b. Their products are secreted into the blood and so stay within the body; they are ductless glands.

_____ c. Examples are glands that secrete sweat, oil, mucus, and digestive enzymes.

_____ d. Examples are glands that secrete hormones.

■ **B14.** Write an example of each of the following. The first one is done for you.

a. Simple coiled tubular gland: _____ **sudoriferous (sweat) gland** _____

b. Simple branched tubular gland: _____

c. Simple branched acinar gland: _____

d. Compound acinar gland: _____

e. Unicellular gland: _____

B15. Complete the table contrasting classes of glands.

Type of Gland	How Secretion is Released	Example
a. Apocrine		
b.	Cell accumulates secretory product in cytoplasm; cell dies; cell and its contents are discharged as secretion.	
c.		Salivary glands

C. Connective tissue   (pages 102–111)

■ **C1.** Write *connective tissue* or *epithelial tissue* next to correct descriptions of these tissues.

a. Consists of many cells with little intercellular substance (matrix):

_____

b. Penetrated by blood vessels (vascular): _____

c. Does not cover body surfaces or line passageways and cavities, but is more

internally located; binds, supports, protects: _____

■ **C2.** List four categories of intercellular materials. (Note that *inter*cellular means "between cells" or "extracellular.") What is the source of these materials?

■ **C3.** The embryonic tissue from which all other connective tissues arise is called

_____. (*Some? None?*) of this type of tissue is found in adults. Another kind of embryonic connective tissue gives support to the umbilical cord. This

tissue is known as _____.

■ **C4.** After you study each adult connective tissue type in the text, refer to diagrams of connective tissues on Figure LG 4-1, G–L. Write the following information on lines provided on the figure.
   a. The name of each tissue
   b. One or more locations of the tissue
   c. One or more functions of the tissue

Now color the following structures on each diagram, using the same colors as you did for diagrams A–F:
○ Nucleus   ○ Intercellular material
○ Cytoplasm   ○ Fat globule (a cellular inclusion), Diagram H

From this activity, you can conclude that connective tissues appear to consist mostly of (*cells? intercellular material [matrix]?*).

■ **C5.** Which diagrams in Fig. LG 4-1 show *connective tissue proper?* _____

C6. List three types of fibers in connective tissue.

Now label fibers in Figure LG 4-1, diagrams G, I, and J. (Note that fibers *are* present in bone, but not visible in diagram K.)

■ **C7.** Identify characteristics of each connective tissue cell type by writing the name of the correct cell type after its description. (Note that several of these cell types are shown in diagrams G, H, and I of Figure LG 4-1.)

Adipocyte	Mast cell
Fibroblast	Melanocyte
Macrophage	Plasma cell

a. Derived from lymphocyte, gives rise to antibodies; helpful in defense

   _____

b. Phagocytic cell that engulfs bacteria and cleans up debris; important during infection; formed from monocyte:

   _____

c. Fat cell: _____

d. Believed to form collagenous and elastic fibers in injured tissue:

   _____

e. Abundant along walls of blood vessels; believed to produce heparin, an anticogulant, as well as histamine, which dilates blood

   vessels: _____

f. Pigment cell: _____

■ **C8.** Which two types of tissue form subcutaneous tissue (superficial fascia)?

C9. Explain what accounts for the "signet ring" (like a class ring with a stone) appearance of adipocytes, as in Figure LG 4-1, H.

■ **C10.** Explain how fat can be helpful.

■ **C11.** Match the common types of dense connective tissue with the decriptions given.

A. Aponeurosis   P. Periosteum
D. Deep fascia   T. Tendon
L. Ligament

_____ a. Connects muscles to bones

_____ b. Holds bones together at joints

_____ c. Flat band or sheet of tissue connecting muscles to each other or to bones

_____ d. Sheet of connective tissue wrapped
around muscle bundles, holding them
in place

_____ e. Covering over bone

■ **C12.** In which of the types of dense connective tissues in Checkpoint C11 are fibers

arranged in an orderly, parallel, _regular arrangement?_ _____

C13. List three locations of each of the following types of connective tissue.
a. Elastic

b. Reticular

■ **C14.** In general, cartilage can endure (_more? less?_) stress than the connective tissues
you have studied so far.

■ **C15.** Match the types of cartilage with the descriptions given.

E. Elastic   F. Fibrous   H. Hyaline

_____ a. Found where strength and rigidity are
needed, as in discs between vertebrae
and in symphysis pubis

_____ b. White, glossy cartilage covering ends
of bones (articular), covering ends of
ribs (costal), and giving strength to
nose, larynx, and trachea

_____ c. Provides strength and flexibility, as in
external part of ear.

■ **C16.** Complete this exercise about bone and blood tissue.

a. Bone tissue is also known as _____ tissue. Compact bone consists of concentric rings, or (*lamellae? canaliculi?*) with bone cells, called (*chondrocytes? osteocytes?*), located in tiny spaces called _____.
Nutrients in blood reach osteocytes by vessels located in the

_____ canal and then via minute, radiating canals called

_____ that extend out to lacunae.

b. Blood, or _____ tissue, consists of a fluid called

_____ containing three types of formed elements. Red blood

cells, or _____, transport oxygen and carbon dioxide; white

cells, also known as _____, provide defense for the body;

thrombocytes assist in the function of blood _____.

C17. Now label the following cell types found in diagrams J, K, or I of Figure LG 4-1: *chondrocytes, osteocytes, erythrocytes, leucocytes, thrombocytes.*

D. Membranes (pages 111–112)

D1. Complete the table about the four types of membranes in the body.

Type of Membrane	Location	Example of Specific Location	Function(s)
a.	Lines body cavities leading to exterior		
b. Serous			Allows organs to glide easily over each other
c.		Lines knee and hip joints	
d.	Covers body surfaces	Skin	

■ **D2.** Check your understanding of membrane types by doing this exercise.

a. The serous membrane covering the heart is known as the

_____, whereas that covering the lungs is called the

_____. The serous membrane over abdominal organs is the

_____.

b. The portion of serous membranes that covers organs (viscera) is called the

_____ layer; that portion lining the cavity is named the

_____ layer. See Figure 20-2, page 574 in your textbook.

c. The _____ layer of a mucous membrane binds epithelium to underlying muscle and serves as a route for oxygen and nutrients to the avascular epithelial layer of the membrane.

d. Another name for skin is the _____ membrane.

e. A _____ membrane secretes a lubricating fluid known as sy-novial fluid. Such a membrane (*does? does not?*) contain epithelium, so (*is? is not?*) classified as an epithelial membrane.

■ **D3.** *A clinical challenge.* Serous membranes are normally relatively free of microor-ganisms, whereas mucous and cutaneous membranes have many microorganisms on them. Based on this information, why may antibiotics be administered to patients undergoing stomach or intestinal surgery?

E. Muscle tissue (pages 112–114)

■ **E1.** These tissues are more (*specific? generalized?*) in their structure and functions than epithelium or connective tissue. (They will be studied in greater detail in Chap-ters 10 and 12.)

■ **E2.** After you study each muscle tissue type in the text, refer to diagrams of muscle tissues in Figure LG 4-1, M–O). Write the following information on lines provided on the figure.

a. The name of each muscle tissue
b. One or more locations of the tissue
c. One or more functions of the tissue

Now color the following structures on each diagram, using the same colors as you did for diagrams A–L.

O Nucleus
O Cytoplasm
O Intercellular material

From this activity, you can conclude that muscle tissues appear to consist mostly of (*cells? intercellular material [matrix]?*).

■ **E3.** Check your understanding of the three muscle types listed in the box by selecting types that best fit descriptions below.

| C. Cardiac   Sm. Smooth   Sk. Skeletal |

_____ a. Tissue forming most of the wall of the heart.

_____ b. Attached to bones

_____ c. Spindle-shaped cells with ends tapering to points

_____ d. Contain intercalated discs

_____ e. Found in walls of intestine, urinary bladder, and blood vessels

_____ f. Cells are multinucleate

F. Nervous Tissue (page 114)

■ **F1.** Refer to diagram P in Figure LG 4-1. Write the following information on lines provided on the figure.

a. The name of the tissue
b. One or more locations of the tissue
c. One or more functions of the tissue

■ **F2.** (*Axons? Dendrites?*) carry nerve impulses towards the cell body of a neuron,

whereas _____ transmit nerve impulses away from the cell body.

G. Tissue repair (pages 114–115)

■ **G1.** Choose the correct term to complete each sentence. Write P for *parenchymal* or S for *stromal*.

_____ a. The part of an organ that consists of functioning cells, such as secreting epithelial cells lining the intestine, is composed of ____ tissue.

_____ b. The connective tissue cells that support the functional cells of the organ are called ____ cells.

_____ c. If only the ____ cells are involved in the repair process, the repair will be close to perfect.

_____ d. If ____ cells are involved in the repair, they will lay down fibrous tissue known as a fibrosis or a scar.

G2. Define each of the following terms, and explain how these conditions occur.

Keloid scar

Adhesions

G3. What is the function of *granulation tissue* in the process of tissue repair?

G4. You have most likely experienced an injury that led to formation of a *scab*. What causes scabs to form?

■ **G5.** List three factors that enhance tissue repair.

■ **G6.** Which of these vitamins performs the following functions in the process of wound healing: A, B, C, D, E, or K?

_____ a. This vitamin is necessary for healing fractures since it enhances calcium absorption from foods in the intestine.
_____ b. This is important in repair of connective tissue and walls of blood vessels.

_____ c. Thiamine, riboflavin, and nicotinic acid are categorized as vitamins of this group. They enhance metabolic processes and therefore help to provide energy for wound repair.
_____ d. This vitamin is helpful in replacement of epithelial tissues, for example, in the lining of the respiratory tract.
_____ e. Believed to promote healing of injured tissues, this vitamin may also help to prevent scarring.
_____ f. This vitamin aids in blood clotting and so helps to prevent excessive blood loss from wounds.

ANSWERS TO SELECTED CHECKPOINTS

B3. (a) Closely packed cells. (b) No.

B4. (a) Basement membrane, extracellular material. (b) Epithelium, collagen, basal lamina. (c) Connective tissue, reticular.

B6. Epithelium of skin and gastrointestinal tract.

B7. Pap smear (Papanicolaou test).

B8. (A) Simple squamous epithelium. (B) Simple cuboidal epithelium. (C) Simple columnar epithelium (unciliated). (D) Simple columnar epithelium (ciliated). (E) Stratified squamous epithelium. (F) Pseudostratified columnar epithelium, ciliated.

B9. (a) Simple columnar, unciliated. (b) Stratified transitional. (c) Stratified squamous. (d) Simple cuboidal. (e) Simple squamous. (f) Pseudostratified columnar. (g) Simple squamous.

B13. (a) EXO. (b) ENDO. (c) EXO. (d) ENDO.

B14. (b) Gastric or uterine glands. (c) Sebaceous (oil) glands. (d) Some salivary glands. (e) Mucus-producing goblet cell.

C1. (a) Epithelial tissue. (b) Connective tissue (except cartilage). (c) Connective tissue.

C2. Fluid, semifluid, gellike, or fibrous; all derived from cells.

C3. Mesenchyme; some; mucous connective tissue (Wharton's jelly).

C4. (G) Loose (areolar) connective tissue. (H) Adipose tissue. (I) Dense connective tissue. (J) Cartilage. (K) Osseous (bone) tissue. (L) Vascular (blood) tissue. Intercellular material (matrix).

C5. G, H, I.

C7. (a) Plasma cell. (b) Macrophage. (c) Adipocyte. (d) Fibroblast. (e) Mast cell. (f) Melanocyte.

C8. Loose (areolar) connective tissue and adipose.

C10. Since fat is a poor conductor of heat, it reduces heat loss from skin. It protects organs, such as kidneys and joints, and serves as an energy source.

C11. (a) T. (b) L. (c) A. (d) D. (e) P.

C12. A, L, T. Tissue is adapted for tension in one direction.

C14. More.

C15. (a) F. (b) H. (c) E.

C16. (a) Osseous, lamellae, osteocytes, lacunae, central (Haversian), canaliculi. (b) Vascular, plasma, erythrocytes, leukocytes, clotting.

D2. (a) Pericardium, pleura, peritoneum. (b) Visceral, parietal. (c) Lamina propria (connective tissue). (d) Cutaneous. (e) Synovial, does not, is not.

D3. Microorganisms from skin or from within the stomach or intestine may be introduced into the peritoneal cavity during surgery; the result may be an inflammation of the serous membrane (peritonitis).

E1. Specific.

E2. (a) Skeletal muscle. (b) Cardiac muscle. (c) Smooth muscle. Cells [known as muscle fibers].

E3. (a) C. (b) Sk. (c) Sm. (d) C. (e) Sm. (f) Sk.

F1. Nervous tissue.

F2. Dendrites, axons *(Hint:* Remember A for Away from cell body).

G1. (a) P. (b) S. (c) P. (d) S.

G5. Nutrition, adequate blood circulation, and (younger) age.

G6. (a) D. (b) C. (c) B. (d) A. (e) E. (f) K.

MASTERY TEST: Chapter 4

Questions 1–5: Circle the letter preceding the one best answer to each question.

1. A surgeon performing abdominal surgery will pass through the skin, then loose connective tissue (subcutaneous), then muscle, to reach the ____ membrane lining the inside wall of the abdomen.

    A. Parietal pleura
    B. Parietal pericardium
    C. Parietal peritoneum
    D. Visceral pleura
    E. Visceral pericardium
    F. Visceral peritoneum

2. The type of tissue that covers body surfaces, lines body cavities, and forms glands is:

    A. Nervous   C. Connective
    B. Muscular   D. Epithelial

3. Neurons are cells that are part of:

    A. Nervous tissue   C. Connective tissue
    B. Muscle tissue   D. Epithelial tissue

4. Which statement about connective tissue is FALSE?

    A. Cells are very closely packed together.
    B. Most connective tissues have an abundant blood supply.
    C. Intercellular substance is present in large amounts.
    D. It is the most abundant tissue in the body.

5. Modified columnar cells that are unicellular glands secreting mucus are known as:

    A. Cilia
    B. Microvilli
    C. Goblet cells
    D. Branched tubular glands

Questions 6–10: Match tissue types with correct descriptions.

Adipose	Simple columnar epithelium
Loose connective tissue	Simple squamous epithelium
Cartilage	Stratified squamous epithelium
Dense connective tissue	

6. Contains lacunae and chondrocytes:

    _____

7. Forms fascia, tendons, and dermis of skin:

    _____

8. Forms thick surface layer of skin on hands and feet, providing extra protection:

    _____

9. Also called areolar connective tissue:

    _____

10. Single layer of flat, scalelike cells:

    _____

Questions 11–20: Circle T (true) or F (false). If the statement is false, change the underlined word or phrase so that the statement is correct.

T F 11. Stratified transitional epithelium lines the <u>urinary bladder.</u>

T F 12. A person who has <u>good nutrition, has good blood supply, and is over 60 years of age</u> will tend to heal rapidly.

T F 13. Hormones are classified as <u>exocrine</u> secretions.

T F 14. Elastic connective tissue, since it provides both stretch and strength, is found in <u>walls of elastic arteries, in the vocal cords, and in some ligaments.</u>

T F 15. Mast cells are formed from <u>lymphocytes, and secrete heparin and antihistamine.</u>

T F 16. Another term for intercellular is <u>intracellular.</u>

T F 17. The surface attachment between epithelium and connective tissue is called <u>basement membrane.</u>

T F 18. <u>Simple</u> squamous epithelium is most likely to line the areas of the body that are subject to wear and tear.

T F 19. Marfan syndrome is a <u>muscle tissue</u> disorder that results in persons being <u>short with short extremities.</u>

T F 20. Appositional growth of cartilage <u>starts later in life than interstitial growth and continues throughout life.</u>

Questions 21–25: fill-ins. Complete each sentence with the word that best fits.

_____ 21. ____ is the study of tissue.

_____ 22. ____ is a dense connective tissue covering over bone.

_____ 23. The kind of tissue which lines alveoli (air sacs) of lungs is ____.

_____ 24. A term that means *without blood vessels* is ____.

_____ 25. All glands are formed of the tissue type named ____.

ANSWERS TO MASTERY TEST: ■ Chapter 4

Multiple Choice

1. C   4. A
2. D   5. C
3. A

Matching

6. Cartilage
7. Dense connective tissue
8. Stratified squamous epithelium
9. Loose connective tissue
10. Simple squamous epithelium

True-False

11. T
12. F. Good nutrition, good blood supply, and is younger
13. F. Endocrine
14. T
15. F. Basophils, and secrete heparin and histamine.
16. F. Extracellular or interstitial
17. T
18. F. Stratified
19. F. Connective tissue; tall with long extremities.
20. T

Fill-ins

21. Histology
22. Periosteum
23. Simple squamous epithelium
24. Avascular
25. Epithelium

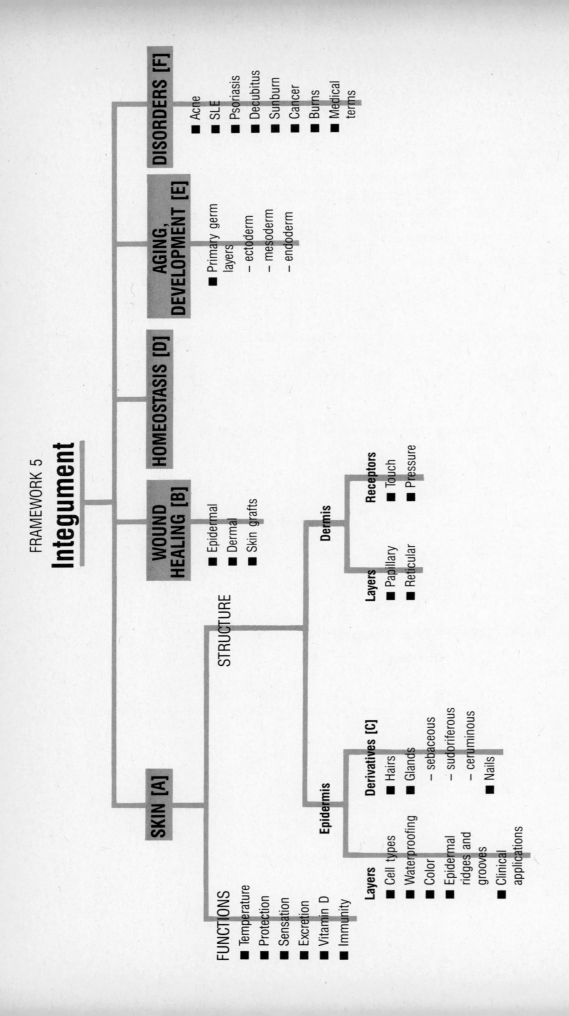

FRAMEWORK 5

Integument

SKIN [A]

WOUND HEALING [B]
- Epidermal
- Dermal
- Skin grafts

HOMEOSTASIS [D]

AGING, DEVELOPMENT [E]
- Primary germ layers
  - ectoderm
  - mesoderm
  - endoderm

DISORDERS [F]
- Acne
- SLE
- Psoriasis
- Decubitus
- Sunburn
- Cancer
- Burns
- Medical terms

FUNCTIONS
- Temperature
- Protection
- Sensation
- Excretion
- Vitamin D
- Immunity

STRUCTURE

Epidermis

Layers
- Cell types
- Waterproofing
- Color
- Epidermal ridges and grooves
- Clinical applications

Derivatives [C]
- Hairs
- Glands
  - sebaceous
  - sudoriferous
  - ceruminous
- Nails

Dermis

Layers
- Papillary
- Reticular

Receptors
- Touch
- Pressure

The Integumentary System

In your human body tour, you have now arrived at your first system, the integumentary system, which envelops the entire body. Composed of the skin, hair, nails and glands, the integument serves as both a barrier and a link with the environment. The many waterproofed layers of cells, as well as the pigmentation of skin, afford protection to the body. Nails, glands, and even hairs offer additional fortification. Sense receptors and blood vessels in skin increase awareness of conditions around the body and facilitate appropriate responses. Even so, this microscopically thin surface cover may be traumatized, invaded, or simply succumb to the passage of time. The integument then demands repair and healing for continued maintenance of homeostasis.

As you begin your study of the integumentary system, first study the Chapter 5 Framework and note key terms associated with each section.

TOPIC OUTLINE AND OBJECTIVES

A. Skin

1. Describe the structure and functions of the skin.
2. Explain the basis for skin color.

B. Skin wound healing

3. Outline the steps involved in epidermal wound healing and deep wound healing.

C. Epidermal derivatives

4. Compare the structure, distribution, and functions of hair, sebaceous (oil), sudoriferous (sweat), and ceruminous glands.

D. Homeostasis

5. Explain the role of the skin in helping to maintain the homeostasis of normal body temperature.

E. Aging and development

6. Describe the effects of aging on the integumentary system.
7. Describe the development of the epidermis, its derivatives, and the dermis.

F. Disorders, medical terminology

8. Describe the causes and effects for the following skin disorders: acne, systemic lupus erythematosus (SLE), psoriasis, decubitus ulcers, sunburn, and skin cancer.
9. Define a burn, classify burns into first, second, and third degree, and describe how to estimate the extent of a burn.
10. Define medical terminology associated with the integumentary system.

WORDBYTES

Now study the following parts of words that may help you better understand the terminology in this chapter.

Wordbyte	Meaning	Example	Wordbyte	Meaning	Example
cera-	wax	*ceru*men	melano-	blackness	*melano*cyte
cut-	skin	*cut*aneous	seb(um)-	tallow, fat	*seb*aceous
derm-	skin	*derm*atologist	sub-	under	*sub*cutaneous
epi-	over	*epi*dermis	sudor-	sweat	*sudor*iferous
-fer	carry	sudori*fer*ous			

CHECKPOINTS

A. Skin (pages 120–124)

■ **A1.** Name the structures included in the integumentary system.

A2. Skin may be one of the most underestimated organs in the body. What functions does your skin perform while it is "just lying there" covering your body? List six functions on the lines provided.

_____   _____

_____   _____

_____   _____

■ **A3.** Answer these questions about the two portions of skin. Label them (bracketed regions) on Figure LG 5-1.

a. The outer layer is named the _____. It is composed of (*connective tissue? epithelium?*).

b. The inner portion of skin, called the _____, is made of (*connective tissue? epithelium?*). The dermis is (*thicker? thinner?*) than the epidermis.

■ **A4.** Most sensory receptors, nerves, blood vessels, and glands are embedded in the (*epidermis? dermis?*). Fill in all label lines on the right side of Figure LG 5-1. Then color those structures and their related color code ovals.

■ **A5.** The tissue underlying skin is called *subcutaneous,* meaning

_____. This layer is also called _____.

It consists of two types of tissue, _____ and

_____. What functions does subcutaneous tissue serve?

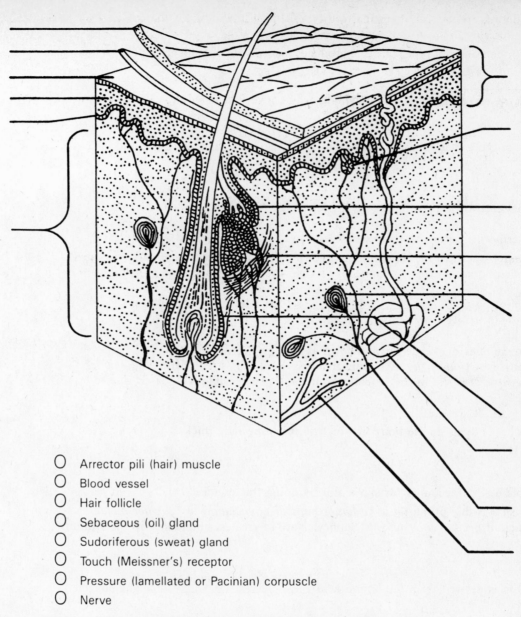

○ Arrector pili (hair) muscle
○ Blood vessel
○ Hair follicle
○ Sebaceous (oil) gland
○ Sudoriferous (sweat) gland
○ Touch (Meissner's) receptor
○ Pressure (lamellated or Pacinian) corpuscle
○ Nerve

Figure LG 5-1 Structure of the skin. Label as directed in Checkpoints A3, A4, and A7.

■ **A6.** Epidermis contains four distinct cell types. Fill in the name of the cell type that fits each description.

  a. Most numerous cell type, this cell produces keratin which helps to waterproof skin:

  _____.

  b. This type of cell produces the pigments which give skin its color:

  _____.

  c. Two cell types that function in immunity: _____ and

  _____.

■ **A7.** Label each of the five layers on the left side of Figure LG 5-1.

■ **A8.** Describe the dermis in this exercise.

  a. The outer one-fifth of the dermis is known as the (*papillary? reticular?*) region. It consists of (*loose? dense?*) connective tissue. Present in finger-like projections

  known as dermal _____ are Meissner's corpuscles, sense receptors sensitive to (*pressure? touch?*).

  b. The remainder of the dermis is known as the _____ region, composed of (*loose? dense?*) connective tissue. Skin is strengthened by (*elastic? collagenous?*) fibers in the reticular layer. Skin is extensible and elastic due to

  _____ fibers.

  c. Name three areas of the body in which skin is especially thick.

■ **A9.** Explain what accounts for skin color by doing this exercise.

  a. Dark skin is due primarily to (*a larger number of melanocytes? greater melanin production per melanocyte?*). Melanocytes are in greatest abundance in the stratum

  _____.

  b. Melanin is derived from the amino acid _____. The enzyme

  which converts tyrosine to melanin is _____. This enzyme

  is activated by _____ light. The hormone

  _____ plays a role in color production in skin of mammals.

  c. The yellowish color of skin of persons of Asian origin is due to the combination

  of the pigments _____ and _____.

  d. What accounts for the pink color of Caucasian skin, especially during blushing and as a cooling mechanism during exercise?

A10. *Clinical application:* explain how the following procedures may help to improve appearance of skin.

a. Collagen implant

b. Chemical exfoliation (skin peel)

B. Skin wound healing (pages 124–126)

■ **B1.** Describe the process of epidermal wound healing in this exercise.

a. State two examples of epidermal wounds.

b. Usually the deepest part of the wound is the (*central? peripheral?*) region.

c. In the process of repair, epidermal cells of the stratum (*corneum? basale?*) break contact from the basement membrane. These are cells at the (*center? periphery?*) of the wound.

d. These basal cells migrate toward the center of the wound, stopping when they meet other similar advancing cells. This cessation of migration is an example of the

phenomenon known as _____. Cancer cells (*do? do not?*) exhibit this characteristic.

e. Both the migrated cells and the remaining epithelial cells at the periphery undergo

_____ to fill in the epithelium up to a normal (or close to normal) level.

B2. The process of deep wound healing involves four phases. List the three or four major events that occur in each phase.

a. Inflammatory

b. Migratory

c. Proliferative

d. Maturation

■ **B3.** *For extra review* of deep wound healing, match the phases in the box with descriptions below.

I. Inflammation	Mig. Migration
Mat. Maturation	P. Proliferation

_____ a. Blood clot temporarily unites edges of wound; blood vessels dilate so neutrophils enter to clean up area.

_____ b. Clot forms a scab; epithelial cells migrate into scab; fibroblasts also migrate to start scar tissue; pink granulation tissue contains delicate new blood vessels.

_____ c. Epithelium and blood vessels grow; fibroblasts lay down many fibers.

_____ d. Scab sloughs off; epidermis grows to normal thickness; collagenous fibers give added strength to healing tissue; blood vessels are more normal.

C. Epidermal derivatives (pages 126–129)

C1. What is the main function of hair? _____

■ **C2.** Complete this exercise about the structure of a hair and its follicle.

a. Arrange the parts of a hair from superficial to deep: ___ ___ ___

A. Shaft   B. Bulb   C. Root

b. Arrange the layers of a hair from outermost to innermost: ___ ___ ___

A. Medulla   B. Cuticle   C. Cortex

c. A hair is composed of (*cells? no cells, but only secretions of cells?*). What accounts for the fact that hairs are waterproofed?

d. Surrounding the root of a hair is the hair _____. The follicle consists of two parts. The (*external? internal?*) root sheath is an extension of deeper layers of the (*epidermis? dermis?*). The internal root sheath consists of cells

derived from the _____.

e. The _____ is the part of a hair follicle where cells undergo mitosis permitting growth of a new hair. What is the function of the papilla of the hair?

f. Which treatment destroys the bulb of a hair so that the hair cannot regrow? (*Use of depilatory? Electrolysis?*)

C3. Describe the relationship between the terms in each pair of terms.

a. Arrector pili muscles/"goose bumps"

b. Sebaceous glands/"blackheads"

C4. In which areas of the body are sebaceous glands most commonly found?

List three functions of these glands.

C5. Describe the composition of perspiration (sweat) and state its functions.

■ **C6.** Check your understanding of skin glands by stating whether the following descriptions refer to *sebaceous, sudoriferous,* or *ceruminous* glands.

a. Sweat glands: _____
b. Simple branched acinar glands leading directly to hair follicle; secrete sebum which

keeps hair and skin from drying out: _____.

c. Line the outer ear canal; secrete ear wax: _____.

C7. Look at one of your own nails and Figure 5-6, page 129 in your text. Identify these parts of your nail: *free edge, nail body, lunula,* and *eponychium* (*cuticle*).

C8. Why does the nail body appear pink, yet the lunula and free edge appear white?

D. Homeostasis: temperature control (pages 129–130)

D1. Humans are homeotherms. What is the meaning of the term *homeotherm?*

D2. Describe how human body temperature is maintained close to 37°C (98.6°F) with the help of sweat glands. Explain why this temperature control mechanism can be called a *negative feedback system.*

E. Effects of aging (page 130)

■ **E1.** Complete the table relating observable changes in aging of the integument to their causes.

Changes	Causes
a. Wrinkles; skin springs back less when gently pinched.	
b.	Macrophages become less efficient.
c.	Loss of subcutaneous fat
d. Dry, easily broken skin	
e.	Decrease in number and size of melanocytes

E2. How is aging of skin related to sunlight?

F. Developmental anatomy (page 130)

F1. Name the three primary germ layers.

■ **F2.** Describe the development of the integument by completing this learning activity.

a. The epidermis is derived from the (*ecto-? meso-? endo-?*) derm. All of its layers are formed by the (*second? fourth?*) month of nine-month human gestation period.

b. The dermis arises from the _____ -derm.

c. Hair, nails, and glands all develop from _____ -derm but grow deeper into the dermis region.

d. At what point do nails begin to develop? _____ month. The nails reach the end of the digits during the (*sixth? ninth?*) month.

e. When does lanugo form? _____ month. Lanugo (*is? is not?*) usually present on a full-term newborn.

f. Which glands develop from the sides of hair follicles? _____

G. Disorders, medical terminology (pages 131–134)

G1. Write a description of these disorders. Include causes, parts of the body most often affected, and groups of people (by age or other factors) most often afflicted.

a. Acne

b. Decubitus ulcers

c. Skin cancer

■ **G2.** Match the name of the disorder with the description given.

A. Acne	L. Lupus (SLE)
D. Decubitus ulcers	M. Melanoma
I. Impetigo	P. Psoriasis

_____ a. Bedsores

_____ b. An autoimmune disease with many symptoms, including "butterfly rash"

_____ c. Staphylococcal or streptococcal infection which may become epidemic in nurseries

_____ d. Inflammation of sebaceous glands especially in chin area, occurs under hormonal influence

_____ e. One kind of skin cancer

_____ f. Chronic disease of 6 to 8 million people in the United States, characterized by reddish plaques or papules, most severe among ages 10 to 50

G3. What reasons would you give if you were advising a person against constant overexposure to sun or the use of suntanning salons?

G4. List four risk factors for skin cancer.

G5. Contrast systemic effects with local effects of burns.

■ **G6.** Identify characteristics of the three different classes of burns by completing this exercise.

a. In a first-degree burn, only the superficial layers of the (*dermis? epidermis?*) are

involved. The tissue appears _____ in color. Blisters (*do? do not?*) form. Give one example of a first-degree burn.

b. Which parts of the skin are injured in a second-degree burn?

   Blisters usually (*do? do not?*) form. Epidermal derivatives, such as hair follicles
   and glands, (*are? are not?*) injured. Healing usually occurs in about three to four
   (*days? weeks?*). For most second-degree burns, grafting (*is? is not?*) required.
c. Third-degree burns are called (*partial? full?*)-thickness burns. Such skin appears
   (*red and blistered? white, brown, or black and dry?*). Such burned areas are usually
   (*painful? not painful?*) since nerve endings are destroyed. Grafting (*is? is not?*)
   required, and scarring (*does? does not?*) result from third-degree burns.

■ **G7.** *A clinical challenge.* Answer these questions about the Lund–Browder method.
a. What does the Lund–Browder method estimate? (*Depth of burn? Amount of
   body surface area burned?*) This method is based upon differences in body (*size?
   proportions?*) of age groups.
b. If an adult and a one-year-old each experience burns over the entire anterior
   surface of both legs, who is more burned? (*Adult? One-year-old?*) If both are
   burned over the entire anterior surface of the head, who is more burned? (*Adult?
   One-year-old?*)

G8. Contrast the following pairs of terms.
a. Subcutaneous/intradermal

b. Hypodermic/topical

c. Eczema/pruritis

■ **G9.** Match the terms in the box with the correct definition.

C. Callus	N. Nevus
E. Erythema	W. Wart
F. Furuncle	

_____ a. Excessive redness of skin due to en-
        larged capillaries in skin
_____ b. Common, contagious, noncancerous
        growth of epithelium due to virus
_____ c. Boil due to infection of hair follicle

_____ d. Area of hardened, thickened skin, as
        in palms and soles, due to pressure
        or friction
_____ e. Mole

ANSWERS TO SELECTED CHECKPOINTS

A1. Skin, its derivatives (hair, nails, and glands), and sense receptors.

A3.

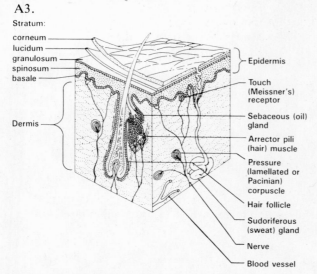

Figure LG 5-1A Structure of the skin.

(a) Epidermis, epithelium. (b) Dermis, connective tissue, thicker.

A4. Dermis. See Figure LG 5-1A above.

A5. Under the skin, superficial fascia or hypodermis, areolar and adipose, anchors skin to underlying tissues and organs.

A6. (a) Keratinocyte. (b) Melanocyte. (c) Langerhan's cell and Granstein cell or nonpigmented granular dendrocytes.

A7. See Figure LG 5-1A above.

A8. (a) Papillary, loose, papillae, touch. (b) Reticular, dense, collagenous, elastic. (c) Palms, soles, and dorsal surface of the body.

A9. (a) Greater melanin production per melanocyte, basale. (b) Tyrosine, tyrosinase, ultraviolet (UV), melanocyte-stimulating hormone (MSH). (c) Carotene and melanin. (d) Vasodilatation of blood vessels.

B1. (a) Skinned knee, first- or second-degree burn. (b) Central. (c) Basale, periphery. (d) Contact inhibition, do not. (e) Mitosis.

B3. (a) I. (b) Mig. (c) P. (d) Mat.

C2. (a) A C B. (b) B C A. (c) Cells, keratin (especially in cuticle). (d) Follicle, external, epidermis, matrix. (e) Matrix (of bulb of hair follicle), contains blood vessels that nourish the hair. (f) Electrolysis.

C6. (a) Sudoriferous. (b) Sebaceous. (c) Ceruminous.

E1.

Changes	Causes
a. Wrinkles; skin springs back less when gently pinched.	Elastic fibers thicken into clumps and fray
b. Increased susceptibility to skin infections and skin breakdown	Macrophages become less efficient.
c. Loss of body heat; increased likelihood of skin breakdown, as in decubitus ulcers	Loss of subcutaneous fat
d. Dry, easily broken skin	Decreased secretion of sebum by sebaceous gland; decreased body fluids
e. Gray or white hair	Decrease in number and size of melanocytes

F2. (a) Ecto-, fourth. (b) Meso. (c) Ecto. (d) Third, ninth. (e) Fifth or sixth, is not. (f) Sebaceous.

G2. (a) D. (b) L. (c) I. (d) A. (e) M. (f) P.

G6. (a) Epidermis, redder, do not, typical sunburn. (b) All of epidermis and upper regions of dermis, do, are not, weeks, is not. (c) Full; white, brown or black and dry; not painful; is; does.

G7. (a) Amount of body surface burned, proportions. (b) Adult, One-year-old.

G9. (a) E. (b) W. (c) F. (d) C. (e) N.

MASTERY TEST: Chapter 5

Questions 1–5: Circle the letter preceding the one best answer to each question.

1. "Goose bumps" occur as a result of:

   A. Contraction of arrector pili muscles
   B. Secretion of sebum
   C. Contraction of elastic fibers in the bulb of the hair follicle
   D. Contraction of papillae

2. Select the one FALSE statement about the stratum basale.

   A. It is the one layer of cells that can undergo cell division.
   B. It consists of a single layer of squamous epithelial cells.
   C. It is part of the stratum germinativum.
   D. It is the deepest layer of the epidermis.

3. Select the one FALSE statement.
   A. Epidermis is composed of epithelium.
   B. Dermis is composed of connective tissue.
   C. Pressure-sensitive Pacinian corpuscles are normally more superficial in location than Meissner's touch receptors.
   D. The amino acid tyrosine is necessary for production of the skin pigment melanin.

4. When older hairs are shed, the cells of the ___ produce new hairs by cell division.
   A. Arrector pili
   B. Connective tissue papilla
   C. Internal sheath
   D. External sheath
   E. Matrix in the base of the bulb

5. Which is derived from mesoderm?
   A. Epidermis   B. Dermis

Questions 6–10: Arrange the answers in correct sequence.

_____ _____ _____ 6. From most serious to least serious type of burn:
   A. First-degree
   B. Second-degree
   C. Third-degree

_____ _____ _____ 7. Deep wound healing involves four phases. List in order the phases following the inflammatory phase.
   A. Migration
   B. Maturation
   C. Proliferation

_____ _____ _____ 8. From outside of hair to inside of hair:
   A. Medulla
   B. Cortex
   C. Cuticle

_____ _____ _____ 9. From most superficial to deepest:
   A. Dermis
   B. Epidermis
   C. Superficial fascia

_____ _____ _____ 10. From most superficial to deepest:
   A. Stratum lucidum
   B. Stratum corneum
   C. Stratum germinativum

Questions 11–20: Circle T (true) or F (false). If the statement is false, change the underlined word or phrase so that the statement is correct.

T F 11. Hairs are noncellular structures composed entirely of nonliving substances secreted by follicle cells.

T F 12. The color of skin is due to a pigment named keratin.

T F 13. The outermost layers of epidermis are composed of dead cells.

T F 14. Eccrine sweat glands are more numerous than apocrine sweat glands, and are especially dense on palms and soles.

T F 15. The dermis consists of two regions; the papillary region is most superficial, and the reticular region is deeper.

T F 16. Temperature regulation is a positive feedback system.

T F 17. The internal root sheath is a downward continuation of the epidermis.

T   F   18. <u>Both epidermis and dermis</u> contain blood vessels (are vascular).

T   F   19. Hair, glands, and nails are all derived from the <u>dermis</u>.

T   F   20. Pacinian corpuscles are <u>pressure</u>-sensitive nerve endings most abundant in the <u>subcutaneous tissue</u>, rather than in the <u>epidermis</u>.

Questions 21–25: fill-ins. Complete each sentence with the word or phrase that best fits.

_____ 21. Skin contains a chemical which, under the influence of ultraviolet radiation, leads to formation of vitamin ____ .

_____ 22. The cells that are sloughed off as skin cells and undergo keratinization are those of the stratum ____ .

_____ 23. Fingerprints are the result of a series of grooves called ____ .

_____ 24. The oily glandular secretion which keeps skin and hairs from drying is called ____ .

_____ 25. When the temperature of the body increases, nerve messages from brain to skin will decrease body temperature by ____ .

ANSWERS TO MASTERY TEST: ■ Chapter 5

Multiple Choice

1. A   4. E
2. B   5. B
3. C

Arrange

6. C B A   9. B A C
7. A C B   10. B A C
8. C B A

True-False

11. F. Composed of different kinds of cells
12. F. Melanin
13. T
14. T
15. T
16. F. Negative
17. F. External root sheath
18. F. Only the dermis
19. F. Epidermis
20. T

Fill-ins

21. D or D_3
22. Corneum
23. Epidermal ridges or grooves
24. Sebum
25. Stimulating sweat glands to secrete and blood vessels to dilate.

UNIT II

Principles of Support and Movement

FRAMEWORK 6

Skeletal Tissue

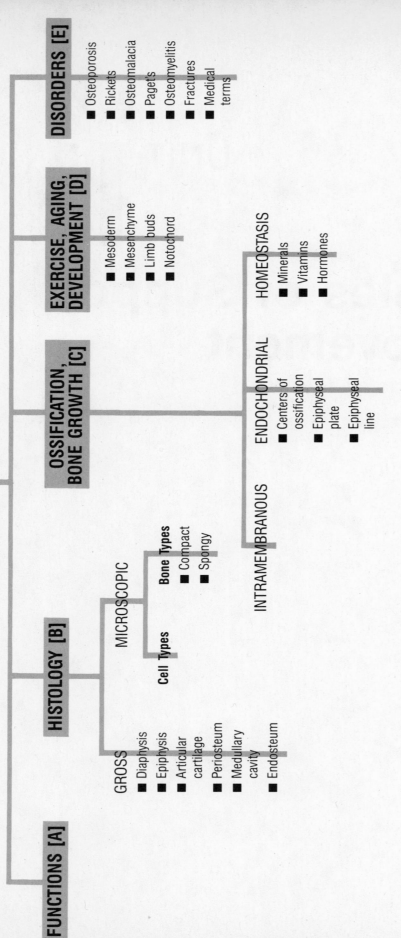

FUNCTIONS [A]

HISTOLOGY [B]

GROSS
- Diaphysis
- Epiphysis
- Articular cartilage
- Periosteum
- Medullary cavity
- Endosteum

MICROSCOPIC

Cell Types

Bone Types
- Compact
- Spongy

OSSIFICATION, BONE GROWTH [C]

INTRAMEMBRANOUS

ENDOCHONDRIAL
- Centers of ossification
- Epiphyseal plate
- Epiphyseal line

HOMEOSTASIS
- Minerals
- Vitamins
- Hormones

EXERCISE, AGING, DEVELOPMENT [D]
- Mesoderm
- Mesenchyme
- Limb buds
- Notochord

DISORDERS [E]
- Osteoporosis
- Rickets
- Osteomalacia
- Paget's
- Osteomyelitis
- Fractures
- Medical terms

Skeletal Tissue

The skeletal system provides the framework for the entire body, affording strength, support, and firm anchorage for the muscles that move the body. In this intial chapter in Unit II, you will explore tissue that forms the skeleton. Bones provide rigid levers covered with connective tissue designed for joint formation and muscle attachment. Microscopically, bones present a variety of cell types intricately arranged in an osseous sea of calcified intercellular material. Bone growth or ossification occurs throughout life. Essential nutrients, hormones, and exercise regulate the growth and maintenance of the skeleton. Fractures and other disorders may result from a lack of such normal regulation.

As you begin your study of the skeletal system, first study the Chapter 6 Framework and read key terms associated with each section.

TOPIC OUTLINE AND OBJECTIVES

A. Functions of skeletal tissue

1. Discuss the components and functions of the skeletal system.

B. Histology of skeletal tissue

2. Describe the histological features of compact and spongy bone tissue.

C. Bone formation, bone growth and homeostasis

3. Contrast the steps involved in intramembranous and endochondral ossification.
4. Describe the processes of bone construction and destruction involved in the homeostasis of bone remodeling.
5. Describe the conditions necessary for normal bone growth and replacement.

D. Effects of aging and exercise on the skeletal system; developmental anatomy of the skeletal system

6. Explain the effects of exercise and aging on the skeletal system.
7. Describe the development of the skeletal system.

E. Disorders, medical terminology

8. Contrast the causes and clinical symptoms associated with osteoporosis, rickets, osteomalacia, Paget's disease, and osteomyelitis.
9. Define a fracture (Fx), describe several kinds of fractures, and describe the sequence of events involved in fracture repair.
10. Define medical terminology associated with skeletal tissue.

WORDBYTES

Now study the following parts of words that may help you better understand terminology in this chapter.

Wordbyte	Meaning	Example	Wordbyte	Meaning	Example
-blast	germ, to form	osteo*blast*	os-, osteo-	bone	*os*sification, *osteo*cyte
-clast	to break	osteo*clast*			
chondr-	cartilage	peri*chondr*ium	peri-	around	*peri*osteum

CHECKPOINTS

A. Functions of skeletal tissue (page 142)

A1. List five functions of the skeletal system.

■ **A2.** What other systems of the body depend on a healthy skeletal system? Explain why in each case.

A3. Name the branch of medical specialty concerned with the bones and joints.

B. Histology of skeletal tissue (pages 142–146)

■ **B1.** The skeletal system consists of two types of tissue. Name these.

■ **B2.** In the following exercise, bone cells are described in the sequence in which they form and later destroy bone. Write the name of the correct bone type after its description.

a. Derived from mesenchyme, these cells can undergo mitosis and differentiate into osteoblasts: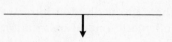

b. Form bone by secreting mineral salts and fibers: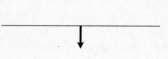

c. Maintain bone tissue; not mitotic: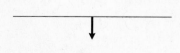

d. Resorb (degrade) bone in bone repair, remodeling, and aging: _____

■ **B3.** Describe the components of bone by doing this exercise.

a. Typical of all connective tissues, bone consists mainly of (*cells? intercellular material?*).

b. The intercellular substance of bone is unique among connective tissues. Collagenous fibers form about (*33? 67?*) percent of the weight of bone, whereas mineral salts account for about _____ percent of the weight of bone.

c. The two main salts present in bone are _____ and _____.

■ **B4.** On Figure LG 6-1, label the *diaphysis, epiphysis, epiphyseal line* and the *medullary* (*marrow*) *cavity*. Then color the parts of a long bone listed on the figure.

■ **B5.** Match the names of parts of a long bone listed in the box with the description below.

> A. Articular cartilage   M. Metaphysis
> E. Endosteum   O. Osteogenic periosteum
> F. Fibrous periosteum

_____ a. Thin layer of hyaline cartilage at end of long bone

_____ b. Region of mature bone where diaphysis joins epiphysis

_____ c. Outer layer of covering over bone into which ligaments and tendons attach

_____ d. Inner layer of covering over bone; osteoblasts here permit increase in diameter of bone

_____ e. Layer of bone cells lining the marrow cavity

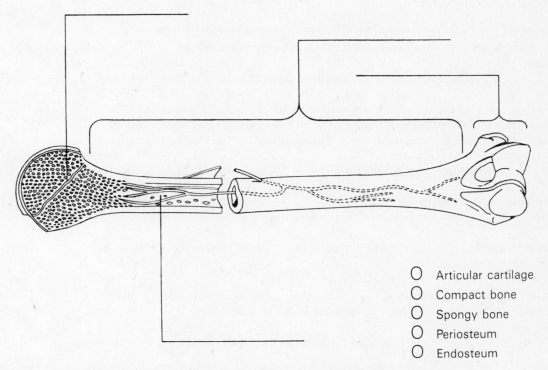

○ Articular cartilage
○ Compact bone
○ Spongy bone
○ Periosteum
○ Endosteum

Figure LG 6-1 Diagram of a long bone that has been partially sectioned lengthwise. Color and label as directed in Checkpoint B4.

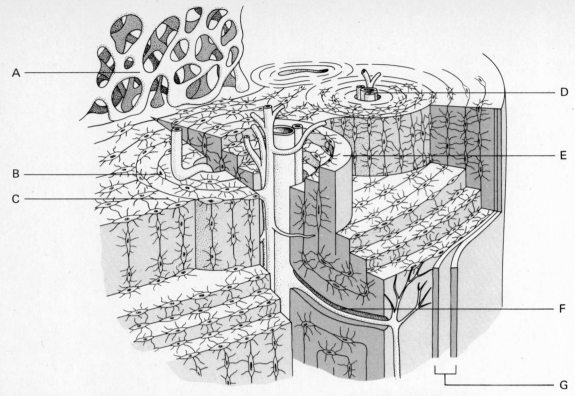

Figure LG 6-2 Osteons (Haversian systems) of compact bone. Identify lettered structures and color as directed in Checkpoint B6.

■ **B6.** Refer to Figure 6-2 and complete this exercise about bone structure.

a. Compact bone is arranged in concentric circle patterns known as

_____. Each individual concentric layer of bone is known

as a _____, labeled with letter _____ in the figure.

b. The osteon pattern of compact bone permits blood vessels and nerves to supply bone cells trapped in hard bone tissue. Blood vessels and nerves penetrate bone

from the periosteum, labeled with letter _____ in the figure. These structures

then pass through horizontal canals, labeled _____, and known as perforating (Volkmann's) canals. These vessels and nerves finally pass into microscopic

channels, labeled _____, in the center of each osteon. Color blood vessels red and blue in the figure.

c. Mature bone cells, known as _____, are located relatively far apart in bone tissue. These are present in "little lakes," or

_____, labeled _____ in the figure. Color ten lacunae green.

d. Minute canals, known as canaliculi, are labeled with letter _____. What is the function of these channels?

C. Bone formation, bone growth, and homeostasis (pages 146–151)

C1. What is mesenchyme? What does it have to do with the skeleton?

■ **C2.** The two main kinds of tissue that compose the "skeleton" of a developing

embryo or fetus are _____ and _____

_____ .

■ **C3.** Which bones of the body form by the process of intramembranous ossification?

■ **C4.** Describe the process of *intramembranous* ossification by filling in the blanks in this exercise. Use all the words in the box.

Calcium salts	Marrow	Spongy bone
Collagenous fibers	Osteoblast	
Compact bone	Ossification	

a. _____ cells cluster in the _____ center in
the fibrous membrane. These cells secrete a matrix composed of

_____ and _____ and form spongy bone.
b. Eventually compact bone forms over the spongy bone. The resulting typical flat
bone is analogous to a jelly sandwich: two firm "slices of bread" comparable to

_____ bone with a "jelly" of _____ bone

that is filled with red _____ .

■ **C5.** Arrange in correct sequence the events of *endochondral* bone formation.

___ ___ ___ ___ ___ ___ ___

A. Secondary ossification centers develop in epiphyses, forming spongy bone there.
B. Meanwhile cartilage cells in the center of the diaphysis enlarge and burst.
C. A cartilage model of future bone is laid down.
D. Cartilage cells die since they are deprived of nutrients; in this way large spaces are formed.
E. Blood vessels penetrate the perichondrium, stimulating periochondrial cells to form osteoblasts. A collar of bone forms and gradually thickens around the diaphysis.
F. Release of alkaline chemicals from these cells causes calcification of cartilage.
G. Blood vessels grow into spaces, forming the marrow cavity.

■ **C6.** The epiphyseal plate consists of four zones. Arrange the letters representing these zones in order starting from the zone closest to the epiphysis.

C. Zone of calcified matrix (dead cartilage cells are replaced by bone cells).
H. Zone of hypertrophic cartilage (cartilage cells die here).
P. Zone of proliferating cartilage (cartilage cells form here).
R. Zone of reserve cartilage (anchors epiphyseal plate to epiphysis).

Epiphysis → _____ → _____ → _____ → _____ → Diaphysis

■ **C7.** Complete this summary statement about bone growth at the epiphyseal plate.

Cartilage cells multiply on the (*epiphysis? diaphysis?*) side of the epiphyseal plate, providing temporary new tissue. But cartilage cells then die and are replaced by bone cells on the (*epiphysis? diaphysis?*) side of the epiphyseal plate.

C8. Contrast the *epiphyseal plate* with the *epiphyseal line.*

C9. What is osteogenic sarcoma? Which age group and which body parts is it most likely to affect?

C10. Defend or dispute this statement: "Once a bone, such as your thighbone, is formed, the bone tissue is never replaced unless the bone is broken."

■ **C11.** Explain the roles of lysosomes and acids in osteoclastic activity.

C12. List factors that appear to be necessary for normal bone growth:

a. Four minerals

b. Four vitamins

■ **C13.** Match the hormones listed in the box with their functions in bone growth and remodeling.

CT. Calcitonin	S. Sex hormones (testosterone,
GH. Growth hormone	estrogens)
PTH. Parathyroid hormone	

_____ a. Produced by the parathyroid gland, it increases osteoclast activity, causing bone destruction.

_____ b. Produced by the thyroid gland, it inhibits osteoclast activity.

_____ c. A pituitary hormone, it enhances bone growth in general; deficiency leads to dwarfism.

_____ d. These hormones stimulate osteoblasts to form new bone and so are responsible for growth spurt at puberty; they can also cause premature closure of epiphyseal plate (short stature) if puberty occurs early.

D. **Effects of exercise and aging on the skeletal system; developmental anatomy of the skeletal system** (pages 151–152)

■ **D1.** Circle correct answers about effects of exercise upon bones.

a. Exercise (*strengthens? weakens?*) bones. A healing bone that is not exercised is likely to become (*stronger? weaker?*) during the period that it is in a cast.

b. The pull of muscles upon bones, as well as the tension on bones as they support body weight during exercise, causes (*increased? decreased?*) production of the protein collagen.

c. The stress of exercise also stimulates (*osteoblasts? osteoclasts?*) and increases production of the hormone calcitonin which inhibits (*osteoblasts? osteoclasts?*).

■ **D2.** Complete this exercise about skeletal changes that occur in the normal aging process.

a. The amount of calcium in bones ____-creases with age. As a result, bones of the elderly are likely to be (*stronger? weaker?*) than bones of younger persons. This change occurs at a younger age in (*men? women?*).

b. Another component of bones that decreases with age is

_____. What is the significance of this change?

■ **D3.** Complete this learning activity about development of the skeleton.

a. Bones and cartilage are formed from (*ecto-, meso-, endo-*)derm which later differentiates into the embryonic connective tissue called

_____. Some of these cells become chondroblasts which eventually form (*bone? cartilage?*), whereas _____-blasts develop into bone.

b. Limb buds appear during the (*fifth? seventh? tenth?*) week of development. At this point the skeleton in these buds consists of (*bone? cartilage?*). Bone begins to form

during week _____.

c. By week (*six? seven? eight?*) the upper extremity is evident, with defined shoulder, arm, elbow, forearm, wrist, and hand. The lower extremity is also developing at a slightly (*faster? slower?*) pace.

d. The notochord develops in the region of the (*head? vertebral column? pelvis?*). Most of the notochord eventually (*forms vertebrae? disappears?*). Parts of it persist

in the _____.

E. **Disorders, medical technology** (pages 153–156)

■ **E1.** Describe osteoporosis in this exercise.

a. Osteoporotic bones exhibit ____-creased bone mass and ____-creased susceptibility to fractures. This disorder is associated with a loss of the hormone

_____. This hormone stimulates osteo-

_____ activity.

b. Circle the category of persons at higher risk for osteoporosis in each pair:

younger persons — older persons

men — women

short, thin persons — tall, large build persons

c. List three factors in daily life that will help to prevent osteoporosis. (*Hint:* Focus on diet and daily activities.)

■ **E2.** Check your understanding of bone disorders by matching the terms in the box with related descriptions.

Osteoma	Osteosarcoma
Osteomalacia	Rickets
Osteomyelitis	

a. Benign bone tumor:

_____

b. Malignant bone tumor:

_____

c. Bone infection, for example, caused by

Staphylococcus: _____

d. Caused by vitamin D deficiency which prevents normal calcium *absorption;* soft bones:

_____

_____

E3. Contrast the terms related to types of fractures in each pair.

a. Simple/compound

b. Partial/complete

c. Pott's/Colles'

E4. *A clinical challenge:* What is the major difference between these two methods of setting a fracture: *closed reduction and open reduction?*

E5. Describe the three main steps involved in repair of a fracture.

a. Fracture hematoma

b. External callus and internal callus

c. Remodeling

ANSWERS TO SELECTED CHECKPOINTS

A2. Essentially all do. For example, muscles need intact bones for movement to occur; bones are site of blood formation; bones provide protection for viscera of nervous, digestive, urinary, reproductive, cardiovascular, respiratory, and endocrine systems; broken bones can injure integument.

B1. Cartilage and bone

B2. (a) Osteoprogenitor. (b) Osteoblast. (c) Osteocyte. (d) Osteoclast.

B3. (a) Intercellular material. (b) 33, 67. (c) Calcium phosphate, calcium carbonate.

B4.

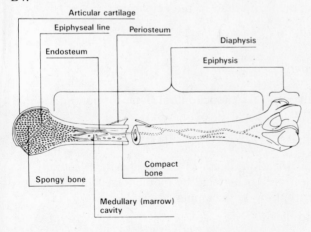

Figure LG 6-1A Diagram of a long bone that has been partially sectioned lengthwise.

B5. (a) A. (b) M. (c) F. (d) O. (e) E.

B6. (a) Osteons (Haversian systems), lamella, E. (b) G, F, D. (c) Osteocytes, lacunae, B. (d) C; contain extensions of osteocytes, and permit passage of nutrients, gases, and wastes between blood and osteocytes.

C2. Hyaline cartilage, fibrous membranes.

C3. Flat bones of the skull and clavicles (collarbones).

C4. (a) Osteoblast, ossification, collagenous fibers, calcium salts. (b) Compact, spongy, marrow.

C5. C, E, B, F, D, G, A

C6. RPHC.

C7. Epiphysis, diaphysis.

C11. Lysosomes release enzymes that may digest the protein collagen whereas acids may cause minerals of bones to dissolve. Both contribute to bone resorption.

C13. (a) PTH. (b) CT. (c) GH. (d) S.

D1. (a) Strengthens, weaker. (b) Increased. (c) Osteoblasts, osteoclasts.

D2. (a) De, weaker, women. (b) Protein, bones are more brittle and vulnerable to fracture.

D3. (a) Meso-, mesenchyme, cartilage, osteo. (b) Fifth, cartilage, six or seven. (c) Eight, slower. (d) Vertebral column, disappears, intervertebral discs.

E1. (a) De-, in-, estrogen, -blast. (b) Older; women; short, thin persons. (c) Adequate intake of calcium, vitamin D, exercise, and not smoking.

E2. (a) Osteoma. (b) Osteosarcoma. (c) Osteomyelitis. (d) Rickets and osteomalacia.

MASTERY TEST: Chapter 6

Questions 1–15: Circle T (true) or F (false). If the statement is false, change the underlined word or phrase so that the statement is correct.

T F 1. Greenstick fractures occur only in <u>adults.</u>

T F 2. Sunlight can help keep bones healthy since sunlight helps <u>the body produce vitamin D.</u>

T F 3. About two-thirds of the weight of bone is due to <u>mineral salts, and one-third is due to bone cells.</u>

T F 4. The epiphyseal plate appears <u>earlier</u> in life than the epiphyseal line.

T F 5. Osteons (haversian systems) are found in <u>compact bone, but not in spongy bone.</u>

T F 6. Another name for the epiphysis is the <u>shaft of the bone.</u>

T F 7. A compound fracture is defined as one in which the bone <u>is broken into many pieces.</u>

T F 8. Haversian canals run longitudinally (lengthwise) through bone, but perforating (Volkmann's) <u>canals run horizontally across bone.</u>

T F 9. Canaliculi are tiny canals containing <u>blood</u> which nourishes bone cells in lacunae.

T F 10. Compact bone that is of intramembranous origin differs <u>structurally</u> from compact bone developed from cartilage.

T F 11. In a long bone the primary ossification center is located in the <u>diaphysis, whereas the secondary center of ossification is in the epiphysis.</u>

T F 12. <u>Osteoblasts</u> are bone-destroying cells.

T F 13. Most bones start out in embryonic life as <u>hyaline cartilage.</u>

T F 14. The layer of compact bone is <u>thicker</u> in the diaphysis than in the epiphysis.

T F 15. <u>Lamellae</u> are small spaces containing bone cells.

Questions 16–20: Arrange the answers in correct sequence.

_____ _____ _____ 16. Steps in marrow formation during endochondral bone formation, from first to last:

    A. Cartilage cells burst, causing intercellular pH to become more alkaline.
    B. pH changes cause calcification, death of cartilage cells; eventually spaces left by dead cells are penetrated by blood vessels.
    C. Cartilage cells hypertrophy with accumulated glycogen.

_____ _____ _____ 17. From most superficial to deepest:

    A. Endosteum   C. Compact bone
    B. Periosteum

_____ _____ _____ 18. Phases in repair of a fracture, in chronological order:

    A. Fracture hematoma formation   C. Callus formation
    B. Remodeling

_____ _____ _____ 19. Phases in formation of bone in embryonic life, in chronological order:

    A. Mesenchyme cells   C. Osteoblasts
    B. Osteocytes

_____ _____ _____ 20. Portions of the epiphyseal plate, from closest to epiphysis to closest to diaphysis:

    A. Zone of reserve cartilage   C. Zone of calcified matrix
    B. Zone of proliferating cartilage

Questions 21–25: fill-ins. Write the word or phrase that best completes the statement.

_____ 21. ____ is a term which refers to the shaft of the bone.

_____ 22. ____ is a disorder primarily of older women associated with decreased estrogen level; it is characterized by weakened bones.

_____ 23. ____ is the connective tissue covering over cartilage in adults and also over embryonic cartilaginous skeleton.

_____ 24. The majority of bones formed by intramembranous ossification are located in the ____ .

_____ 25. Vitamin ____ is a vitamin that is necessary for absorption of calcium from the gastrointestinal tract, and so it is important for bone growth and maintenance.

ANSWERS TO MASTERY TEST: ■ Chapter 6

True-False

1. F. Children
2. T
3. F. Mineral salts, and one-third is due to collagenous fibers (The very few cells contribute little to bone weight.)
4. T
5. T
6. F. End of the bone (often bulbous)
7. F. Protrudes through the skin
8. T
9. F. Parts of osteocytes and fluid from blood vessels in haversian canals, but not blood itself
10. F. Only in origin
11. T
12. F. Osteoclasts
13. T
14. T
15. F. Lacunae

Arrange

16. C A B
17. B C A
18. A C B
19. A C B
20. A B C

Fill-ins

21. Diaphysis
22. Osteoporosis
23. Perichondrium
24. Skull; also the clavicles
25. D (or D_3)

FRAMEWORK 7

Axial Skeleton

INTRODUCTION

TYPES OF
BONES

SURFACE
MARKINGS

DIVISIONS
- Axial
- Appendicular

BONES OF AXIAL SKELETON [80]

SKULL [B]

Markings
- Sutures
- Fontanels
- Sinuses
- Foramina

Skull Bones (28)
- Cranial (8)
- Facial (14)
- Middle ear (6)

HYOID [C]

VERTEBRAE (26) [C]

- Divisions
  - cervical (17)
  - thoracic (12)
  - lumbar (5)
  - sacrum (1)
  - coccyx (1)
- Curves
- Typical vertebra

THORAX (25) [D]

- Sternum (1)
- Ribs (24)
  - true
  - false
  - floating

DISORDERS, TERMS [E]

- Herniated disc
- Abnormal curves
- Spina bifida
- Fractures

The Skeletal System: The Axial Skeleton

CHAPTER 7

The 206 bones of the human skeleton are classified into two divisions—axial and appendicular—based on their locations. The axial skeleton is composed of the portion of the skeleton immediately surrounding the axis of the skeleton, primarily the skull bones, vertebral column, and bones surrounding the thorax. Bones of the appendages, or extremities, comprise the appendicular skeleton. Chapter 7 begins with an introduction to types and surface characteristics (markings) of all bones and then focuses on the 80 bones of the axial skeleton. Disorders of the vertebral column are also included.

First study the Chapter 7 Framework and note key terms associated with each section.

TOPIC OUTLINE AND OBJECTIVES

A. Introduction: types of bones, surface markings, divisions of the skeletal system

1. Classify the four principal types of bones on the basis of shape and location.
2. Describe the various markings on the surfaces of bones.

B. Skull

3. Identify the bones of the skull and the major markings associated with each.
4. Identify the principal sutures, fontanels, paranasal sinuses, and foramina of the skull.

C. Hyoid bone and vertebral column

5. Identify the bones of the vertebral column and their principal markings.

D. Thorax

6. Identify the bones of the thorax and their principal markings.

E. Disorders

7. Contrast herniated (slipped) disc, abnormal curves, spina bifida, and fractures of the vertebral column as disorders associated with the skeletal system.

WORDBYTES

Now study the following parts of words that may help you better understand terminology in this chapter.

Wordbyte	Meaning	Example	Wordbyte	Meaning	Example
annulus	ring	*annulus* fibrosus	ethm-	sieve	*ethm*oid bone
costa	rib	*costa*l cartilage	lambd-	L-shaped	*lambd*oidal suture
cribr-	sieve	*cribr*iform plate	pulp-	(soft) flesh	nucleus *pulp*osus
crist-	crest	*crist*a galli			

CHECKPOINTS

A. **Introduction: types of bones, surface markings, divisions of the skeletal system** (pages 160–161)

A1. Complete the table about the four major and two minor types of bones.

Type of Bone	Structural Features	Examples
a. Long	Slightly curved to absorb stress better	
b.		Wrist, ankle bones
c.	Composed of two thin plates of bone	
d. Irregular		
e. Sutural (or Wormian)		
f.	Small bones in tendons	

A2. In general, what are functions of surface markings of bones?

A3. Contrast the bone markings in each of the following pairs.
a. Tubercle/tuberosity

b. Crest/line

c. Fossa/foramen

d. Condyle/epicondyle

■ **A4.** Write the name of the correct answer from the following list of specific bone markings on the line next to its description. The first one has been done for you.

Articular *facet* on vertebra	Maxillary *sinus*
External auditory *meatus*	Optic *foramen*
Greater *trochanter*	Styloid *process*
Head of humerus	Superior orbital *fissure*

a. Air-filled cavity within a bone, connected to nasal cavity: ___**Maxillary sinus**___

b. Narrow, cleftlike opening between adjacent parts of bone; passageway for blood vessels and nerves: _____

c. Rounded hole, passageway for blood vessels and nerves: _____

d. Tubelike passageway through bone: _____

e. Large, rounded projection above constricted neck: _____

f. Large, blunt projection on the femur: _____

g. Sharp slender projection: _____

h. Smooth, flat surface: _____

■ **A5.** Describe the bones in the two principal divisions of the skeletal system by completing this exercise. It may help to refer to Figure LG 8-1, page 133.

　　a. Bones that lie along the axis of the body are included in the (*axial? appendicular?*) skeleton.

　　b. The axial skeleton includes the following groups of bones. Indicate how many bones are in each category.

　　　　_____ Skull (cranium, face)　　　　_____ Vertebrae

　　　　_____ Earbones　　　　　　　　　　_____ Sternum

　　　　_____ Hyoid　　　　　　　　　　　　_____ Ribs

　　c. The total number of bones in the axial skeleton is _____.

　　d. The appendicular skeleton consists of bones in which parts of the body?

　　e. Write the number of bones in each category. Note that you are counting bones on one side of the body only.

　　　　_____ Left shoulder girdle　　　　　_____ Left hipbone

　　　　_____ Left upper extremity (arm, forearm,　　_____ Left lower extremity (thigh, kneecap, leg,
　　　　　　　　wrist, hand)　　　　　　　　　　　　　　　　　foot)

　　f. There are _____ bones in the appendicular skeleton.

　　g. In the entire human body there are _____ bones.

B. Skull (pages 162–175)

■ **B1.** Color the skull bones on Figure LG 7-1a, b, and c. Be sure to color the corresponding color code oval for each bone listed on the figure.

O　Ethmoid bone
O　Frontal bone
O　Lacrimal bone
O　Mandible bone
O　Maxilla bone
O　Nasal bone
O　Occipital bone
O　Parietal bone
O　Sphenoid bone
O　Temporal bone
O　Zygomatic bone

Figure LG 7-1 Skull bones. (a) Skull viewed from right side.

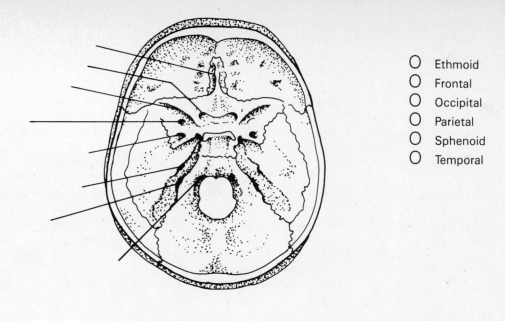

O Ethmoid
O Frontal
O Occipital
O Parietal
O Sphenoid
O Temporal

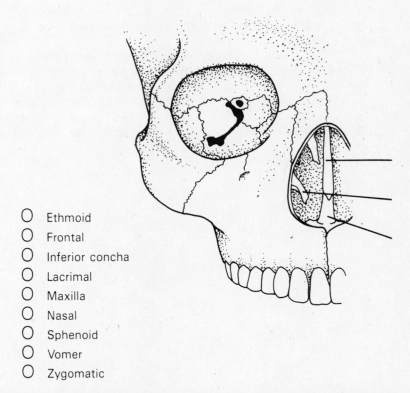

O Ethmoid
O Frontal
O Inferior concha
O Lacrimal
O Maxilla
O Nasal
O Sphenoid
O Vomer
O Zygomatic

Figure LG 7-1 Skull bones. (b) Floor of the cranial cavity. (c) Anterior view of bones that form the right orbit and the nose. Color and label as directed.

B2. Check your understanding of locations and functions of skull bones by writing the name of each skull bone next to its description.

_____ a. This bone forms the lower jaw, including the chin.

_____ b. These are the cheek bones; they also form lateral walls of the orbit of the eye.

_____ c. Tears pass through tiny foramina in these bones; they are the smallest bones in the face.

_____ d. The bridge of the nose is formed by these bones.

_____ e. Organs of hearing (internal part of ears) are located in and protected by these bones.

_____ f. This bone sits directly over the spinal column; it contains the hole through which the spinal cord connects to the brain.

_____ g. The name means "wall." The bones form most of the roof and much of the side walls of the skull.

_____ h. These bones form most of the roof of the mouth (hard palate) and contain the sockets into which upper teeth are set.

_____ i. L-shaped bones form the posterior parts of the hard palate and nose.

_____ j. Commonly called the forehead, it provides protection for the anterior portion of the brain.

_____ k. A fragile bone, it forms much of the roof and internal structure of the nose.

_____ l. It serves as a "keystone," since it binds together many of the other bones of the skull. It is shaped like a bat, with the wings forming part of the sides of the skull and the legs at the back of the nose.

_____ m. This bone forms the inferior part of the septum dividing the nose into two nostrils.

_____ n. Two delicate bones form the lower parts of the side walls of the nose.

B3. Answer these questions about the cranium.

a. What is the main function of the cranium?

b. Which bones make up the cranium (rather than the face)?

c. Define *sutures*.

Between which two bones is the sagittal suture located? _____

On Figure 7-1a, label the following sutures: *coronal, squamosal, lambdoidal.*

d. "Soft spots" of a newborn baby's head are known as _____

What is the location of the largest one? _____

B4. List three functions of the *fontanels.*

■ **B5.** Complete the table describing major markings of the skull.

Marking	Bone	Function
a. Greater wings	Sphenoid	
b.		Forms superior portion of septum of nose
c.		Site of pituitary gland
d. Petrous portion		
e.		Largest hole in skull; passageway for spinal cord
f.	(2)	Bony sockets for teeth
g.		Passageway for sound waves to enter ear
h. Condylar process	Mandible	

■ **B6.** Now label each of the markings listed in the above table on Figure LG 7-1.

■ **B7.** *For extra review.* The 12 pairs of nerves attached to the brain are called cranial nerves (Figure 14-3, page 390 of your text). Holes in the skull permit passage of these nerves to and from the brain. These nerves are numbered according to the order in which they attach to the brain (and leave the cranium) from I (most anterior) to XII (most posterior). To help you visualize their sequence, label the foramina for cranial nerves in order on the left side of Figure LG 7-1b. Complete the table summarizing these foramina. The first one is done for you.

Number and Name of Cranial Nerve	Location of Opening for Nerve
a. I Olfactory	Cribriform plate of ethmoid bone
b. II Optic	
c. III Oculomotor IV Trochlear V Trigeminal (ophthalmic branch) VI Abducens	
d. V Trigeminal (maxillary branch)	
e. V Trigeminal (mandibular branch)	
f. VII Facial VIII Vestibulocochlear	
g. IX Glossopharyngeal X Vagus XI Accessory	
h. XII Hypoglossal	

■ **B8.** Do this exercise about bony structures related to the nose.

a. Name four bones that contain paranasal sinuses.

_____   _____   _____   _____

   Practice identifying their locations the next time you have a cold!

b. List three functions of paranasal sinuses.

c. Most internal portions of the nose are formed by a bone that includes these markings: paranasal sinuses, conchae, and part of the nasal septum. Name the bone:

_____ .

d. What functions do nasal conchae perform?

e. On Figure LG 7-1c, label the bone that forms the inferior portion of the nasal septum. The anterior portion of the nasal septum consists of flexible

_____ .

■ **B9.** *A clinical challenge. TMJ syndrome* refers to disorders involving the only movable joint in the skull, the joint between the _____ and

_____ bones. List three signs or symptoms of this problem.

■ **B10.** A good way to test your ability to visualize locations of important skull bones is to try to identify bones that form the orbit of the eye. On Figure LG 7-1c, locate six bones that form the orbit. (Note: the tiny portion of the seventh bone, the superior tip of the palatine bone, is not visible in the figure.)

C. Hyoid bone and vertebral column   (pages 175–184)

■ **C1.** Identify the location of the hyoid bone on yourself. Place your hand on your throat and swallow. Feel your larynx move upward? The hyoid sits just (*superior? inferior?*) to the larynx at the level of the mandible.

■ **C2.** In what way is this bone unique among all the bones of the body?

■ **C3.** The five regions of the vertebral column are grouped (A–E) in Figure LG 7-2. Color vertebrae in each region and be sure to select the same color for the corresponding color code oval. Now write on lines next to A–E the number of vertebrae in each region. One is done for you. Also label the first two cervical vertebrae.

■ **C4.** Name structures that provide flexibility to the vertebral column:

_____.

■ **C5.** Note which regions of the vertebral column on Figure LG 7-2 normally retain an

anteriorly concave curvature in the adult: _____ and

_____. These are considered (*primary? secondary?*) curvatures. This classification is based upon the fact that these curves (*were present originally during fetal life? are more important?*).

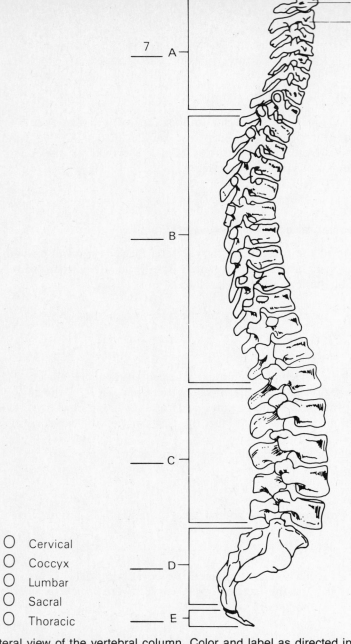

7 —— A

B

C

○ Cervical
○ Coccyx
○ Lumbar
○ Sacral
○ Thoracic

D

E

Figure LG 7-2 Right lateral view of the vertebral column. Color and label as directed in Checkpoint C3.

■ **C6.** On Figure LG 7-3, color the parts of vertebrae and corresponding color code ovals. As you do this, notice differences in size and shape among the three vertebral types in the figure.

Spinal cord

(a)

(b)

(c)

(d)

○ Body
○ Facet and demifacet for articulation with rib
○ Intervertebral disc
○ Intervertebral foramen
○ Lamina
○ Pedicle

○ Spinous process (spine)
○ Superior articular process
○ Transverse foramen
○ Transverse process
○ Vertebral foramen

Figure LG 7-3 Typical vertebrae. (a) Thoracic vertebra, superior view. (b) Thoracic vertebrae, right lateral view. (c) Cervical vertebra, superior view. (d) Lumbar vertebra, superior view. Color and label as directed in Checkpoint C6.

■ **C7.** Identify distinctive features of vertebrae in each region.

C. Cervical	S. Sacral
L. Lumbar	T. Thoracic

_____ a. Small body, foramina for vertebral blood vessels in transverse processes

_____ b. Only vertebrae that articulate with ribs

_____ c. Massive body, blunt spinous process and articular processes directed medially or laterally

_____ d. Long spinous processes that point inferiorly

_____ e. Articulate with the two hipbones

121

C8. *A clinical challenge.* State the clinical significance of these sacral markings.

a. Sacral hiatus

b. Sacral promontory

D. Thorax (pages 185–186)

D1. Name the structures that compose the thoracic cage. (Why is it called a "cage," the thoracic cage?)

■ **D2.** On Figure LG 7-4, color the three parts of the sternum.

D3. *A clinical challenge.* State one reason why a sternal puncture may be performed.

■ **D4.** Complete this exercise about ribs. Refer again to Figure LG 7-4.

a. There are a total of _____ ribs (_____ pairs) in the human skeleton.
b. Ribs slant in such a way that the anterior portion of the rib is (*superior? inferior?*) to the posterior end of the rib.

c. Posteriorly, all ribs articulate with _____. Ribs 1 to 10 also pass (*anterior? posterior?*) to and articulate with transverse processes of vertebrae. (Note this on a skeleton or Figure 7-17, page 185 of the text.) Of what functional advantage is such an arrangement?

d. Anteriorly, ribs numbered _____ to _____ attach to the sternum directly by

means of strips of hyaline cartilage, called _____ cartilage. These ribs are called (*true? false?*) ribs. Color those ribs on Figure LG 7-4, leaving the costal cartilages white.

e. Ribs 8 to 10 are called _____. Select a different color and color them again leaving costal cartilages white. Do these ribs attach to the sternum? (*Yes? No?*) If so, in what manner?

f. Ribs _____ and _____ are called "floating ribs." Now color them a different color. Why are these ribs so named?

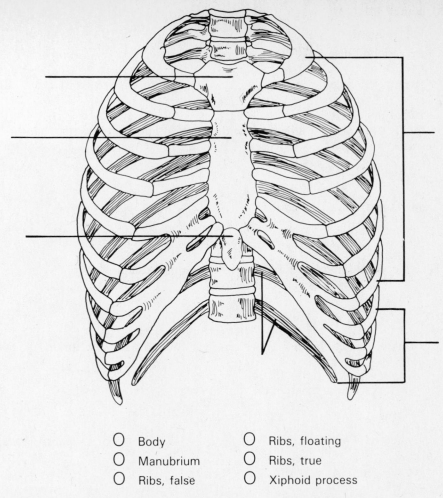

○ Body   ○ Ribs, floating

○ Manubrium   ○ Ribs, true

○ Ribs, false   ○ Xiphoid process

Figure LG 7-4 Anterior view of the thorax. Color as directed in Checkpoints D2 and D4.

g. What function is served by the costal groove?

h. What occupies intercostal spaces?

■ **D5.** At what point are ribs most commonly fractured?

E. Disorders (pages 187–188)

■ **E1.** Complete this exercise about slipped discs.

a. The normal intervertebral disc consists of two parts: an outer ring of (*hyaline?*

elastic? fibro-?) cartilage called _____ and a soft, elastic,

inner portion called the _____.

b. Ligaments normally keep discs in alignment with vertebral bodies. What may happen if these ligaments weaken?

c. Why might pain result from a slipped disc?

d. In what part of the vertebral column are slipped discs most common? What symptoms may result from a slipped disc in this region?

■ **E2.** Match types of abnormal curvatures of the vertebral column with descriptions below.

| K. Kyphosis | L. Lordosis | S. Scoliosis |

_____ a. Exaggerated lumbar curvature; "swayback"

_____ b. Exaggerated thoracic curvature; "hunchback"

_____ c. S- or C-shaped lateral bending

E3. Imperfect union of the vertebral arches at the midline is the condition known as

_____. Why is it crucial that the vertebral foramen be completely surrounded by bone? What problems may result from incomplete closure?

ANSWERS TO SELECTED CHECKPOINTS

A4. (b) Superior orbital *fissure.* (c) Optic *foramen.* (d) External auditory *meatus.* (e) *Head* of humerus. (f) Greater *trochanter.* (g) Styloid *process.* (h) Articular *facet* on vertebra.

A5. (a) Axial. (b) 22 skull, 6 earbones (studied in Chapter 17), 1 hyoid, 26 vertebrae, 1 sternum, 24 ribs. (c) 80. (d) Shoulder girdles, upper extremities, hipbones, lower extremities. (e) 2 left shoulder girdle, 30 left upper extremity, 1 left hipbone, 30 left lower extremity. (f) 126. (g) 206.

B1. Bone labels are capitalized; markings are in lowercase with related bone initial in parentheses.

B2.

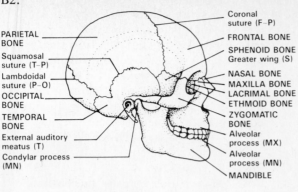

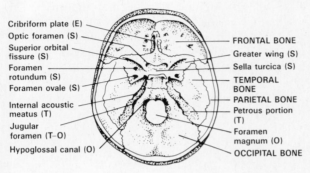

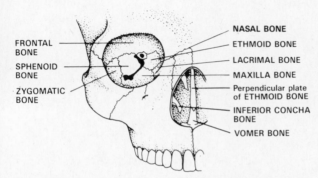

Figure LG 7-1A Skull bones. (a) Skull viewed from right side. (b) Floor of the cranial cavity. (c) Anterior view of bones that form the right orbit and the nose.

(a) Mandible. (b) Zygomatic (malar). (c) Lacrimal. (d) Nasal. (e) Temporal. (f) Occipital. (g) Parietal. (h) Maxilla. (i) Palatine. (j) Frontal. (k) Ethmoid. (l) Sphenoid. (m) Vomer. (n) Inferior concha.

B3. (a) Protects the brain. (b) Frontal, parietals, occipital, temporals, ethmoid, sphenoid. (c) Immovable, fibrous joints between skull bones; between parietal bones; see Figure LG 7-1A at Checkpoint B1 above. (d) Fontanels; between frontal and parietal bones.

B5.

Marking	Bone	Function
a. Greater wings	Sphenoid	**Form part of side walls of skull**
b. **Perpendicular plate**	**Ethmoid**	Forms superior portion of septum of nose
c. **Sella turcica**	**Sphenoid**	Site of pituitary gland
d. Petrous portion	**Temporal**	**Houses middle ear and inner ear**
e. **Foramen magnum**	**Occipital**	Largest hole in skull; passageway for spinal cord
f. **Alveolar processes**	(2) **Maxillae and mandible**	Bony sockets for teeth
g. **External auditory meatus**	**Temporal**	Passageway for sound waves to enter ear
h. Condylar process	Mandible	**Articulates with temporal bone (in TMJ)**

B6. See Figure LG 7-1A.

B7. See Figure LG 7-1A.

Number and Name of Cranial Nerve		Location of Opening for Nerve
a. I	Olfactory	Cribriform plate of ethmoid bone
b. II	Optic	**Optic foramen of sphenoid bone**
c. III	Oculomotor	**Superior orbital fissure of sphenoid bone**
IV	Trochlear	
V	Trigeminal (ophthalmic branch)	
VI	Abducens	
d. V	Trigeminal (maxillary branch)	**Foramen rotundum of sphenoid bone**
e. V	Trigeminal (mandibular branch)	**Foramen ovale of sphenoid bone**
f. VII	Facial	**Internal acoustic meatus**
VIII	Vestibulocochlear	
g. IX	Glossopharyngeal	**Jugular foramen**
X	Vagus	
XI	Accessory	
h. XII	Hypoglossal	**Hypoglossal canal**

B8. (a) Frontal, ethmoid, sphenoid, maxilla. (b) Warm and humidify air since air sinuses are lined with mucous membrane; serve as resonant chambers for speech and other sounds; make the skull lighter weight. (c) Ethmoid. (d) Like paranasal sinuses, they are covered with mucous membrane, which warms, humidifies, and cleanses air entering the nose. (e) Vomer, cartilage.

B9. Temporal, mandible; pain, clicking noise, or limitation of movement associated with the joint.

B10. See on Figure LG 7-1A above: frontal, zygomatic, maxilla, lacrimal, ethmoid, sphenoid.

C1. Superior.

C2. It articulates (forms a joint) with no other bones.

C3. A, cervical (7); B, thoracic (12); C, lumbar (5); D, sacrum (1); E, coccyx (1); C1, atlas; C2, axis.

C4. Intervertebral discs and ligaments.

C5. B (thoracic) and D (sacral); primary; were present originally during fetal life.

C6.

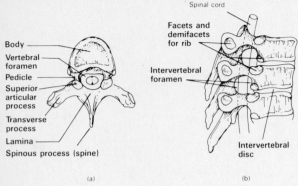

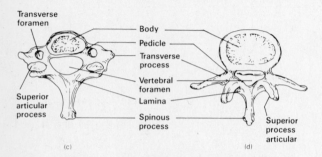

Figure LG 7-3A Typical vertebrae. (a) Thoracic vertebra, superior view. (b) Thoracic vertebrae, right lateral view. (c) Cervical vertebra, superior view. (d) Lumbar vertebra, superior view.

C7. (a) C. (b) T. (c) L. (d) T. (e) S.
D2.

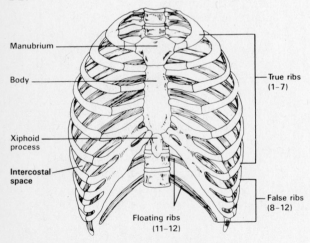

Figure LG 7-4A Anterior view of the thorax.

D4. (a) 24, 12. (b) Inferior. (c) Bodies of thoracic vertebrae, anterior; prevents ribs from slipping posteriorward. (d) 1, 7, costal, true. (e) False, yes; they attach to sternum indirectly via seventh costal cartilage. (f) 11, 12; they have no anterior attachment to sternum (g) Provides protective channel for intercostal nerve, artery, and vein. (h) Intercostal muscles. Refer also to Figure LG 7-4A.

D5. Just anterior to the costal angle, especially involving ribs 3–10.

E1. (a) Fibro-, annulus fibrosus, nucleus pulposus. (b) Annulus fibrosus may rupture, allowing the nucleus pulposus to protrude (herniate), causing a "slipped disc." (c) Disc tissue may press on spinal nerves. (d) L4–L5 or L5–sacrum; pain in posterior of thigh and leg(s) due to pressure on sciatic nerve, which attaches to spinal cord at L4 to S3.

E2. (a) L. (b) K. (c) S.

MASTERY TEST: Chapter 7

Questions 1–10: Circle the letter preceding the one best answer to each question.

1. All of these bones contain paranasal sinuses EXCEPT:

   A. Frontal   D. Sphenoid
   B. Maxilla   E. Ethmoid
   C. Nasal

2. Choose the one FALSE statement.

   A. There are seven vertebrae in the cervical region.
   B. The cervical region normally exhibits a curve that is slightly concave anteriorly.
   C. The lumbar vertebrae are superior to the sacrum.
   D. Intervertebral discs are located between bodies of vertebrae.

3. The hard palate is composed of ____ bones.

   A. Two maxilla and two mandible
   B. Two maxilla and two palatine
   C. Two maxilla
   D. Two palatine
   E. Vomer, ethmoid, and two temporal

4. The lateral wall of the orbit is formed mostly by which two bones?

   A. Zygomatic and maxilla
   B. Zygomatic and sphenoid
   C. Sphenoid and ethmoid
   D. Lacrimal and ethmoid
   E. Zygomatic and ethmoid

5. Choose the one TRUE statement.

   A. All of the ribs articulate anteriorly with the sternum.
   B. Ribs 8 to 10 are called true ribs.
   C. There are 23 ribs in the male skeleton and 24 in the female skeleton.
   D. Rib 7 is larger than rib 3.
   E. Cartilage discs between vertebrae are called costal cartilages.

6. Which is the largest fontanel?

   A. Frontal   C. Sphenoid
   B. Occipital   D. Mastoid

7. Immovable joints of the skull are called:

   A. Wormian bones   D. Sinuses
   B. Sutures   E. Fontanels
   C. Conchae

8. ____ articulate with every bone of the face except the mandible.

   A. Lacrimal bones   D. Sphenoid bones
   B. Zygomatic bones   E. Ethmoid bones
   C. Maxillae

9. All of these markings are parts of the sphenoid bone EXCEPT:

   A. Lesser wings
   B. Optic foramen
   C. Crista galli
   D. Sella turcica
   E. Pterygoid processes

10. All of these bones are included in the axial skeleton EXCEPT:

    A. Rib   D. Hyoid
    B. Sternum   E. Ethmoid
    C. Clavicle

Questions 11–15: Arrange the answers in correct sequence.

_____ _____ _____ 11. From anterior to posterior:

    A. Ethmoid bone
    B. Sphenoid bone
    C. Occipital bone

_____ _____ _____ 12. Parts of vertebra from anterior to posterior:

    A. Vertebral foramen
    B. Body
    C. Lamina

_____ _____ _____ 13. Vertebral regions, from superior to inferior:

    A. Lumbar
    B. Thoracic
    C. Cervical

_____ _____ _____ 14. From superior to inferior:

    A. Atlas
    B. Axis
    C. Occipital bone

_____ _____ _____ 15. From superior to inferior:

    A. Atlas
    B. Manubrium of sternum
    C. Hyoid

Questions 16–20: Circle T (true) or F (false). If the statement is false, change the underlined word or phrase so that the statement is correct.

T F 16. The space between two ribs is called the costal groove.

T F 17. The thoracic and sacral curves are called primary curves, meaning that they retain the original curve of the fetal vertebral column.

T F 18. The jugular vein passes through the same foramen as cranial nerves V, VI, and VII.

T F 19. In general, foramina, meati, and fissures serve as openings in the skull for nerves and blood vessels.

T F 20. The annulus fibrosus portion of an intervertebral disc is a firm ring of fibrocartilage surrounding the nucleus pulposus.

Questions 21–25: fill-ins. Write the word or phrase that best completes the statement.

_____ 21. A finger- or toothlike projection called the dens is part of the ____ bone.

_____ 22. The ramus, angle, mental foramen, and alveolar processes are all markings on the ____ bone.

_____ 23. The squamous, petrous, and zygomatic portions are markings on the ____ bone.

_____ 24. The perpendicular plate, crista galli, and superior and middle conchae are markings found on the ____ bone.

_____ 25. The sternum is often used for a marrow biopsy because ____.

ANSWERS TO MASTERY TEST: ■ Chapter 7

Multiple Choice

1. C	6. A
2. B	7. B
3. B	8. C
4. B	9. C
5. D	10. C

Arrange

11. A B C	14. C A B
12. B A C	15. A C B
13. C B A	

True-False

16. F. Intercostal space
17. T
18. F. IX, X, and XI
19. T
20. T

Fill-ins

21. Axis (second cervical vertebra)
22. Mandible
23. Temporal
24. Ethmoid
25. It contains red bone marrow and it is readily accessible.

FRAMEWORK 8

The Appendicular Skeleton

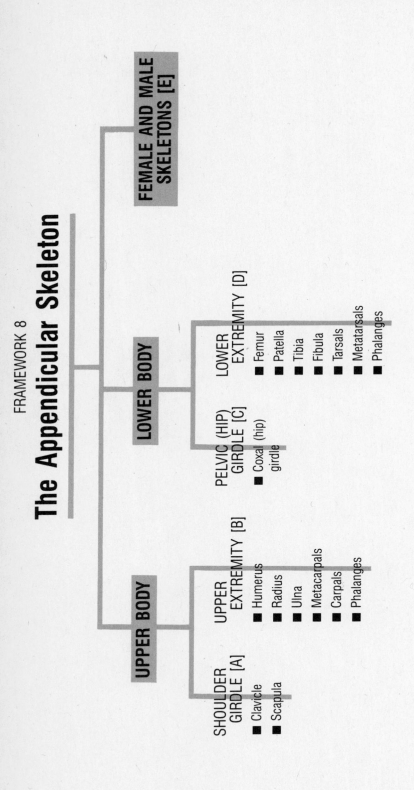

The Skeletal System: The Appendicular Skeleton

The appendicular skeleton includes the bones of the limbs, or extremities, as well as the supportive bones of the shoulder and hip girdles. Male and female skeletons exhibit some differences in skeletal structure, noted particularly in bones of the pelvis.

To begin your study of the appendicular skeleton, first refer to the Chapter 8 Framework and study key terms associated with each section.

TOPIC OUTLINE AND OBJECTIVES

A. Pectoral (shoulder) girdles

1. Identify the bones of the pectoral (shoulder) girdle and their major markings.

B. Upper extremities

2. Identify the upper extremity, its component bones, and their markings.

C. Pelvic (hip) girdle

3. Identify the components of the pelvic (hip) girdle and their principal markings.

D. Lower extremities

4. Identify the lower extremity, its component bones, and their markings.

5. Define the structural features and importance of the arches of the foot.

E. Male and female skeletons

6. Compare the principal structural differences between female and male skeletons, especially those that pertain to the pelvis.

Now study the following parts of words that may help you better understand terminology in this chapter.

Wordbyte	Meaning	Example		Wordbyte	Meaning	Example
acro-	tip	*acro*mion process		-physis	to grow	pubic sym*physis*
cap-	head	*cap*itulum		semi-	half	*semi*lunar notch
meta-	beyond	*meta*tarsal		sym-	together	pubic *sym*physis

CHECKPOINTS

A. Pectoral (shoulder) girdles (page 191)

■ **A1.** Do this exercise about the pectoral girdle.

a. Which bones form the pectoral girdle?

b. Do these bones articulate (form a joint) with vertebrae or ribs? _____

c. The pectoral girdle is part of the (*axial? appendicular?*) skeleton. Identify the point (marked by *) on Figure LG 8-1 at which the shoulder girdle articulates with the axial skeleton. Name the two bones forming that joint. Palpate (press and feel) the

bones at this joint on yourself. _____ _____

■ **A2.** Write the name of the bone that articulates with each of these markings on the scapula.

a. Acromion process: _____

b. Glenoid cavity: _____

c. Coracoid process: _____

■ **A3.** Study a scapula carefully, using a skeleton or Figure 8-3, page 193, in your text. Then match the markings in the box with the descriptions given.

A. Axillary border	M. Medial border
I. Infraspinatus fossa	S. Spine

_____ a. Sharp ridge on the posterior suface

_____ b. Despression inferior to the spine; location of infraspinatus muscle

_____ c. Edge closest to the vertebral column

_____ d. Thick edge closest to the arm

B. Upper extremities (pages 191–195)

■ **B1.** List the bone (or groups of bones) in the upper extremity from proximal to distal. Indicate how many of each bone there are. Two are done for you. Refer to Figure LG 8-1 to check your answers.

a. **Humerus** _____ (1)

b. _____ ( )

c. _____ ( )

d. _____ ( )

e. **Metacarpals** _____ (5)

f. _____ ( )

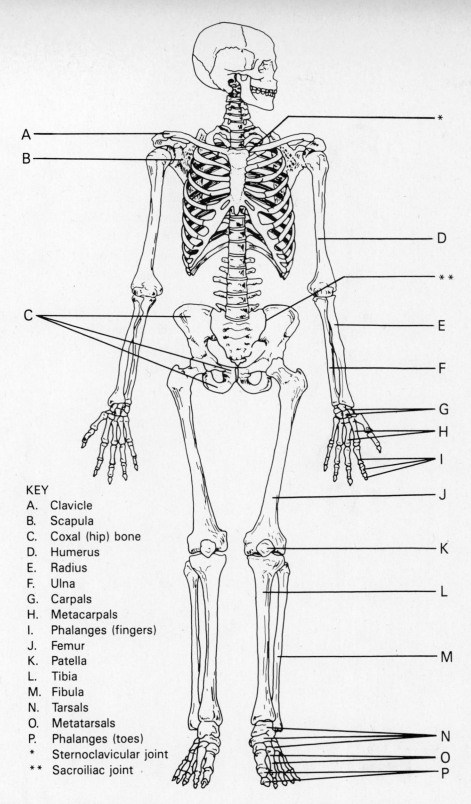

A

B

*

D

**

C

E

F

G

H

I

J

K

L

M

N

O

P

KEY
A. Clavicle
B. Scapula
C. Coxal (hip) bone
D. Humerus
E. Radius
F. Ulna
G. Carpals
H. Metacarpals
I. Phalanges (fingers)
J. Femur
K. Patella
L. Tibia
M. Fibula
N. Tarsals
O. Metatarsals
P. Phalanges (toes)
* Sternoclavicular joint
** Sacroiliac joint

Figure LG 8-1 Anterior view of the skeleton. Color, label, and answer questions as directed in Checkpoints A1, B1, C1, D1, D2, and D7.

■ **B2.** On Figure LG 8-2, select different colors and color each of the markings indicated by O. Where possible, color markings on both the anterior and posterior views.

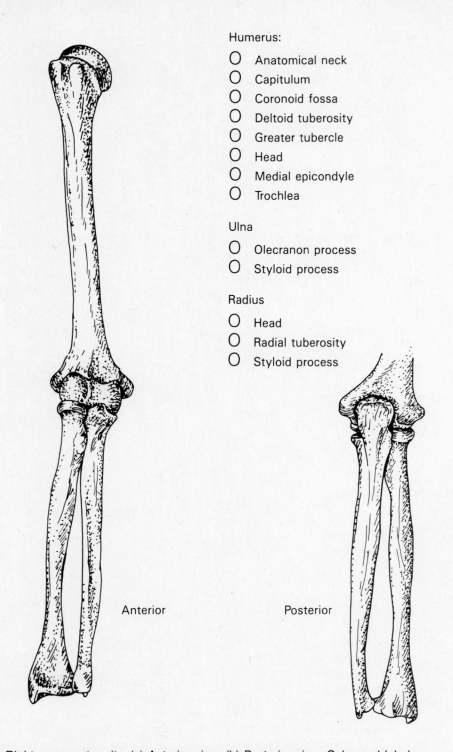

Humerus:

O   Anatomical neck
O   Capitulum
O   Coronoid fossa
O   Deltoid tuberosity
O   Greater tubercle
O   Head
O   Medial epicondyle
O   Trochlea

Ulna

O   Olecranon process
O   Styloid process

Radius

O   Head
O   Radial tuberosity
O   Styloid process

Anterior

Posterior

Figure LG 8-2   Right upper extremity. (a) Anterior view. (b) Posterior view. Color and label as directed in Checkpoint B2.

■ **B3.** Name the marking that fits each description. Write *H, U,* or *R* also to indicate if the marking is part of the humerus (H), ulna (U), or radius (R). One has been done for you.

a. Articulates with glenoid fossa: _____**head**_____ (__**H**__).

b. Rounded head that articulates with radius: _____ (_____)

c. Posterior depression that receives the olecranon process: _____ (_____)

d. Half-moon-shaped curved area that articulates with trochlea: _____ (_____)

e. Slight depression in which the head of the radius pivots: _____ (_____)

f. Biceps brachii muscle attaches here: _____ (_____)

■ **B4.** Answer these questions about wrist bones.

a. Wrist bones are called _____. There are (*5? 7? 8? 14?*) of them in each wrist.

b. The wrist bone most subject to fracture is the bone named

_____, located just distal to the (*radius? ulna?*).

B5. Trace an outline of your hand. Draw in and label all bones.

C. Pelvic (hip) girdle (pages 195–198)

■ **C1.** Describe the pelvic bones in this exercise.

a. Name the bones that form the pelvic girdle. _____

   Which bones form the pelvis? _____
b. Which of these bones is/are part of the axial skeleton?

c. Locate the point (**) on Figure LG 8-1 at which the pelvic girdle portion of the appendicular skeleton articulates with the axial skeleton. Name the bones involved

   in that joint: _____ and _____.

C2. Contrast the two principal parts of the pelvis. Describe their locations and name the structures which compose them.
a. Greater (false) pelvis

b. Lesser (true) pelvis

■ **C3.** Refer to Figure LG 8-3 and answer these questions about pelvic markings.
a. Trace with your pencil, and then label, the *brim of the pelvis* on the figure. It is a relatively (*smooth? irregular?*) line that demarcates the (*superior? inferior?*) border of the lesser (true) pelvis. The brim of the pelvis encircles the pelvic (*inlet? outlet?*)
b. The bony border of the pelvic outlet is (*smooth? irregular?*). Why is the pelvic outlet so named?

■ **C4.** Answer the following questions about coxal bones.
a. Each coxal (hip) bone originates as three bones which fuse early in life. These

   bones are the _____, _____, and

   _____. At what location do the bones fuse?

b. The largest of the three bones is the _____. A ridge along the superior border is called the iliac crest. Locate this on yourself.

c. The iliac crest ends anteriorly as the _____ spine. The crest

   ends posteriorly as the _____. This marking causes a dimpling of the skin just lateral to the sacrum, which can be used as a landmark for administering hip injections accurately.

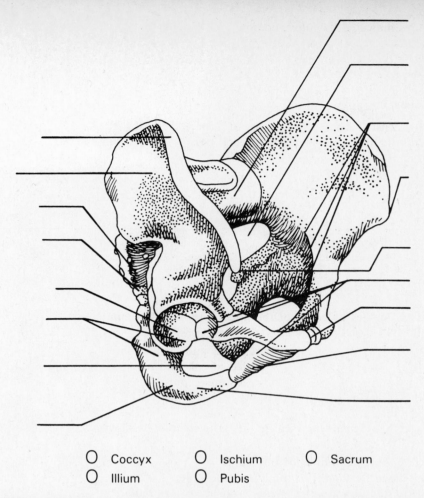

○ Coccyx   ○ Ischium   ○ Sacrum

○ Illium   ○ Pubis

Figure LG 8-3 Right anterolateral view of the pelvis. Color, label, and answer questions as directed in Checkpoints C3 and C6.

■ **C5.** Complete the table about markings of the coxal bones.

Marking	Location on coxal bone	Function
a. Greater sciatic notch		
b.		Supports most of body weight in sitting position
c.		Fibrocartilaginous joint between two coxal bones
d.		Socket for head of femur
e. Obturator foramen	Large foramen surrounded by pubic and ischial rami and acetabulum	

C6. Color the pelvic structures indicated by color code ovals on Figure LG 8-3. Also label the following markings: *acetabulum, anterior superior iliac spine, greater sciatic notch, iliac crest, inferior pubic ramus, ischial tuberosity, obturator foramen, sacral promontory,* and *symphysis pubis.*

D. Lower extremities (pages 198–203)

■ **D1.** Refer to Figure LG 8-1 and list the bones (or groups of bones) in the lower extremity from proximal to distal. Indicate how many of each bone there are. One is done for you.

a. **Femur** (1) e. _____ ( )

b. _____ ( ) f. _____ ( )

c. _____ ( ) g. _____ ( )

d. _____ ( )

D2. Contrast size, location, and names of bones of the upper and lower extremities by coloring the bones as follows. Color on one side of Figure LG 8-1 only.

Humerus and femur (red)   Carpals and tarsals (blue)
Patella (brown)   Metacarpals and metatarsals (orange)
Ulna and tibia (green)   Phalanges (purple)
Radius and fibula (yellow)

■ **D3.** Circle the term that correctly indicates the location of these parts of the lower extremity.

a. The head is the (*proximal? distal?*) epiphysis of the femur.
b. The greater trochanter is (*lateral? medial?*) to the lesser trochanter.
c. The intercondylar fossa is on the (*anterior? posterior?*) surface of the femur.
d. The tibial condyles are more (*concave? convex?*) than the femoral condyles.
e. The lateral condyle of the femur articulates with the (*fibula? lateral condyle of the tibia?*).
f. The tibial tuberosity is (*superior? inferior?*) to the patella.
g. The tibia is (*medial? lateral?*) to the fibula.
h. The outer portion of the ankle is the (*lateral? medial?*) malleolus which is part of the (*tibia? fibula?*).

■ **D4.** The tarsal bone that is most superior in location (and that articulates with the tibia and fibula) is the _____. The largest and strongest of the tarsals is the _____.

D5. Answer these questions about the arch of the foot.
a. How is the foot maintained in an arched position?

b. Locate each of these arches on your own foot. Refer to Figure 8-14, page 203 of your text. (*For extra review:* List the bones that form each arch.)

Longitudinal: medial side

Longitudinal: lateral side

Transverse

c. What causes flatfoot?

■ **D6.** Now that you have seen all of the bones of the appendicular skeleton, complete this table relating to common and anatomical names of bones.

Common Name	Anatomical Name
a. Shoulder blade	
b.	Pollex
c. Collarbone	
d. Heel bone	
e.	Olecranon process
f. Kneecap	
g.	Tibial crest
h. Toes	
i. Palm of hand	
j. Wrist bones	

D7. *For extra review.* Label all bones marked with label lines on Figure LG 8-1.

E. Female and male skeletons (pages 204–205)

E1. State several characteristics of the female pelvis that make it more suitable for childbirth than the male pelvis.

E2. Identify specific differences in pelvic structure in the two sexes by placing M before characteristics of the male pelvis and F before structural descriptions of the female pelvis.

_____ a. Shallow greater pelvis

_____ b. Heart-shaped inlet

_____ c. Pubic arch greater than 90° angle

_____ d. Ischial spine turned inward

_____ e. Pelvic outlet comparatively small

_____ f. General structure light and thin

ANSWERS TO SELECTED CHECKPOINTS

A1. (a) Two clavicles and two scapulas. (b) No. (c) Appendicular, clavicles and manubrium of sternum.

A2. (a) Clavicle. (b) Humerus. (c) None; muscles and ligaments attach here.

A3. (a) S. (b) I. (c) M. (d) A.

B1. Humerus—1, ulna—1, radius—1, carpals—8, metacarpals—5, phalanges—14.

B2.

B3. (b) Capitulum (H). (c) Olecranon fossa (H). (d) Semilunar (trochlear) notch (U). (e) Radial notch (U). (f) Radial tuberosity (R).

B4. (a) Carpals, 8. (b) Scaphoid, radius.

C1. (a) Two os coxa (hip) bones; hip bones plus sacrum and coccyx. (b) Sacrum and coccyx. (c) Sacrum and iliac portion of hip bone (sacroiliac).

C3.

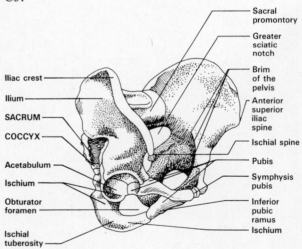

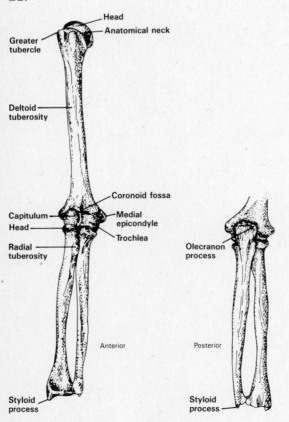

Figure LG 8-2A Right upper extremity. (a) Anterior view. (b) Posterior view.

See Figure LG 8-3A. Note that bone labels are capitalized; markings are in lowercase. (a) Smooth, superior, inlet. (b) Irregular; feces, urine, semen, menstrual flow, and baby (at birth) exit via openings in the muscular floor attached to the bony outlet.

C4. (a) Ilium, ischium, pubis; acetabulum. (b) Ilium. (c) Anterior superior iliac; posterior superior iliac spine.

C5.

Marking	Location on coxal bone	Function
a. **Greater sciatic notch**	**Inferior to posterior inferior iliac spine**	**Sciatic nerve passes inferior to notch**
b. **Ischial tuberosity**	**Posterior and inferior to obturator foramen**	Supports most of body weight in sitting position
c. **Symphysis pubis**	**Most anterior portion of pelvis**	Fibrocartilaginous joint between two coxal bones
d. **Acetabulum**	**At junction of ilium, ischium, and pubis**	Socket for head of femur
e. Obturator foramen	Large foramen surrounded by pubic and ischial rami and acetabulum	**Blood vessels and nerves pass through**

C6. See Figure LG 8-3A above.

D1. (a) Femur, 1. (b) Patella, 1. (c) Tibia, 1. (d) Fibula, 1. (e) Tarsals, 7. (f) Metatarsals, 5. (g) Phalanges, 14.

D3. (a) Proximal. (b) Lateral. (c) Posterior. (d) Concave. (e) Lateral condyle of the tibia. (f) Inferior. (g) Medial. (h) Lateral, fibula.

D4. Talus, calcaneus.

D6. (a) Scapula. (b) Thumb. (c) Clavicle. (d) Calcaneus. (e) Elbow. (f) Patella. (g) Shinbone. (h) Phalanges. (i) Metacarpals. (j) Carpals.

E2. (a) F. (b) M. (c) F. (d) M. (e) M. (f) F.

MASTERY TEST: Chapter 8

Questions 1–5: Circle the letter preceding the one best answer to each question.

1. The point at which the upper part of the appendicular skeleton is joined to (articulates with) the axial skeleton is at the joint between:

    A. Sternum and ribs
    B. Humerus and clavicle
    C. Scapula and clavicle
    D. Scapula and humerus
    E. Sternum and clavicle

2. All of following are markings on the femur EXCEPT:

    A. Acetabulum
    B. Head
    C. Condyles
    D. Greater trochanter
    E. Intercondylar fossa

3. All of the following are bones in the lower extremity EXCEPT:

    A. Talus   D. Ulna
    B. Tibia   E. Fibula
    C. Calcaneus

4. Which structures are on the posterior surface of the upper extremity (in anatomical position)?

    A. Radial tuberosity and lesser tubercle
    B. Trochlea and capitulum
    C. Coronoid process and coronoid fossa
    D. Olecranon process and olecranon fossa

5. The humerus articulates with all of these bones EXCEPT:

    A. Ulna   C. Clavicle
    B. Radius   D. Scapula

6. Choose the FALSE statement.

    A. The capitulum articulates with the head of the radius.
    B. The medial and lateral epicondyles are located at the distal ends of the tibia and fibula.
    C. The coronoid fossa articulates with the ulna when the forearm is flexed.
    D. The trochlea articulates with the trochlear notch of the ulna.

_____ _____ _____ 7. According to size of the bones, from largest to smallest:

A. Femur
B. Ulna
C. Humerus

_____ _____ _____ 8. Parts of the humerus, from proximal to distal:

A. Anatomical neck
B. Surgical neck
C. Head

_____ _____ _____ 9. From proximal to distal:

A. Phalanges
B. Metacarpals
C. Carpals

_____ _____ _____ 10. From superior to inferior:

A. True pelvis
B. False pelvis
C. Pelvic brim

_____ _____ _____ 11. Markings on coxal (hip) bones in anatomical position, from superior to inferior:

A. Acetabulum
B. Ischial tuberosity
C. Iliac crest

Questions 12–20: Circle T (true) or F (false). If the statement is false, change the underlined word or phrase so that the statement is correct.

T F 12. Another name for the true pelvis is the greater pelvis.

T F 13. The scapulae do articulate with the vertebrae.

T F 14. The olecranon process is a marking on the ulna, and the olecranon fossa is a marking on the humerus.

T F 15. The female pelvis is deeper and more heart-shaped than the male pelvis.

T F 16. The greater tubercle of the humerus is lateral to the lesser tubercle.

T F 17. There are 14 phalanges in each hand and also in each foot.

T F 18. The fibula articulates with the femur, tibia, talus, and calcaneus.

T F 19. The organs contained within the right iliac, hypogastric, and left iliac portions (ninths) of the abdomen are located in the true pelvis.

T F 20. The total number of bones in one upper extremity (including arm, forearm, wrist, hand, fingers, but excluding shoulder girdle) is 29.

Questions 21–25: fill-ins. Write the word or phrase that best completes the statement.

_____ 21. In about three-fourths of all carpal bone fractures, only the ____ bone is involved.

_____ 22. A fracture of the distal end of the fibula with injury to the tibial articulation is known as a ____ fracture.

_____ 23. The ____ is the thinnest bone in the body compared to its length.

_____ 24. The point of fusion of the three bones forming the coxal bone is the ____.

_____ 25. The kneecap is the common name for the ____.

ANSWERS TO MASTERY TEST: ■ Chapter 8

Multiple Choice

1. E   4. D
2. A   5. C
3. D   6. B

Arrange

7. A C B   10. B C A
8. C A B   11. C A B
9. C B A

True-False

12. F. Lesser
13. F. Do not
14. T
15. F. Shallower and more oval
16. T
17. T
18. F. Only tibia and talus
19. F. False
20. F. 30

Fill-ins

21. Scaphoid
22. Pott's
23. Fibula
24. Acetabulum
25. Patella

FRAMEWORK 9

Articulations

FUNCTIONAL
- Synarthroses
- Amphiarthroses
- Diarthroses

STRUCTURAL

Fibrous
- Suture
- Syndesmosis
- Gomphosis

Cartilaginous
- Synchondrosis
- Symphysis

Synovial

Structure
- Articular cartilage
- Articular capsule
  - fibrous capsule
  - synovial membrane
- Synovial cavity
- Accessory ligaments

Movements [B]
- Gliding
- Angular
  - flexion
  - extension
  - hyperextension
  - abduction
  - adduction
- Rotation
- Circumduction
- Special

Types [B]
- Gliding
- Hinge
- Pivot
- Ellipsoidal
- Saddle
- Ball-and-socket

**Selected Articulations [C]
(capsule, ligaments, bursae)**
- Shoulder
  - labrum
- Hip
  - labrum
- Knee
  - menisci

DISORDERS, TERMS [D]

Articulations

CHAPTER

9

In the past three chapters you have learned a great deal about the 206 bones in the body. Separated, or disarticulated, these bones would constitute a pile as disorganized as the rubble of a ravaged city deprived of its structural integrity. Fortunately bones are arranged in precise order and held together in specific conformations — articulations or joints — that permit bones to function effectively. Joints may be classified by function (how movable) or by structure (the type of tissue forming the joint). Synovial joints are most common; their structure, movements, and types will be discussed. A detailed examination of three important synovial joints (the shoulder, hip, and knee) is also included. Joint disorders such as arthritis are usually not life-threatening, but plague much of the population, particularly the elderly. An introduction to joint disorders completes this chapter.

Be sure to look over the Chapter 9 Framework with its key terms at the start and finish of your study of the chapter.

TOPIC OUTLINE AND OBJECTIVES

A. Classification of joints

1. Define an articulation and identify the factors that determine the types and degree (range) of movement at a joint.
2. Contrast the structure, kind of movement, and location of fibrous, cartilaginous, and synovial joints.

B. Movements and types of synovial (diarthrotic) joints

3. Discuss and compare the movements possible at various synovial joints.

C. Selected articulations of the body

4. Describe selected articulations of the body with respect to the bones that enter into their formation, structural classification, and anatomical components.

D. Disorders, medical terminology

5. Describe the causes and symptoms of common joint disorders, including rheumatism, rheumatoid arthritis (RA), osteoarthritis, gouty arthritis, bursitis, dislocation, and sprain.
6. Define medical terminology associated with articulations.

Now study the following parts of words that may help you better understand terminology in this chapter.

Wordbyte	Meaning	Example	Wordbyte	Meaning	Example
amphi-	both	*amphi*arthrotic	-osis	condition of	syndesm*osis*
arthr-	joint	osteo*arthr*itis	rheum-	water discharge	*rheum*atoid arthritis
articulat-	joint	*articulat*ion	syn-	together	*syn*arthrotic
cruci-	cross	*cruci*ate			
-itis	inflammation	arthr*itis*			

CHECKPOINTS

A. Classification of joints   (pages 208–211)

A1. Define the term *articulation* (*joint*).

■ **A2.** Fill in the blanks below to name three classes of joints according to the amount of movement they permit.

a. Synarthroses: _____

b. _____: slightly movable

c. _____: freely movable

■ **A3.** Name three classes of joints based on structure.

■ **A4.** Describe fibrous joints by completing this exercise.

a. Fibrous joints (*have? lack?*) a joint cavity. They are held together by

_____ connective tissue.

b. One type of fibrous joint is a _____ found between skull

bones. Such joints are (*freely? slightly? im-?*) movable or _____-arthrotic.

c. The distal end of the tibia/fibula joint is a fibrous joint. It is (*more? less?*) mobile

than a suture and is therefore _____-arthrotic.

■ **A5.** Contrast two types of cartilaginous joints in this exercise.

a. Synchondroses involve (*hyaline? fibrous?*) cartilage between bone. An example is

the _____ cartilage between diaphysis and epiphysis of a growing bone. This cartilage (*persists through life? is replaced by bone during adult life?*). Synchondroses are (*somewhat movable? immovable?*).

b. Fibrocartilage is present in the type of joint known as _____.

These joints permit some movement, so are called _____

-arthrotic. Two locations of symphyses are _____ and

_____.

■ **A6.** What structural features of synovial joints make them more freely movable than fibrous or cartilaginous joints?

■ **A7.** On Figure LG 9-1, color the indicated structures.

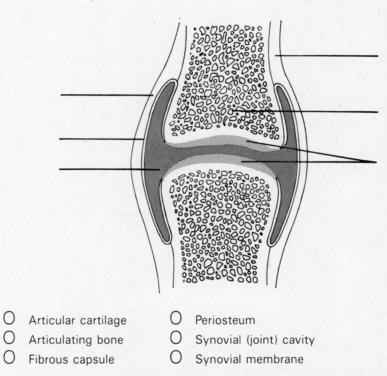

  ○ Articular cartilage ○ Periosteum
  ○ Articulating bone ○ Synovial (joint) cavity
  ○ Fibrous capsule ○ Synovial membrane

Figure LG 9-1 Structure of a generalized synovial joint. Color as directed in Checkpoint A7.

■ **A8.** Select the parts of a synovial joint (listed in the box) that fit the descriptions below.

A. Articular cartilage	SF. Synovial fluid
F. Fibrous capsule	SM. Synovial membrane
L. Ligaments	

_____ a. Hyaline cartilage that covers ends of articulating bones, but does not bind them together

_____ b. With the consistency of uncooked egg white, it lubricates the joint and nourishes articular cartilage.

_____ c. Connective tissue membrane that lines synovial cavity and secretes synovial fluid

_____ d. Parallel fibers in some fibrous capsules; bind bones together

_____ e. Together these form the articular capsule (2 answers)

A9. Explain how joints can produce noises, as in "cracking your knuckles."

A10. *A clinical challenge.* Contrast the following procedures and tell how each can be useful for patients with joint problems:
arthrocentesis/arthroscopy

A11. Describe the structure, function, and location of the following structures that may be associated with synovial joints:

a. Articular discs (menisci)

b. Bursae

B. Movements and types of synovial joints   (pages 211–217)

B1. The design of synovial joints permits free movement of bones. However, if bones moved too freely, they could move right out of their joint cavities (dislocation). List and briefly describe four factors that account for limitation of movement at synovial joints.

■ **B2.** From the terms listed in the box, choose the one that fits the type of movement in each case. Not all answers will be used.

Abd. Abduction	E. Extension	I. Inversion
Add. Adduction	F. Flexion	P. Plantar flexion
C. Circumduction	G. Gliding	R. Rotation
D. Dorsiflexion		

_____ a. Decrease in angle between anterior surfaces of bones (or between posterior surfaces at knee and toe joints)

_____ b. Simplest kind of movement that can occur at a joint; no angular or rotary motion involved; example: ribs moving against vertebrae

_____ c. State of entire body when it is in anatomical position

_____ d. Movement away from the midline of the body

_____ e. Movement of a bone around its own axis

_____ f. Position of foot when heel is on the floor and rest of foot is raised

■ **B3.** Perform the action described. Then write in the name of the type of movement.

a. Describe a cone with your arm, as if you are winding up to pitch a ball. The

movement is called _____ .

b. Stand in anatomical position (palms forward). Turn your palms backward. This

action is called _____ .

c. Move your fingers from "fingers together" to "fingers apart" position. This action

is _____ of fingers.

d. Raise your shoulders, as if to shrug them. This movement is called

_____ of the shoulders.

e. Stand on your toes. This action at the ankle joint is called

_____ .

f. Grasp a ball in your hand. Your fingers are performing the type of movement

called _____ .

■ **B4.** Identify the kinds of movements shown in Figure LG 9-2. Write the name of the movement below each figure. Use the following terms: *abduction, adduction, extension, flexion,* and *hyperextension.*

■ **B5.** Check your understanding of types of synovial joints by choosing the type of joint that fits the description. (Answers may be used more than once.)

B. Ball-and-socket	H. Hinge
E. Ellipsoidal	P. Pivot
G. Gliding	S. Saddle

_____ a. Monaxial joint; only rotation possible

_____ b. Joint between carpal and metacarpal of the thumb joint

_____ c. Shoulder and hip joints

_____ d. Spoollike surface articulated with concave surface

_____ e. Monaxial joint; only flexion and extension possible

_____ f. Biaxial joints (two answers)

C. Selected articulations of the body   (pages 217–224)

■ **C1.** Complete this exercise describing the shoulder, hip, and knee joints. Consult Figures 9-6 to 9-8 on pages 219–222 of your text.

a. The articular capsule is strengthened by ligaments, such as the

_____-humeral ligament between the coracoid process of the scapula and the (*greater? lesser?*) tubercle of the humerus. Name one other ligament that forms part of this joint.

b. The shoulder joint is further strengthened and stabilized by the glenoid

_____, which is a rim (or liplike structure) of fibrocartilage; it increases the concavity of the glenoid cavity, providing the head of the humerus with a more stable socket.

c. List three bursae associated with the humeroscapular joint.

d. The articular capsule of the coxal, or hip, joint includes circular fibers, called the

_____, which form a protective collar around the head of the femur. Name the three ligaments that reinforce the articular capsule.

e. The stability of the hip joint is further enhanced by the ligament of the head of the femur and the transverse ligament of the acetabulum. Both of these are (*intra? extra?*)-capsular ligaments. The hip joint (*does? does not?*) have a labrum.

f. The knee (tibiofemoral) joint (*does? does not?*) include a complete capsule uniting the two bones. This factor contributes to the relative (*strength? weakness?*) of this joint.

g. A number of ligaments do provide some strength and support. Name two extra-capsular ligaments:

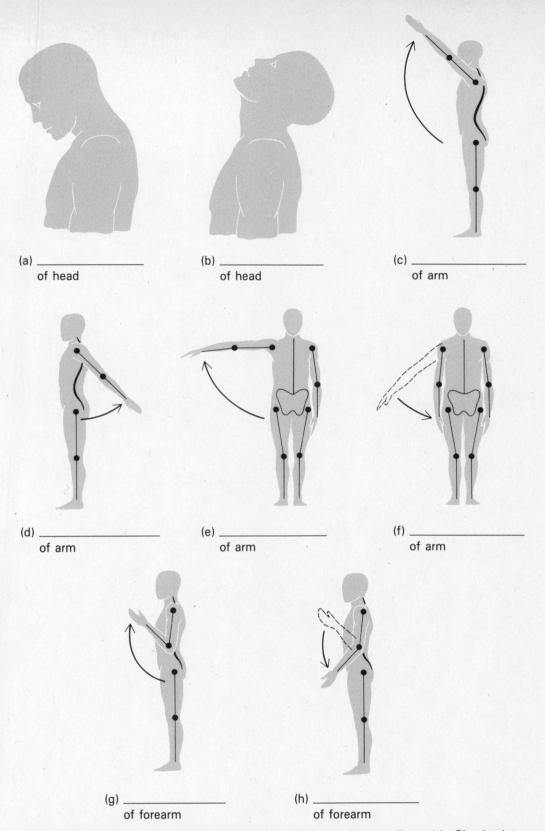

(a) _____
  of head

(b) _____
  of head

(c) _____
  of arm

(d) _____
  of arm

(e) _____
  of arm

(f) _____
  of arm

(g) _____
  of forearm

(h) _____
  of forearm

Figure LG 9-2 Movements at synovial joints. Answer questions as directed in Checkpoint B4 and in Chapter 11.

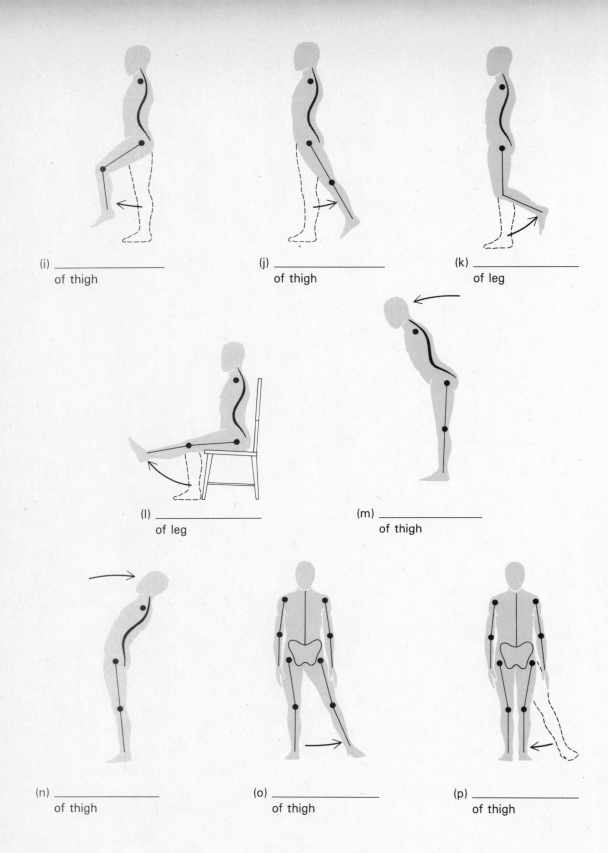

(i) _____ of thigh

(j) _____ of thigh

(k) _____ of leg

(l) _____ of leg

(m) _____ of thigh

(n) _____ of thigh

(o) _____ of thigh

(p) _____ of thigh

h. Name two intracapsular ligaments of the knee joint.

Which one is more often involved in serious knee injuries?
i. The knee joint (*does? does not?*) include a labrum. Instead it has two liplike car-

tilage discs called _____, which add stability to the joint.
j. There are (*no? three? over a dozen?*) bursae associated with the knee joint.

■ **C2.** Make a final comparison of the shoulder, hip, and knee joints by completing this exercise.
a. Which of these joints has the widest range of motion? _____

b. Which has the most limited range of motion? _____

c. Which is most stable and so is rarely dislocated? _____

d. Which is least stable? _____

■ **C3.** *A clinical challenge.* Which three knee structures beginning with C's are examined closely when a knee is injured?

D. Disorders, medical terminology (pages 225–227)

■ **D1.** How are *arthritis* and *rheumatism* related? Choose the correct answer.
   A. Arthritis is a form of rheumatism.
   B. Rheumatism is a form of arthritis.

■ **D2.** List several forms of arthritis. Name one symptom common to all forms of this ailment.

■ **D3.** Contrast *rheumatoid arthritis* with *osteoarthritis*. Write *More, Less, Yes* or *No* for answers.

	Rheumatoid Arthritis	Osteoarthritis
a. Which is more common?		
b. Which is usually more damaging?		
c. Is this an inflammatory autoimmune condition?		
d. Is synovial membrane affected?		
e. Does articular cartilage degenerate?		
f. Does fibrous tissue join bone ends?		
g. Is movement limited?		

■ **D4.** Briefly describe each of these disorders involving articulations by doing this exercise.

a. Gouty arthritis is a condition due to an excess of _____ in the

blood leading to deposit of _____ in joints. This condition is more common among (*males? females?*).

b. An acute chronic inflammation of a bursa is called _____.
Inflammation of certain bursae associated with the knee joint is known as

_____ knee.

c. *Luxation,* or _____, is displacement of a bone from its joint with tearing of ligaments. Three most common sites of dislocations are the

_____, _____, and

_____ joints.

d. Pain in a joint is known as _____.

ANSWERS TO SELECTED CHECKPOINTS

A2. (a) Immovable. (b) Amphiarthrotic. (c) Diarthrotic.
A3. Fibrous, cartilage, synovial.
A4. (a) Lack, fibrous. (b) Suture, im-, syn. (c) More, amphi-.

A5. (a) Hyaline, epiphyseal, is replaced by bone during adult life, immovable. (b) Symphysis amphi-, discs between vertebrae, symphysis pubis between coxal bones.

A6. The space between the articulating bones and the absence of tissue between those bones (which might restrict movement) make the joints more freely movable.

A7.

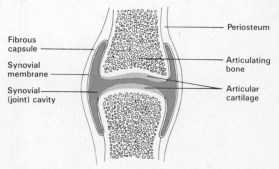

Figure LG 9-1A Structure of a generalized synovial joint.

A8. (a) A. (b) SF. (c) SM. (d) L. (e) SM and F.

B2. (a) F. (b) G. (c) E. (d) Abd. (e) R. (f) D.

B3. (a) Circumduction. (b) Pronation. (c) Abduction. (d) Elevation. (e) Plantar flexion. (f) Flexion.

B4. (a) Flexion; (b) Hyperextension. (c) Flexion. (d) Hyperextension. (e) Abduction. (f) Adduction. (g) Flexion. (h) Extension. (i) Flexion, while leg also slightly flexed. (j) Hyperextension, while leg extended. (k) Flexion, while thigh extended. (l) Extension, with thigh flexed. (m) Flexion. (n) Hyperextension. (o) Abduction. (p) Adduction.

B5. (a) P. (b) S. (c) B. (d) H. (e) H. (f) E and S.

C1. (a) Coraco-, greater, glenohumeral or transverse humeral ligament. (b) Labrum. (c) Subscapular, subdeltoid, subacromial or subcoracoid. (d) Zona orbicularis; iliofemoral, ischiofemoral, pubofemoral. (e) Intra, does. (f) Does not, weakness. (g) Tibial collateral, fibular collateral, patellar, oblique popliteal, arcuate popliteal. (h) Anterior and posterior cruciates; anterior cruciate. (i) Does not, menisci. (j) Over a dozen.

C2. (a) Shoulder. (b) Knee. (c) Hip. (d) Knee.

C3. Tibial collateral ligament, anterior cruciate ligament, and medial meniscus (cartilage).

D1. A.

D2. Rheumatoid arthritis (RA), osteoarthritis, and gouty arthritis; pain.

D3.

	Rheumatoid Arthritis	Osteoarthritis
a. Which is more common?	Less	More
b. Which is usually more damaging?	More	Less
c. Is this an inflammatory autoimmune condition?	Yes	No
d. Is synovial membrane affected?	Yes	No
e. Does articular cartilage degenerate?	Yes	Yes
f. Does fibrous tissue join bone ends?	Yes	No
g. Is movement limited?	Yes	Yes

D4. (a) Uric acid, uric acid and sodium urate, males. (b) Bursitis, housemaid's or carpet layer's. (c) Dislocation; finger, thumb, shoulder. (d) Arthralgia.

MASTERY TEST: Chapter 9

Questions 1–14: Circle T (true) or F (false). If the statement is false, change the underlined word or phrase so that the statement is correct.

T F 1. A fibrous joint is one in which there is <u>no joint cavity and bones are held together by fibrous connective tissue</u>.

T F 2. <u>Sutures, syndesmoses, and symphyses</u> are kinds of fibrous joints.

T F 3. Osteoarthritis is <u>less common and usually more damaging than</u> rheumatoid arthritis.

T F 4. <u>Ball-and-socket, gliding, pivot, and ellipsoidal joints</u> are all diarthrotic joints.

T F 5. All fibrous joints are <u>synarthrotic and all cartilaginous joints are amphiarthrotic</u>.

T F 6. In synovial joints synovial membranes <u>cover the surfaces of articular cartilages</u>.

T F 7. Bursae are <u>saclike structures that reduce friction</u> at joints.

T F 8. Synovial fluid becomes <u>more</u> viscous when there is friction at a joint.

T F 9. When your arm is in the supine position, your radius and ulna are <u>parallel (not crossed)</u>.

T F 10. When you touch your toes, the major action you perform at your hip joint is called <u>hyperextension</u>.

T F 11. The <u>only type of joint that is triaxial</u> is the ball-and-socket.

T F 12. Abduction is movement <u>away from</u> the midline of the body.

T F 13. The <u>elbow, knee, and ankle</u> joints are all hinge joints.

T F 14. Joints that are very stable tend to have <u>more</u> mobility than joints that are unstable.

Questions 15–16: Arrange the answers in correct sequence.

_____ _____ _____ 15. From most mobile to least mobile:

A. Amphiarthrotic
B. Diarthrotic
C. Synarthrotic

_____ _____ _____ 16. Stages in rheumatoid arthritis, in chronological order:

A. Articular cartilage is destroyed and fibrous tissues joins exposed bone.
B. The synovial membrane produces pannus which adheres to articular cartilage.
C. Synovial membrane becomes inflamed and thickened, and synovial fluid accumulates.

Questions 17–20: Circle the letter preceding the one best answer to each question.

17. Which structure is extracapsular in location?

A. Acetabular labrum
B. Posterior cruciate ligament
C. Ligament of the head of the femur
D. Tibial collateral ligament

18. All of these structures are associated with the knee joint EXCEPT:

A. Glenoid labrum
B. Patellar ligament
C. Infrapatellar bursa
D. Medial meniscus
E. Tibial collateral ligament

19. "Housemaid's knee" is:

A. Sprained knee
B. Tendinitis
C. Inflammation of a bursa of the knee
D. Torn cartilage

20. A suture is found between:

A. The two pubic bones
B. The two parietal bones
C. Radius and ulna
D. Diaphysis and epiphysis
E. Tibia and fibula (distal ends)

Questions 21–25: fill-ins. Write the word or phrase that best completes the statement.

_____ 21. ____ are saclike structures which reduce friction at joints.

_____ 22. ____ is the forcible wrenching or twisting of a joint with partial rupture of it, but without dislocation.

_____ 23. The action of pulling the jaw back from a thrust-out position so that it becomes in line with the upper jaw is the movement called ____.

_____ 24. Another name for a freely movable joint is ____.

_____ 25. The type of joint between the atlas and axis and also between proximal ends of the radius and ulna is a ____ joint.

ANSWERS TO MASTERY TEST: ■ Chapter 9

True-False

1. T
2. F. Sutures and syndesmoses
3. F. More common and usually less damaging than
4. T
5. F. Either synarthrotic or amphiarthrotic, and the same is true of cartilaginous
6. F. Do not cover surfaces of articular cartilages
7. T
8. F. Less
9. T
10. F. Flexion
11. T
12. T
13. T
14. F. Less

Arrange

15. B A C
16. C B A

Multiple Choice

17. D   19. C
18. A   20. B

Fill-ins

21. Bursae
22. Sprain
23. Retraction
24. Diarthrotic
25. Pivot (or synovial or diarthrotic)

FRAMEWORK 10

Muscle Tissue

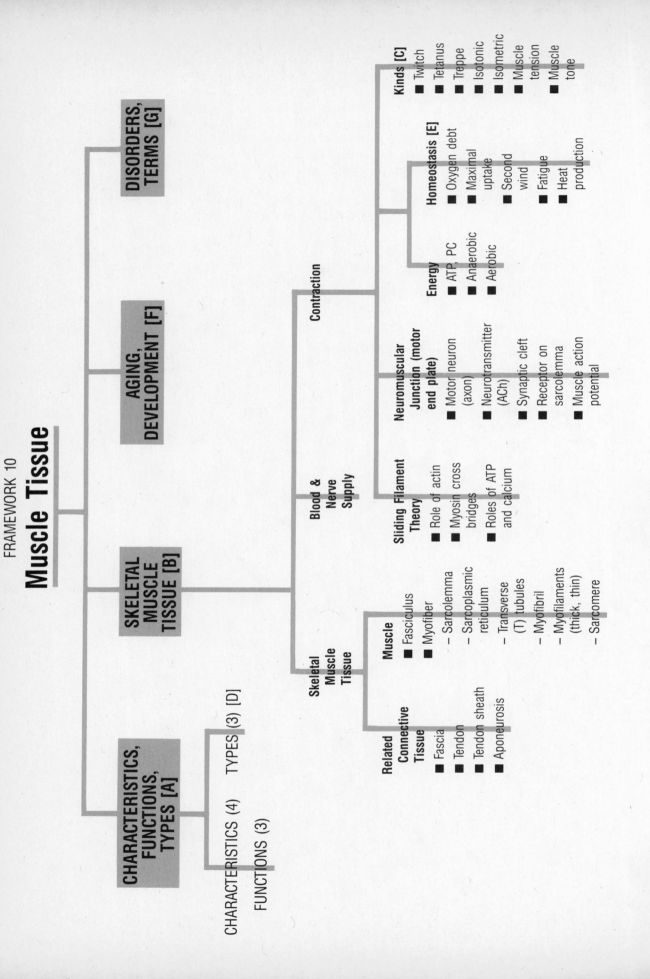

CHARACTERISTICS, FUNCTIONS, TYPES [A]

CHARACTERISTICS (4)   TYPES (3) [D]

FUNCTIONS (3)

SKELETAL MUSCLE TISSUE [B]

AGING, DEVELOPMENT [F]

DISORDERS, TERMS [G]

Skeletal Muscle Tissue

Related Connective Tissue
- Fascia
- Tendon
- Tendon sheath
- Aponeurosis

Muscle
- Fasciculus
- Myofiber
  – Sarcolemma
  – Sarcoplasmic reticulum
  – Transverse (T) tubules
  – Myofibril
  – Myofilaments (thick, thin)
  – Sarcomere

Blood & Nerve Supply

Contraction

Sliding Filament Theory
- Role of actin
- Myosin cross bridges
- Roles of ATP and calcium

Neuromuscular Junction (motor end plate)
- Motor neuron (axon)
- Neurotransmitter (ACh)
- Synaptic cleft
- Receptor on sarcolemma
- Muscle action potential

Energy
- ATP, PC
- Anaerobic
- Aerobic

Homeostasis [E]
- Oxygen debt
- Maximal uptake
- Second wind
- Fatigue
- Heat production

Kinds [C]
- Twitch
- Tetanus
- Treppe
- Isotonic
- Isometric
- Muscle tension
- Muscle tone

Muscle Tissue

Muscles are making it possible for you read this paragraph: facilitating movement of your eyes and head, maintaining your posture, producing heat to keep you comfortable. Muscles do not work in isolation, however. Connective tissue binds muscle cells into bundles and attaches muscle to bones. Blood delivers the oxygen and nutrients for muscle work. And nerves initiate a cascade of events that culminate in muscle contraction.

A tour of the mechanisms of muscle action finds the visitor caught in the web of muscle protein filaments sliding back and forth as the muscle contracts and relaxes. Neurotransmitters, as well as calcium and power-packing ATP, serve as chief regulators. Muscles are observed contracting in a multitude of modes, from twitch to treppe and isotonic to isometric. It may be surprising to hear a tour guide's announcement that most muscle cells present now were actually around at birth. Very few muscle cells can multiply, but fortunately muscle cells can greatly enlarge, as with exercise. With aging, cell numbers decline and so do muscle strength and endurance. Certain disorders decrease muscle function and debilitate even in very early years of life.

To organize your study of muscle tissue, glance over the Chapter 10 Framework now. Be sure to refer to the Framework frequently and note relationships among key terms in each section.

TOPIC OUTLINE AND OBJECTIVES

A. Characteristics, functions, types of muscle tissue

1. List the characteristics and functions of muscle tissue.
2. Compare the location, microscopic appearance, nervous control, functions, and regenerative capacities of the three kinds of muscle tissue.

B. Skeletal muscle tissue and contraction

3. Describe the principal events associated with the sliding-filament theory of muscle contraction.
4. Describe the structure and importance of a neuromuscular junction (motor end plate) and a motor unit.
5. Identify the source of energy for muscle contraction.
6. Define the all-or-none principle of muscular contraction.

C. Kinds of contractions; types of skeletal muscle fibers

7. Describe different types of normal contractions performed by skeletal muscles.

D. Cardiac muscle tissue, smooth muscle tissue

E. Homeostasis

8. Compare oxygen debt, fatigue, and heat production as examples of muscle homeostasis.

F. Aging and development of muscles

9. Explain the effects of aging on muscle tissue.
10. Describe the development of the muscular system.

G. Disorders, medical terminology

11. Define such common disorders as fibrosis, "charleyhorse," muscular dystrophy, myasthenia gravis (MG), spasms, fibrillation, and tics.

12. Define medical terminology associated with the muscular system.

WORDBYTES

Now study the following parts of words that may help you better understand terminology in this chapter.

Wordbyte	Meaning	Example	Wordbyte	Meaning	Example
a-	not	*a*trophy	graph-	to write	electromyo*graphy*
-algia	pain	fibromy*algia*	-lemma	rind, skin	sarco*lemma*
apo-	from	*apo*neurosis	myo-	muscle	*myo*fiber, *myo*sin
dys-	bad	muscular *dys*trophy	mys-	muscle	epi*mys*ium
-gen	to produce	phospha*gen* system	sarco-	flesh, muscle	*sarco*lemma
			troph-	nourishment	muscular dys*trophy*

CHECKPOINTS

A. Characteristics, functions, types of muscle tissue (page 230)

A1. List and define four characteristics of muscle tissue.

a. _____

b. _____

c. _____

d. _____

■ **A2.** List three functions of muscle tissue that are important for maintenance of homeostasis.

■ **A3.** Match the muscle types listed in the box with descriptions below.

| C. Cardiac   Sk. Skeletal   Sm. Smooth |

_____ a. Involuntary muscle found in blood vessels and intestine

_____ b. Involuntary striated muscle

_____ c. Striated voluntary muscle attached to bones

B. Skeletal muscle tissue and contraction (pages 230–241)

■ **B1.** Contrast two kinds of fascia by indicating which of the following are characteristics of superficial fascia (S) or deep fascia (D).

_____ a. Immediately under the skin

_____ b. Composed of dense connective tissue

_____ c. Contains much fat, so provides insulation and protection

_____ d. Also called subcutaneous layer

B2. Contrast *tendon* and *aponeurosis*.

■ **B3.** Arrange the following terms in correct order from largest to smallest in size: *fasciculus, myofilament* (*thick or thin*), *myofiber* (*muscle cell*), *myofibril*

_____ → _____ → _____ → _____

■ **B4.** Refer to Figure LG 10-1 and complete this activity about muscle structure.
  a. Color the structures indicated by color code ovals.
  b. Using lead lines provided, label the following structures on Figures LG 10-1b or
     10-1c: *A band, I band, sarcomere, triad, Z line*

■ **B5.** *For extra review.* Match the correct term from the list at right with its description.

> | | |
> |---|---|
> | A band | Sarcoplasm |
> | I band | Triad |
> | Sarcomere | |

a. Cytoplasm of muscle cell

_____

b. A transverse tubule, along with sarcoplasmic
   reticulum on either side

_____

c. Extends from Z line to Z line

_____

d. Dark area in striated muscle; contains thick

   and thin filaments _____

e. Light area on either side of Z line; location of

   thin filaments only _____

■ **B6.** After you study Figure 10-4, page 235 in the text, contrast thick and thin fila-
     ments in this exercise:

  a. Thin filaments are composed mostly of molecules of _____.
     Each actin molecule contains a binding site for (*myosin? actin?*).

  b. Two other proteins in thin filaments are _____ arranged in

     helical strands, and _____ located at regular intervals along
     thin filaments. Troponin contains binding sites for actin, tropomyosin, and (*ATP?
     calcium?*).
  c. (*Thick? Thin?*) myofilaments are composed mostly of myosin. Myosin molecules
     are shaped much like (*basketballs? golf clubs?*). The "head" of the club is called a

     cross bridge and contains sites for attachment with _____ and

     _____.

  B7. State the desired effects and disadvantages of uses of anabolic steroids.

■ **B8.** Summarize one theory of muscle contraction in this Checkpoint.
  a. In order to effect muscle shortening (or contraction), heads (cross bridges) of

     _____ myofilaments act like oars pulling on

     _____ molecules of thin filaments.
  b. As a result, (*thick? thin?*) myofilaments move toward the center of the sarcomere.
     Since the thick and thin myofilaments (*shorten? slide?*) to decrease the length of
     the sarcomere, this theory of muscle contraction is known as the

     _____ theory.

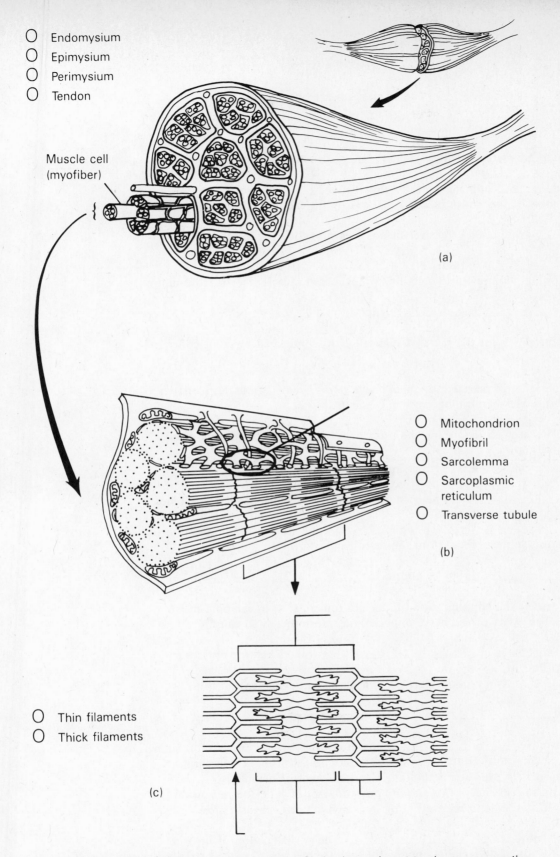

○ Endomysium
○ Epimysium
○ Perimysium
○ Tendon

Muscle cell
(myofiber)

(a)

○ Mitochondrion
○ Myofibril
○ Sarcolemma
○ Sarcoplasmic
   reticulum
○ Transverse tubule

(b)

○ Thin filaments
○ Thick filaments

(c)

Figure LG 10-1 Diagram of skeletal muscle. (a) Skeletal muscle cut to show cross section and longitudinal section with connective tissues. (b) Section of one muscle cell (myofiber). (c) Detail of sarcomere of muscle cell. Color and label as indicated in Checkpoint B4.

■ **B9.** Refer to Figure LG 10-2 and identify structures involved in nerve stimulation of muscle in this exercise.

    a. A nerve impulse travels along (*an axon? a dendrite?*) of a nerve cell (at letter *A*) toward a muscle (letter _____). The axon is enlarged at its end to form a synaptic _____ (at letter *B*).

    b. Inside the bulb are synaptic vesicles (letter _____), which release chemical transmitter (letter _____) into the synaptic cleft (letter _____).

    c. The effectiveness of the transmitter is enhanced by the fact that the sarcolemma (letter _____) exhibits an extensive surface area providing the infolded _____ (letter *G*).

    d. The axon terminal along with the closely approximated muscle fiber sarcolemma is known as a _____ (letter *I*).

B10. For *extra review*. Color the parts of Figure LG 10-2 indicated by color code ovals.

■ **B11.** Muscles that control precise movements have (*more? fewer?*) muscle fibers per motor unit than muscles controlling gross movements.

■ **B12.** Complete this exercise describing the principal events that occur during muscle contraction and relaxation.

    a. A nerve impulse causes release of a chemical transmitter named _____ at a neuromuscular junction. This chemical combines with _____ on the muscle fiber membrane and initiates an impulse on the muscle membrane, from the sarcolemma to the _____ tubules and to the _____ reticulum.

    b. In a relaxed muscle the concentration of calcium ions (Ca^{2+}) in sarcoplasm is (*high? low?*). The effect of the nerve impulse and transmitter is to cause Ca^{2+} to pass from storage areas in _____ out to the sarcoplasm surrounding myofilments.

    c. In relaxed muscle myosin cross bridges are not attached to the actin in thin filaments and _____ is bound to myosin cross bridges, while the tropomyosin-_____ complex blocks binding sites on actin.

    d. The released calcium ions attach to (*myosin? troponin?*), causing a structural change which leads to exposure of binding sites on (*myosin? actin?*).

    e. The nerve impulse also causes (*formation? breakdown?*) of ATP located on myosin cross bridges. This results in release of energy which activates myosin cross bridges to bind to and move _____. The oarlike action of myosin cross bridges (heads of golf clubs) upon actin is called a _____ stroke.

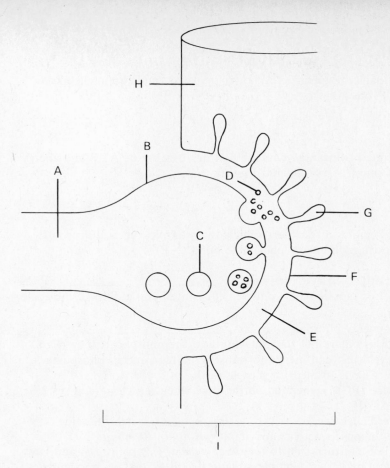

○ Motor neuron (axon with end bulb) ○ Neurotransmitter
○ Synaptic vesicle ○ Skeletal muscle myofiber (cell)

Figure LG 10-2 Diagram of a neuromuscular junction (motor end plate). Refer to Checkpoints B9 and B10.

f. Repeated power strokes slide actin filaments (*toward or even across? away from?*) the H zone and M line, and so shorten the sarcomere (and entire muscle).

g. Relaxation of a muscle occurs when an enzyme named _____ destroys ACh. This terminates impulse conduction over the muscle. Calcium ions

are then moved from sarcoplasm back into _____.

h. With a low level of Ca^{2+} now in the sarcoplasm surrounding myofilaments,

_____ reforms from ADP on myosin cross bridges, while tropomyosin-troponin complex once again blocks binding sites on

_____. As a result, thick and thin filaments detach, slip back

into normal position, and the muscle is said to _____.

i. The movement of Ca^{2+} back into sarcoplasmic reticulum (B12g) is (*an active? a passive?*) process. After death, a supply of ATP (*is? is not?*) available, so an active transport process cannot occur. Explain why the condition of rigor mortis results.

■ **B13.** Complete this exercise about energy sources for muscle contraction.

a. Breakdown of (*ADP? ATP?*) provides the energy muscles use for contraction. Recall from Checkpoint B12e that ATP is attached to (*actin? myosin?*) cross bridges and so is available to energize the power stroke. Complete the chemical reaction showing ATP breakdown.

   ATP →

b. ATP must be regenerated constantly. One method involves use of ADP and energy from food sources. Complete that chemical reaction.

   ADP+

   In essence ADP is serving as a transport vehicle that can pick up and drop off energy stored in an extra high energy phosphate bond (~P).

c. But ATP is used for other cell activities such as _____. To assure adequate energy for muscle work, muscle cells contain an additional molecule

   for transporting high energy phosphate; this is _____.
   Complete the reaction in Figure LG 10-3 showing how the phosphocreatine (PC) and ADP transport "vehicles" can meet and transfer the high energy phosphate "trailer" so that more ATP is formed for muscle work.

d. How is phosphocreatine (PC) regenerated during time when muscles are at rest? Show this on Figure LG 10-3.

e. ATP and PC, together called the _____ system, provide only enough energy to power muscle activity for about (*an hour? 10 minutes? 15 seconds?*). After that, muscles turn first to (*aerobic? anaerobic?*) pathways, and later to (*aerobic? anaerobic?*) pathways. Complete Figure LG 10-4 to show how muscles get energy via these pathways.

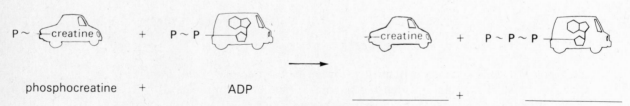

phosphocreatine   +   ADP   _____ +   _____

Figure LG 10-3   High energy molecules of the phosphagen system: the (~P) "trailer" tradeoff. Complete figure as indicated in Checkpoint B13c, d.

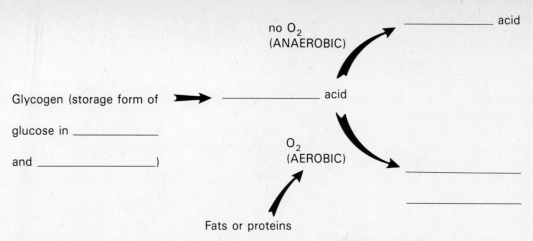

Glycogen (storage form of

glucose in _____

and _____)

no O₂
(ANAEROBIC)

_____ acid

_____ acid

O₂
(AEROBIC)

_____

_____

Fats or proteins

Figure LG 10-4 Anaerobic and aerobic energy sources for muscles. Complete figure as indicated in Checkpoint B13e.

B14. State the all-or-none principle.

■ **B15.** This principle applies to (*individual motor units? an entire muscle such as the biceps?*).

■ **B16.** List three factors than can decrease the strength of a muscle contraction.

B17. Briefly discuss electromyography (EMG). Why and how is this procedure performed?

C. Kinds of contractions; types of skeletal muscle fibers   (pages 242–247)

■ **C1.** Match the terms in the box with the definitions below.

L. Latent period	R. Refractory period
M. Myogram	T. Twitch

_____ a. Rapid, jerky response to a single
       stimulus

_____ b. Recording of a muscle contraction

_____ c. Period between application of a
       stimulus and start of a contraction

_____ d. Period when a muscle is not respon-
       sive to a stimulus

■ **C2.** Choose the type of contraction that fits each descriptive phrase.

Muscle tone	Treppe	Tetanus

a. Sustained partial contraction of some portion of skeletal muscle; some fibers contracted,

   others not: _____
b. Sustained contraction due to stimulation at a rate of 30 stimuli per second:

   _____
c. More forceful contraction of skeletal muscle in response to same strength stimuli after muscle has contracted several times:

   _____

d. Phenomenon that is the principle behind athletic warmups: _____
e. Voluntary contraction of muscles, such as

   biceps: _____

■ **C3.** Contrast isometric and isotonic contractions by doing this exercise.
   a. A contraction in which a muscle shortens while tension (tone) of the muscle remains constant is known as an (*isometric? isotonic?*) contraction.
   b. In an isometric contraction the muscle length (*shortens? stays about the same?*), and tension of the muscle (*increases? stays the same?*).
   c. *A clinical challenge.* (*Isometric? Isotonic?*) exercise, such as carrying heavy objects, weight lifting, or waterskiing may be dangerous for cardiac patients since blood pressure may increase dramatically during the exercise period.

C4. Define *wave summation*. Explain how summation is involved in tetanic contractions.

C5. Explain why a muscle must be stretched adequately but not excessively in order to produce an optimal contraction.

C6. *A clinical challenge.* Complete this exercise about clinical aspects of muscle change.
   a. (*Spastic? Flaccid?*) is a term that refers to muscles with less than normal muscle tone. State a cause of flaccid muscles.

   b. _____ is a term that means decrease in size or wasting away of muscles. Explain how transcutaneous muscle stimulations (TMS) may prevent atrophy.

   c. Hypertrophy means an increase in (*number? size?*) of muscle fibers, as in development of muscles through exercise.

■ **C7.** The substance in muscle that stores oxygen until oxygen is needed by mitochondria is _____. This protein is structurally somewhat like the

_____-globin molecule in blood that also binds to and stores

oxygen. Both of these molecules have a _____ color that accounts for the color of blood and also of red muscle. White muscle fibers have a relatively (*high? low?*) amount of myoglobin.

■ **C8.** Refer to the table below and contrast the three types of skeletal muscle by doing this activity.

Type and Rate of Contraction	Color	Myoglobin Concentration	Mitochondria and Blood Vessels	Source of ATP	Fatigue Easily?
I. Slow	Red	High	Many	Aerobic	No
IIA. Fast	Red	Very high	Very high	Aerobic	Moderate
IIB. Fast	White	Low	Few	Anaeorbic	Yes

a. Postural muscles (as in neck and back) are used (*constantly? mostly for short bursts of energy?*). Therefore it is appropriate that they contract at a (*fast? slow?*) rate. Such muscle tissue consists mostly of slow red fibers that (*do? do not?*) fatigue easily. These appear red since they have (*much? little?*) myoglobin and (*many? few?*) blood vessels. They have (*many? few?*) mitochondria and therefore can depend on (*aerobic? anaerobic?*) metabolism for energy.

b. Muscles of the arms are used (*constantly? mostly for short bursts of energy?*) as in lifting and throwing. Therefore they must contract at a (*fast? slow?*) rate. The arms consist mainly of fast twitch white fibers. These respond rapidly to nerve impulses and contract (*fast? slowly?*). They (*do? do not?*) fatigue easily since they are designed for the relatively inefficient processes of (*aerobic? anaerobic?*) metabolism.

c. Endurance exercises tend to enhance development of the more efficient (aerobic)

_____ fibers, whereas weight lifting, which requires short

bursts of energy, tends to develop _____ fibers.

D. Cardiac muscle tissue, smooth muscle tissue (pages 247–251)

■ **D1.** Compare the structure and function of skeletal and cardiac muscle by completing this table. State the significance of characteristics that have asterisks (*).

Characteristic	Skeletal Muscle	Cardiac Muscle
a. Number of nuclei per myofiber	Several (multinucleate)	
b. Number of mitochondria per myofiber (More or Fewer)*		
c. Striated appearance due to alternated actin and myosin myofilaments (Yes or No)	Yes	
d. Arrangement of muscle fibers (Parallel or Branching)		
e. Nerve stimulation required for contraction (Yes or No)*		
f. Length of refractory period (Long or Short)*		

■ **D2.** Contrast smooth muscle with the muscle tissue you have already studied by circling the correct answers in this paragraph.

Smooth muscle fibers are (*cylinder? spindle?*)-shaped with (*several nuclei? one nucleus?*) per cell. They (*do? do not?*) contain actin and myosin. However, due to the irregular arrangement of these filaments, smooth muscle tissue appears (*striated? nonstriated or "smooth"?*). In general, smooth muscle contracts and relaxes more (*rapidly? slowly?*) than skeletal muscle does, and smooth muscle holds the contraction for a (*shorter? longer?*) period of time than skeletal muscle does.

D3. Compare the two types of smooth muscle.

Muscle Type	Structure	Spread of Stimulus	Locations
a. Visceral		Impulse spreads and causes contraction of adjacent fibers	
b.			Blood vessels, iris of eye

■ **D4.** Smooth muscle (*is? is not?*) normally under voluntary control. List three chemicals released in the body that can also lead to smooth muscle contraction.

■ **D5.** Arrange the three types of your muscle tissue in correct sequence according to ability (from most to least) to regenerate during your lifetime.

_____ _____ _____ A. Heart muscle
B. Biceps muscle
C. Muscle of an artery or intestine

D6. *For extra review* of the three muscle types. Review Chapter 4, Checkpoints E1-E3, and Figure LG 4-1, M–O, pages LG 70 and 76–77.

E. Homeostasis (pages 252–253)

■ **E1.** Review Figure LG 10-4 and complete this exercise about muscle contraction in the presence or absence of oxygen.

a. During strenuous exercise an adequate supply of oxygen may not be available to

muscles. They must convert pyruvic acid to _____ acid by an (*aerobic? anaerobic?*) process. Excessive amounts of lactic acid in muscle tissue

may account for muscle _____.

b. After exercising, a person breathes rapidly for a while. During this period sufficient

oxygen enters the body to "pay back the _____ debt," that is, to provide oxygen for the complete catabolism of stored lactic acid to

_____, _____, and much energy in the

form of _____.

E2. Relate these terms to training of athletes:

a. Maximal oxygen uptake

b. "Second wind"

■ **E3.** Indicate what percent of energy used during muscular contraction may go toward each of these purposes:

a. Mechanical work (contraction): _____ percent

b. Heat production (to maintain body temperature): _____ percent

F. Aging and development of muscles (pages 253–254)

F1. List three changes in skeletal muscles that occur with normal aging.

■ **F2.** Complete this exercise about the development of muscles.

a. Which of the three types of muscle tissue develop from mesoderm? (*Skeletal? Cardiac? Smooth?*).

b. Part of the mesoderm forms columns on either side of the developing nervous system. This tissue segments into blocks of tissue called _____. The first pair of somites forms on day (*10? 20? 30?*) of gestation. By day 30 a total of _____ pairs of somites are present.

c. Which part of a somite develops into vertebrae? (*Myo-? Derma-? Sclero-?*) derm. The dermis of the skin and other connective tissues are formed from

_____-tomes, while most skeletal muscles develop from

_____-tomes.

G. Disorders, medical terminology (pages 254–255)

■ **G1.** Write the correct medical term related to muscles after its description.

a. Muscle or tendon pain and stiffness, such as "charleyhorse":

_____

b. Formation of fibrous tissue where it does not belong, for example, replacing skeletal or cardiac muscle: _____

c. Inherited, muscle-destroying disease causing atrophy of muscles:

_____

d. A muscle tumor: _____

e. Any disease of muscle: _____

f. Loss or impairment of motor or muscular function due to nerve or muscle

disorder: _____

■ **G2.** Describe myasthenia gravis in this exercise.

a. In order for skeletal muscle to contract, a nerve must release the chemical

_____ at the myoneural junction. Normally ACh binds to

_____ on the muscle fiber membrane.

b. It is believed that a person with myasthenia gravis produces

_____ that bind to these receptors, making them unavailable for ACh binding. Therefore, ACh (*can? cannot?*) stimulate the muscle, and it is weakened.

c. One treatment for this condition employs _____ drugs, which enhance muscle contraction by permitting the ACh molecules that *do* bind to act longer (and not be destroyed by AChE).

d. Other treatments include _____-suppressants, which decrease the patient's antibody production, or _____, which segregates the patient's harmful antibodies.

G3. Define each of these types of abnormal muscle contractions.

a. Spasm

c. Tic

b. Fibrillation

A2. Motion, heat maintenance, and posture.
A3. (a) Sm. (b) C. (c) Sk.
B1. (a) S. (b) D. (c) S. (d) S.
B3. Fasciculus, myofiber (muscle cell), myofibril, myofilament (thick or thin)
B4.

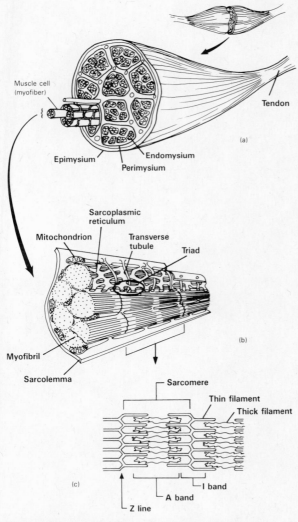

Figure LG 10-1A Diagram of skeletal muscle. (a) Skeletal muscle cut to show cross section and longitudinal section with connective tissues. (b) Section of one muscle cell (myofiber). (c) Detail of sarcomere of muscle cell.

B5. (a) Sarcoplasm. (b) Triad. (c) Sarcomere. (d) A band. (e) I band.
B6. (a) Actin, myosin. (b) Tropomyosin, troponin, calcium. (c) Thick, golf clubs, actin and ATP.
B8. (a) Myosin, actin. (b) Thin, slide, sliding filament.
B9. (a) An axon, H, bulb. (b) *C, D. E.* (c) *F,* subneural clefts. (d) Neuromuscular junction, myoneural junction (MNJ), or motor end plate.
B11. Fewer.
B12. (a) Acetylcholine (ACh), receptors, transverse, sarcoplasmic. (b) Low, sarcoplasmic reticulum. (c) ATP, troponin. (d) Troponin, actin. (e) Breakdown, actin, power. (f) Toward or even across. (g) Acetylcholinesterase (AChE), sarcoplasmic reticulum. (h) ATP, actin (or thin filament), relax (or lengthen). (i) An active, is not; myosin cross bridges stay attached to actin and the muscles remain in a state of partial contraction (rigor mortis) for 1–2 days.
B13. (a) ATP, myosin, ATP → ADP + P + energy. (b) ADP + P + energy (from foods) → ATP. (c) Active transport; creatine that combines with phosphate to form phosphocreatine (PC); phosphocreatine (PC) + ADP → creatine + ATP. (d) PC + ADP ← C + ATP. (Draw arrow to LEFT in Figure LG 10-3.) (e) Phosphagen, 15 seconds, anaerobic, aerobic.

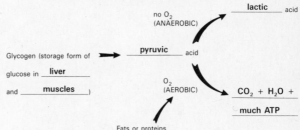

Figure LG 10-4A Anaerobic and aerobic energy sources for muscles.

B15. Individual motor units.
B16. Fatigue, lack of nutrients, lack of oxygen.

C1. (a) T. (b) M. (c) L. (d) R.
C2. (a) Muscle tone. (b) Tetanus. (c) Treppe. (d) Treppe. (e) Tetanus.
C3. (a) Isotonic. (b) Stays about the same, increases. (c) Isometric.
C7. Myoglobin, hemo-, red, low.
C8. (a) Constantly, slow, do not, much, many, many, aerobic. (b) Mostly for short bursts of energy, fast, fast, do, anaerobic. (c) Fast twitch red, fast twitch white.
D1.

Characteristic	Skeletal Muscle	Cardiac Muscle
a. Number of nuclei per myofiber	Several (multinucleate)	One
b. Number of mitochondria per myofiber (More or Fewer)*	Fewer	More, since heart muscle requres constant generation of energy
c. Striated appearance due to alternated actin and myosin myofilaments (Yes or No)	Yes	Yes
d. Arrangement of muscle fibers (Parallel or Branching)	Parallel	Branching
e. Nerve stimulation required for contraction (Yes or No)*	Yes	No, so heart can contract without nerve stimulation, but nerves can increase or decrease heart rate
f. Length of refractory period (Long or Short)*	Short	Long, to allow heart time to relax between beats

D2. Spindle, one nucleus, do, nonstriated or "smooth," slowly, longer.
D4. Is not; hormones, pH changes, O_2 and CO_2 levels, and certain ions.
D5. C B A
E1. (a) Lactic, anaerobic, fatigue. (b) Oxygen; carbon dioxide, water, ATP.
E3. (a) 15. (b) 85.
F2. (a) All three: skeletal, cardiac, and smooth. (b) Somites, 20, 44. (c) Sclero-, derma-, myo-.
G1. (a) Fibromyalgia. (b) Fibrosis. (c) Muscular dystrophy. (d) Myoma. (e) Myopathy. (f) Paralysis.
G2. (a) Acetylcholine (ACh), receptors. (b) Antibodies, cannot. (c) Anticholinesterase (anti-AChE). (d) Immuno-, plasmapheresis.

Questions 1–11: Circle T (true) or F (false). If the statement is false, change the underlined word or phrase so that the statement is correct.

T F 1. Tendons are <u>cords of connective tissue, whereas aponeuroses are broad, flat bands</u> of connective tissue.

T F 2. During contraction of muscle <u>both A bands and I bands</u> get shorter.

T F 3. Fascia is <u>one type of skeletal muscle tissue.</u>

T F 4. <u>Elasticity</u> is the ability of a muscle to be stretched or extended.

T F 5. Myasthenia gravis is caused by <u>an excess of acetylcholine production at the myoneural junction.</u>

T F 6. Muscle fibers remain relaxed if there are <u>few calcium ions in the sarcoplasm.</u>

T F 7. <u>Atrophy</u> means decrease in muscle mass.

T F 8. Muscles that are used mostly for quick bursts of energy (such as those in the arms) contain large numbers of <u>fast twitch white fibers.</u>

T F 9. Most of the energy released during muscle contraction is used for <u>heat production.</u>

T F 10. In a relaxed muscle fiber <u>thin and thick myofilaments overlap to form the A band.</u>

T F 11. As a result of aging, muscle tissue is largely <u>replaced by fat,</u> and muscle strength <u>decreases.</u>

Questions 12–13: Arrange the answers in correct sequence.

_____ _____ _____ 12. According to the amount of muscle tissue they surround, from most to least:

    A. Perimysium
    B. Endomysium
    C. Epimysium

_____ _____ _____ 13. From largest to smallest:

    A. Myofibril
    B. Myofilament
    C. Muscle fiber (myofiber or cell)

_____ 14. All of the following molecules are parts of thin filaments EXCEPT:

A. Actin   C. Tropomyosin
B. Myosin   D. Troponin

_____ 15. Choose the one statement that is FALSE:

A. A band refers to the anisotropic band.
B. The A band is darker than the I band.
C. Thick myofilaments reach the Z line in relaxed muscle.
D. The H zone contains thick myofilaments, but not thin ones.
E. Thick myofilaments are made of myosin.

_____ 16. Which of the following statements about the role of calcium in muscle contraction is FALSE?

A. Calcium ions are not necessary for muscle contraction.
B. Calcium ions are released from sarcoplasmic reticulum to sarcoplasm as a result of nerve stimulation.
C. Calcium ions are actively transported into the sarcoplasmic reticulum by calsequestrin and calcium ATPase.

_____ 17. Which statement about muscle physiology in the relaxed state is FALSE?

A. Myosin cross bridges are bound to ATP.
B. Calcium ions are stored in sarcoplasmic reticulum.
C. Myosin cross bridges are bound to actin.
D. Tropomyosin-troponin complex is bound to actin.

_____ 18. The staircase phenomenon refers to ____ contractions.

A. Tetanic   C. Tonic
B. Treppe   D. Isotonic

_____ 19. Choose the FALSE statement about cardiac muscle.

A. Cardiac muscle has a long refractory period.
B. Cardiac muscle has one centrally located nucleus per fiber.
C. Cardiac muscle cells are called cardiac muscle fibers.
D. Cardiac fibers are separated by intercalated discs.
E. Cardiac fibers are spindle-shaped with no striations.

_____ 20. Most voluntary movements of the body are results of ____ contractions.

A. Isometric   D. Tetanic
B. Fibrillation E. Spasm
C. Twitch

_____ 21. Muscle cells are nonstriated and spindle-shaped with one nucleus per cell.

_____ 22. Cardiac muscle remains contracted longer than skeletal muscle does because ___ is slower in cardiac than in skeletal muscle.

_____ 23. It is during the ___ period of a muscle contraction that calcium ions are released from sarcoplasmic reticulum and myosin cross bridge activity begins to occur.

_____ 24. ___ is the transmitter released from synaptic vesicles of axons supplying skeletal muscle.

_____ 25. ADP + phosphocreatine → ___. (Write the products.)

ANSWERS TO MASTERY TEST: ■ Chapter 10

True-False

1. T
2. F. A bands but not I bands
3. F. Dense connective tissue (which may surround skeletal muscle)
4. F. Extensibility
5. F. An autoimmune response in which antibodies are produced which bind onto receptors on the sarcolemma

Arrange

12. C A B
13. C A B

Multiple Choice

14. B   18. B
15. C   19. E
16. A   20. D
17. C

Fill-ins

21. Smooth
22. Passage of calcium ions from the extracellular fluid through sarcolemma
23. Latent
24. Acetylcholine
25. ATP + creatine

True-False

6. T   9. T
7. T   10. T
8. T   11. T

FRAMEWORK 11

The Muscular Systems

SKELETAL MUSCLES, MOVEMENT [A]
- Origin, insertion
- Lever, fulcrum
- Arrangements of fasciculi
- Group actions

NAMING SKELETAL MUSCLES [B]
- Direction
- Location
- Size
- No. of origins
- Shape
- Site of origin, insertion
- Action

PRINCIPAL SKELETAL MUSCLES

HEAD, NECK [C]
- Facial expression
- Mastication
- Eyeball movement
- Tongue movement
- Soft palate
- Pharynx
- Oral cavity
- Larynx
- Head movement

TRUNK MUSCLES [D]
- Abdominal wall
- Breathing
- Pelvic floor
- Perineum

UPPER BODY [E]
- Shoulder girdle
- Arm
- Forearm
- Wrist, hand, fingers

VERTEBRAL COLUMN [F]

LOWER BODY [G]
- Thigh
- Leg
- Foot & toes

INTRAMUSCULAR INJECTIONS, RUNNING INJURIES [H]

The Muscular System

More than 600 different muscles attach to the bones of the skeleton. Muscles are arranged in groups that work much as a symphony, with certain muscles quiet while others are performing. The results are smooth, harmonious movements, rather than erratic, haphazard discord.

Much information can be gained by a careful initial examination of two aspects of each muscle: its name and location. A muscle name may offer such clues as size or shape, direction of fibers, or points of attachment of the muscle. A look at the precise location of the muscle and the joint it crosses—combined with logic—can usually lead to a correct conclusion of action(s) of that muscle.

Chapter 11 focuses on all of the principal skeletal muscles of the body. First study the Chapter 11 Framework and the key terms associated with each section.

TOPIC OUTLINE AND OBJECTIVES

A. How skeletal muscles produce movement

1. Describe the relationship between bones and skeletal muscles in producing body movements.
2. Define a lever and fulcrum and compare the three classes of levers on the basis of placement of the fulcrum, effort, and resistance.
3. Identify the various arrangements of muscle fibers in a skeletal muscle and relate the arrangements to the strength of contraction and range of movement.
4. Discuss most body movements as activities of groups of muscles by explaining the roles of the prime mover, antagonist, synergist, and fixator.

B. Naming skeletal muscles

5. Define the criteria employed in naming skeletal muscles.

C. Principal skeletal muscles of the head and neck

D. Principal skeletal muscles that act on the abdominal wall, muscles used in breathing, muscles of the pelvic floor and perineum

E. Principal skeletal muscles that move the shoulder girdle and upper extremity

F. Principal skeletal muscles that move the vertebral column

G. Principal skeletal muscles that move the lower extremity

6. Identify the principal skeletal muscles in different regions of the body by name, origin, insertion, action, and innervation.

H. Intramuscular (IM) injections; running injuries

7. Discuss the administration of drugs by intramuscular (IM) injection.
8. Describe several injuries related to running.

WORDBYTES

Now study the following parts of words that may help you better understand terminology in this chapter.

Wordbyte	Meaning	Example	Wordbyte	Meaning	Example
bi-	two	*bi*ceps	grac-	slender	*grac*ilis
brachi-	arm	*brachi*alis	-issimus	the most	long*issimus* capitis
brev-	short	flexor digitorum *brev*is	lat-	broad	*lat*issimus dorsi
			maxi-	large	gluteus *maxi*mus
			mini-	small	gluteus *mini*mus
bucc-	mouth, cheek	*bucc*inator	or-	mouth	orbicularis *or*is
cap-, -ceps	head	*cap*itis, tri*ceps*	rect-	straight	*rect*us femoris
-cnem-	leg	gastro*cnem*ius	sartor-	tailor	*sartor*ius
delt-	Greek D (Δ)	*delt*oid	serra-	toothed, notched	*serra*tus anterior
gastro-	stomach, belly	*gastro*cnemius			
genio-	chin	*genio*glossus	teres-	round	*teres* major
glossus	tongue	*gloss*ary, stylo*glossus*	tri-	three	*tri*ceps femoris
			vast-	large	*vast*us lateralis
glute-	buttock	*glute*us medius			

CHECKPOINTS

A. How skeletal muscles produce movement   (pages 260–263)

■ **A1.** What structures constitute the *muscular system*?

■ **A2.** Refer to Figure LG 11-1 and consider flexion of your own forearm as you do this learning activity.

a. In flexion your forearm serves as a rigid rod, or _____, which

moves about a fixed point, called a _____ (your elbow joint, in this case).

b. Hold a weight in your hand as you flex your forearm. The weight plus your forearm serve as the (*effort? fulcrum? resistance?*) during this movement.

c. The effort to move this resistance is provided by contraction of a

_____. Note that if you held a heavy telephone book in your

hand while your forearm is flexed, much more _____ by your arm muscles would be required.

d. In Figure LG 11-1 identify the exact point at which the muscle causing flexion attaches to the forearm. It is the (*proximal? distal?*) end of the (*humerus? radius? ulna?*). Write an E and an I on the two lines next to the arrow at that point in the figure. This indicates that this is the site where the muscle exerts its effort (E) in the lever system, and it is also the insertion (I) end of the muscle. (More about insertions in a minute.)

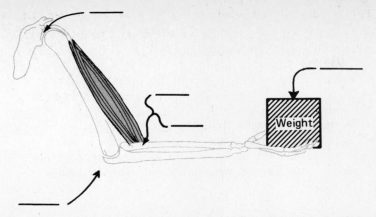

Figure LG 11-1 The lever-fulcrum principle is illustrated by flexion of the forearm. Complete the figure as directed in Checkpoint A2.

e. Each skeletal muscle is attached to at least two bones. As the muscle shortens, one bone stays in place and so is called the (*origin? insertion?*) end of the muscle. What bone in the figure appears to serve as the origin bone?

   Write O on that line at that point in the figure.
f. Now label the remaining arrows in Figure LG 11-1: F at fulcrum and R at resistance. This is an example of a (*first? second? third?*) class lever.

   _____

■ **A3.** Describe the roles of lever systems in the body by completing this exercise.

a. There are (*2? 3? 5? 7?*) classes of lever systems. Levers are categorized according to relative postions of effort (E), fulcrum (F), and resistance (R).
b. Now hyperextend your head as if to look at the sky (and refer to Figure 11-2a, page 262 of your text). The weight of your face and jaw serves as (*E? F? R?*), while your neck muscles provide (*E? F? R?*). The fulcrum is the joint between the

   _____ and the _____ bones. This is an

   example of a _____-class lever.

■ **A4.** Correlate fascicular arrangement with muscle power and range of motion of muscles.

a. A muscle with (*many? long?*) fibers will tend to have great strength. An example is the (*parallel? pennate?*) arrangement.
b. A muscle with (*many? long?*) fibers will tend to have great range of motion. An example is the (*parallel? pennate?*) arrangement.

■ **A5.** Refer again to Figure LG 11-1 and do this exercise about how muscles of the body work in groups.

a. The muscle that contracts to cause flexion of the forearm is called a

_____. An example of a prime mover in this action would be

the _____ muscle.

b. The triceps brachii must relax as the biceps brachii flexes the forearm. The triceps is an extensor. Since its action is opposite to that of the biceps, the triceps is called (*a synergist? an agonist? an antagonist?*) of the biceps.

c. What would happen if the flexors of your forearm were functional, but not the antagonistic extensors?

d. What action would occur if both the flexors and extensors contracted simultaneously?

e. Muscles that assist or cooperate with the prime mover to cause a given action are

known as _____, whereas muscles that stabilize a bone (such as the scapula) so that prime movers and synergists can move another bone (such

as the humerus) are called _____.

B. Naming skeletal muscles (pages 263)

■ **B1.** Review the Wordbyte section above. Match the names of the following muscles with their meanings.

> A. Large muscle of the buttock region
> B. Belly-shaped muscle in leg
> C. Thigh muscle with four origins
> D. The broadest muscle of the back
> E. Large muscle in medial thigh area
> F. Muscle that raises the upper lip

_____ a. Latissimus dorsi

_____ b. Vastus medialis

_____ c. Gluteus maximus

_____ d. Quadriceps femoris

_____ e. Gastrocnemius

_____ f. Levator labii superioris

■ **B2.** As you study the names of muscles, you will find that most of them provide a good description of the muscle. For each of the following, indicate the type of clue that each part of the name gives. The first one is done for you.

A. Action	N. Number of heads or origins
D. Direction of fibers	P. Points of attachment of origin and insertion
L. Location	S. Size or shape

__DL__ a. Rectus abdominis

_____ b. Flexor carpi ulnaris

_____ c. Biceps brachii

_____ d. Sternocleidomastoid

_____ e. Adductor longus

C. Principal skeletal muscles of the head and neck (pages 263–282)

■ **C1.** After studying Exhibit 11-2 (pages 267–268), in your text, check your understanding of the muscles of facial expression. Write the name of the muscle that answers each description. Locate muscles in Figure LG 11-2a, O–R. Cover the key and write the name of each facial muscle next to its lettered leader line.

a. Allows you to show surprise by raising your eyebrows and forming horizontal

forehead wrinkle: _____

b. Muscle surrounding opening of your mouth; allows you to use your lips in kissing

and in speech: _____

c. Muscle for smiling and laughing since it draws the outer portion of the mouth

upward and outward: _____

d. Circular muscle around eye; closes eye: _____

■ **C2.** Essentially all of the muscles controlling facial expression receive nerve impulses

via the _____ nerve, which is cranial nerve (_III? V? VII?_).

■ **C3.** Place your index finger and thumb on the origin and insertion of each of the muscles that move your lower jaw. (Refer to Exhibit 11-3 and Figure 11-5, pages 270–271 in the text, for help.) Then do this learning activity.

a. Two large muscles help you to close your mouth forcefully, as in chewing. Both of these act by (_lowering the maxilla? elevating the mandible?_). The

_____ covers your temple and the _____ covers the ramus of the mandible.

b. Most of the muscles involved in chewing are ones that help you to (_open? close?_) your mouth. Think about this the next time you go to the dentist and try to hold your mouth open for a long time, with only your _____ muscles to force your mouth wide open.

c. The muscles that move the lower jaw, and so aid in chewing, are innervated by

cranial nerve (_III? V? VII?_), known as the _____ nerve.

d. Refer to Figure LG 11-2a. Muscles _A_ and _B_ are used primarily for (_facial expressions? chewing?_), whereas muscles _O, P, Q,_ and _R_ are used mainly for (_facial expressions? chewing?_).

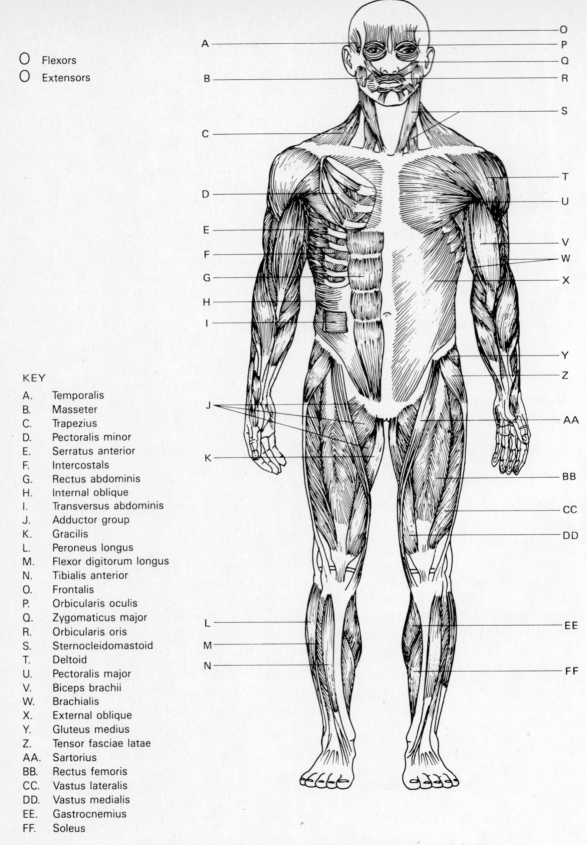

○ Flexors
○ Extensors

KEY

A. Temporalis
B. Masseter
C. Trapezius
D. Pectoralis minor
E. Serratus anterior
F. Intercostals
G. Rectus abdominis
H. Internal oblique
I. Transversus abdominis
J. Adductor group
K. Gracilis
L. Peroneus longus
M. Flexor digitorum longus
N. Tibialis anterior
O. Frontalis
P. Orbicularis oculis
Q. Zygomaticus major
R. Orbicularis oris
S. Sternocleidomastoid
T. Deltoid
U. Pectoralis major
V. Biceps brachii
W. Brachialis
X. External oblique
Y. Gluteus medius
Z. Tensor fasciae latae
AA. Sartorius
BB. Rectus femoris
CC. Vastus lateralis
DD. Vastus medialis
EE. Gastrocnemius
FF. Soleus

Figure LG 11-2 Major muscles of the body. Some deep muscles are shown on the left side of the figure. All muscles on the right side are superficial. Label and color as directed. (a) Anterior view. (b) Posterior view.

O Flexors
O Extensors

A
B
C
D
E
F
G
H
I
J
K

L
M
N
O
P
Q
R
S
T

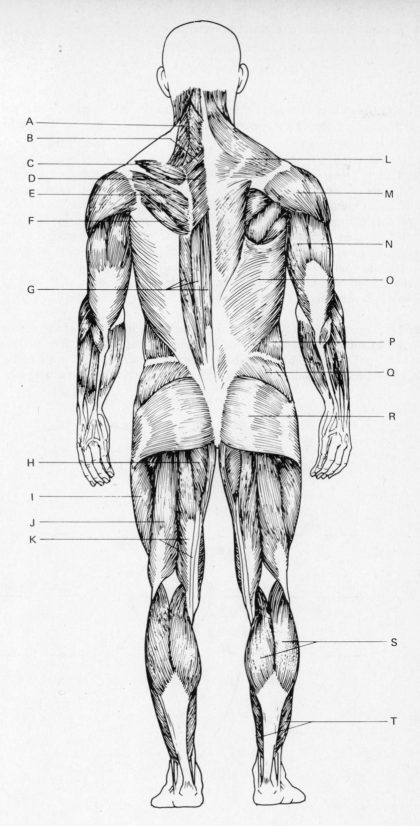

KEY

A. Levator scapulae
B. Rhomboideus minor
C. Supraspinatus
D. Rhomboideus major
E. Infraspinatus
F. Teres major
G. Sacrospinalis
H. Adductors
I. Vastus lateralis
J. Biceps femoris
K. Semimembranosus and
       semitendinosus
L. Trapezius
M. Deltoid
N. Triceps brachii
O. Latissimus dorsi
P. External oblique
Q. Gluteus medius
R. Gluteus maximus
S. Gastrocnemius
T. Soleus

■ **C4.** Complete this exercise about muscles that move the eyeballs.

a. Why are these muscles called *extrinsic* eyeball muscles?

b. Label Figure LG 11-3 with initials of the names of the six extrinsic eye muscles to indicate *direction of movement* of each muscle. One has been done for you.

IO. Inferior oblique
IR. Inferior rectus
LR. Lateral rectus

MR. Medial rectus
SO. Superior oblique
SR. Superior rectus

c. Now write next to each muscle the number of the cranial nerve innervating that muscle. One has been done for you.

d. Note that most of the extrinsic eye muscles are innervated by cranial nerve (*III? IV? VI?*), as indicated by the name of that nerve (*oculomotor*). Cranial nerve VI, the *abducens,* supplies the muscles that causes most (*medial? lateral?*) movement, or abduction of the eye. (*Hint:* Remember the chemical symbol for sulfate [SO_4] to remind yourself that the remaining muscle, superior oblique [SO] is supplied by the fourth cranial nerve.)

■ **C5.** *For extra review* of actions of eye muscles, work with a study partner. One person moves the eyes in a particular direction; the partner then names the eye muscles used for that action. *Note:* Will both eyes use muscles of the same name? For example, as you look to your right, will both eyes contract the lateral rectus?

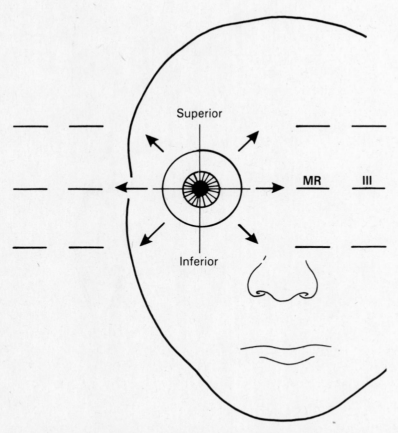

Figure LG 11-3 Right eyeball with arrows indicating directions of eye movements produced by the extrinsic eye muscles. Label as directed in Checkpoint C4.

■ **C6.** Actions such as opening your mouth for eating or speaking and also swallowing require integrated action of a number of muscles. You have already studied some that allow you to open your mouth (such as masseter and temporalis). We will now consider muscles of the tongue, floor of the oral cavity, pharynx, and larynx.

Identify locations of three bony structures relative to your tongue. Tell whether each is anterior or posterior and superior or inferior to your tongue. (For help refer to Figure 11-7, page 274 of the text.)

a. Styloid process of temporal bone _____

b. Hyoid bone _____

c. Chin (anterior portion) _____

■ **C7.** Remembering the locations of these bony points and that muscles move a structure (tongue) by pulling on it, determine the direction that the tongue is pulled by each of these muscles.

a. Styloglossus _____

b. Hyoglossus _____

c. Genioglossus _____

■ **C8.** The three muscles listed in Checkpoint C7 are innervated by cranial nerve

_____; the name _____ nerve indicates that this pair of cranial nerves supplies the region "under the tongue."

■ **C9.** Match groups of muscles in each answer with descriptions below.

> A. Styloglossus, hyoglossus, and genioglossus
> B. Superior, middle, and inferior constrictors
> C. Stylopharyngeus, salpingopharyngeus, and palatopharyngeus
> D. Stylohyoid, mylohyoid, and geniohyoid
> E. Thyrohyoid, sternohyoid, and omohyoid

_____ a. Located superior to the hyoid (suprahyoid), these form the floor of the oral cavity. (Recall that the hyoid is directly posterior to the inferior border of the U-shaped mandible.)

_____ b. These muscles cover the anterior of the larynx and trachea. These permit you to elevate your thyroid cartilage ("Adam's apple") during swallowing to prevent food from entering your larynx. (Try it.)

_____ c. Forming part of the walls of the pharynx, these muscles also elevate the larynx during swallowing. And they close off the nasopharynx (so that food does not back up into the nasal cavity) during swallowing. Finally one of these muscles opens your auditory tube (leading to the middle ear) during swallowing, an action that helps to "pop" your ears while you are descending from a high altitude.

_____ d. Three muscles that squeeze the pharynx to propel a bolus of food into the esophagus during swallowing.

_____ e. Muscles permitting tongue movements.

■ **C10.** Using a mirror, find the origin and insertion of your left sternocleidomastoid muscle. (See Figure 11-4, pages 268–269 in your text.) The muscle contracts when you pull your chin down and to the right; this diagonal muscle of your neck will then be readily located. Note that the left sternocleidomastoid pulls your face toward the (*same? opposite?*) side. It also (*flexes? extends?*) the head.

■ **C11.** Alternately flex and extend vertebrae of your neck by looking at your toes and then toward the sky. As you look at the sky, you are (*flexing? extending and then hyperextending?*) your head and cervical vertebrae. On which surface of the neck would you expect to find extensors of the head and neck? (*Anterior? Posterior?*) Note that these muscles are the most superior muscles of the columns of extensors of the vertebrae. Find them on Figure 11-21 (page 307) in your text. Now name three of these muscles.

D. Principal skeletal muscles that act on the abdominal wall, muscles used in breathing, muscles of the pelvic floor and perineum (pages 283–289)

■ **D1.** Each half of the abdominal wall is composed of (*two? three? four?*) muscles. Describe these in the exercise below.

a. Just lateral to the midline is the rectus abdominis muscle. Its fibers are (*vertical? horizontal?*), attached inferiorly to the _____ and superiorly to the _____. Contraction of this muscle permits (*flexion? extension?*) of the vertebral column.

b. List the remaining abdominal muscles that form the sides of the abdominal wall from most superficial to deepest.

_____ → _____ → _____
   most superficial   deepest

c. Do all three of these muscles have fibers running in the same direction? (*Yes? No?*) Of what advantage is this?

■ **D2.** Answer these questions about muscles used for breathing.

a. The diaphragm is _____-shaped. Its oval origin is located

_____. Its insertion is not into bone, but rather into dense connective tissue forming the roof of the diaphragm; this tissue is called the

_____.

b. Contraction of the diaphragm flattens the dome, causing the size of the thorax to (*increase? decrease?*), as occurs during (*inspiration? expiration?*).

c. The name *intercostals* indicates that these muscles are located

_____. Which set is used during expiration? (*Internal? External?*)

D3. *For extra review.* Cover the key to Figure LG 11-2a and write labels for muscles *F, G, H, I,* and *X.*

D4. Look at the inferior of the human pelvic bones on a skeleton (or refer to Figure 8-7, page 197 in your text). Note that a gaping hole (outlet) is present. Pelvic floor muscles attach to the bony pelvic outlet. Name these muscles and state functions of the pelvic floor.

■ **D5.** Fill in the blanks in this paragraph. *Diaphragm* means literally

_____ (*dia-*) _____ (*-phragm*). You are

familiar with the diaphragm that separates the _____ from the abdomen. The *pelvic diaphragm* consists of all of the muscles of the

_____ floor plus their fasciae. It is a "wall" or barrier between the inside and outside of the body.

E. Principal skeletal muscles that move the shoulder girdle and upper extremity (pages 290–304)

■ **E1.** From the list in the box, select the names of muscles that move the shoulder girdle as indicated.

> LS. Levator scapulae   PM. Pectoralis minor
> RMM. Rhomboideus   SA. Serratus anterior
>   major and minor   T. Trapezius

_____ a. Superiorly (elevation)   _____ c. Towards vertebrae (adduction)

_____ b. Inferiorly (depression)   _____ d. Away from vertebrae (abduction)

■ **E2.** Note on Figure LG 11-2a and 11-2b that the only 2 muscles listed in Checkpoint

E1 that are superficial are the _____ and a small portion of the

_____ .

■ **E3.** On Figure LG 11-2a and 11-2b identify the pectoralis major, deltoid, and latissimus dorsi muscles. All three of these muscles are (*superficial? deep?*). They are all directly involved with movement of the (*shoulder girdle? humerus? radius/ulna?*).

■ **E4.** Now write next to each muscle name listed below the letters of ALL *points of origin and insertion* and all *actions* that apply.

Points of Origin or Insertion	Actions (of Humerus)
C. Clavicle	Ab. Abducts
H. Humerus	Ad. Adducts
Sc. Scapula	EH. Extension, hyperextension
St. Sternum	F. Flexion
VS. Vertebrae and sacrum	

Muscles

a. Pectoralis major _____ _____

b. Deltoid _____ _____

c. Latissimus dorsi _____ _____

■ **E5.** Combine your knowledge of muscle actions with your knowledge of movements at joints from Chapter 9. Return to Figure 9-2, page 151, and write the name(s) of one or two muscles that produce each of the actions (a–h). Write muscle names next to each figure.

■ **E6.** Complete the table describing three muscles that move the forearm.

Muscle Name	Origin	Insertion	Action on Forearm
a.		Radial tuberosity	
b. Brachialis			
c.			Extension

■ **E7.** Complete this exercise about muscles that move the wrist and fingers.
   a. Examine your own forearm, palm, and fingers. There is more muscle mass on the (*anterior? posterior?*) surface. You therefore have more muscles that can (*flex? extend?*) your wrist and fingers.
   b. Locate the flexor carpi ulnaris muscle on Figure 11-19, page 300 in your text. What

   action does it have other than flexion of the wrist? _____
   What muscles would you expect to abduct the wrist?

c. What is the difference in location between flexor digitorum superficialis and flexor digitorum profundus?

d. What muscle helps you to point (extend) your index finger?

■ **E8.** *For extra review* of muscles that move the upper extremitites, write the name of one or more muscles that fit these descriptions.

a. Covers most of the posterior of the humerus _____

b. Turns your hand from palm down to palm up position _____

c. Originates from upper eight or nine ribs; inserts on scapula; moves scapula laterally

_____

d. Used when a baseball is grasped _____

e. Antagonist to serratus anterior_____
f. Largest muscle of the chest region; used to throw a ball in the air (flex humerus) and

to adduct arm_____
g. Raises or lowers scapula, depending on which portion of the muscle contracts

_____

h. Controls action at the elbow for a movement such as the downstroke in hammering

a nail _____
i. Hyperextends the humerus, as in doing the "crawl" stroke in swimming or exerting

a downward blow; also adducts the humerus _____

F. Principal skeletal muscles that move the vertebral column (pages 305–308)

■ **F1.** Describe the muscles that comprise the sacrospinalis. Locate and label on Figure LG 11-2b.

a. The sacrospinalis muscle is also called the _____.

b. The muscle consists of three groups: _____ (lateral),

_____ (intermediate), and _____
(medial).

c. In general, these muscles have attachments between _____.
d. They are (*flexors? extensors?*) of the vertebral column, and so are (*synergists? antagonists?*) of the rectus abdominis muscles.

■ **F2.** Explain why it is common for women in their final weeks of pregnancy to experience frequent back pains.

■ **F3.** Choose the correct origin and insertion of the scalene muscles:
  A. Ribs—iliac crest
  B. Cervical vertebrae—occipital and temporal bones
  C. Cervical vertebrae—first two ribs
  D. Thoracic vertebrae—sacrum

G. Principal skeletal muscles that move the lower extremity (pages 309–321)

G1. Cover the key in Figure LG 11-2a and 11-2b and identify by size, shape and location the major muscles that move the lower extremity.

■ **G2.** Now match muscle names in the box with their descriptions below.

Ad.	Adductor group	Ham.	Hamstrings
Gas.	Gastrocnemius	Il.	Iliopsoas
GMax.	Gluteus maximus	QF.	Quadriceps femoris
GMed.	Gluteus medius	Sar.	Sartorius

_____ a. Consists of four heads: rectus femoris and three vastus muscles (lateralis, medialis, and intermedius).

_____ b. This muscle mass lies in the posterior (flexor) compartment of the thigh; antagonist to quadriceps femoris.

_____ c. Attached to lumbar vertebrae, anterior of ilium, and lesser trochanter, it crosses anterior to hip joint.

_____ d. Large muscle mass of the buttocks; antagonist to the iliopsoas.

_____ e. The only one of these muscles located in the leg (between knee and ankle), it forms the "calf." Attaches to calcaneus by "Achilles tendon."

_____ f. Forms the medial compartment of the thigh; moves femur medially.

_____ g. Crossing the femur obliquely, it moves lower extremity into "tailor position."

_____ h. Located posterior to upper, outer portion of the ilium, it forms a preferred site for intramuscular (IM) injections.

■ **G3.** Complete the table by marking *X* next to each action produced by contraction of the muscles listed. (Some muscles will have two answers.)

Key to actions in table: Ab. Abduct; Ad. Adduct; EH, extend or hyperextend; F, flex

	Movements of Thigh (hip joint)				Movement of Leg (knee)	
	Ab	Ad	EH	F	E	F
a. Iliopsoas						
b. Gluteus maximus						
c. Adductor mass						
d. Tensor fasciae latae						
e. Quadriceps femoris						
f. Hamstrings						
g. Gracilis						
h. Sartorius						

■ **G4.** Refer to Figure LG 11-4.

  a. Locate the major muscle groups of the right thigh in this cross section. Color the groups as indicated by color code ovals on the figure.

  b. Cover the key to Figure LG 11-4 and identify each muscle. Relate position of muscles in this figure to views of muscles in Figures LG 11-2a and 11-2b.

■ **G5.** What muscles cause the actions i–p shown in Figure LG 9-2 (page LG 152)? Write the muscle names next to each diagram.

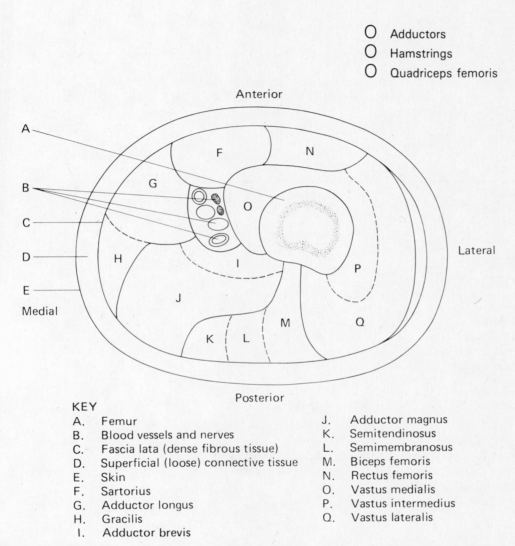

○ Adductors
○ Hamstrings
○ Quadriceps femoris

KEY

A. Femur
B. Blood vessels and nerves
C. Fascia lata (dense fibrous tissue)
D. Superficial (loose) connective tissue
E. Skin
F. Sartorius
G. Adductor longus
H. Gracilis
I. Adductor brevis

J. Adductor magnus
K. Semitendinosus
L. Semimembranosus
M. Biceps femoris
N. Rectus femoris
O. Vastus medialis
P. Vastus intermedius
Q. Vastus lateralis

Figure LG 11-4 Cross section of the right thigh midway between hip and knee joints. Structures *A* to *E* are nonmuscle; *F* to *Q* are skeletal muscles of the thigh. Color as directed in Checkpoint G4.

■ **G6.** Perform these actions of your foot and toes. Feel which muscles are contracting. Then match names of actions with descriptions.

DF. Dorsiflex	F. Flex toes
Ev. Evert foot	In. Invert foot
Ex. Extend toes	PF. Plantar flex

_____ a. Jump, as if to touch ceiling   _____ d. Lift toes upward away from floor

_____ b. Walk around on your heels   _____ e. Move sole of foot medially

_____ c. Curl toes down   _____ f. Move sole of foot laterally

■ **G7.** To review details of leg muscles that move the foot, complete this table. Note that muscles within a compartment tend to have similar functions. Muscles with * flex or extend only the great toe (not all toes). Use the same key for foot and toe actions as in the box for Checkpoint G6.

	DF PF	In Ev	F Ex
Posterior compartment: a. Gastrocnemius and soleus b. Tibialis posterior c. Flexor digitorum longus d. Flexor hallucis longus*			
Lateral compartment: e. Peroneus (longus and brevis)			
Anterior compartment: f. Tibialis anterior g. Extensor digitorum longus h. Extensor hallicis longus*			

■ **G8.** Do this Checkpoint about additional muscles of the foot.

a. Muscles whose main muscle mass lies within the foot (rather than the leg) are known as (*intrinsic? extrinsic?*) muscles of the foot.

b. Most intrinsic foot muscles are on the (*dorsal? plantar?*) surface of the foot. (*Hint:* Feel your own foot.) These muscles are arranged in (*two? four?*) layers. The first layer is most (*superficial? deep?*).

c. The muscle that moves the great toe medially is known as the

_____

■ **G9.** *For extra review* of all muscles, color flexors and extensors using color code ovals on Figures LG 11-2a and 11-2b. This activity will allow you to see on which sides of the body most muscles with those actions are located. Omit plantar flexors and dorsiflexors. Note that some muscles are flexors *and* extensors—at different joints. (See Mastery Test question 23 also.)

H. Intramuscular (IM) injections; running injuries (pages 322–325)

H1. State three reasons why intramuscular (IM) injections may be the method of choice for administration of drugs.

■ **H2.** *A clinical challenge.* Why is the gluteus medius considered a safer site for intramuscular injections than the gluteus maximus?

■ **H3.** Two other muscles commonly used for intramuscular injections are the

_____ in the thigh and the _____ in the upper extremity.

■ **H4.** Answer these questions about running injuries.

a. The most common site of injury for runners is the (*calcaneal tendon? groin? hip? knee?*).

b. *Patellofemoral stress syndrome* is a technical term for _____
Briefly describe this problem.

c. *Shinsplint syndrome* refers to soreness along the (*patella? tibia? fibula?*).
d. Write several suggestions you might make to a beginning runner to help to avoid runners' injuries.

ANSWERS TO SELECTED CHECKPOINTS

A1. Skeletal muscle tissues and connective tissues.
A2. (a) Lever, fulcrum. (b) Resistance. (c) Muscle, effort. (d) Proximal, radius. (e) Origin, scapula. (f) See Figure LG 11-1A; third.

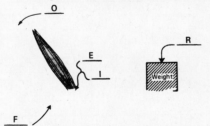

Figure LG 11-1A The lever-fulcrum principle is illustrated by flexion of the forearm.

A3. (a) 3. *R, E*, atlas, occipital, first.
A4. (a) Many, pennate. (b) Long, parallel.

A5. (a) Prime mover (agonist), biceps brachii. (b) An antagonist. (c) Your forearm would stay in the flexed position. (d) None: each opposing muscle would negate the action of the other. (e) Synergists, fixators.
B1. (a) D. (b) E. (c) A. (d) C. (e) B. (f) F.
B2. (b) A, P, L. (c) N, L. (d) P. (e) A, S.
C1. (a) Frontalis portion of epicranius. (b) Orbicularis oris. (c) Zygomaticus major. (d) Orbicularis oculi.
C2. Facial, VII.
C3. (a) Elevating the mandible, temporalis, masseter. (b) Close, lateral pterygoid. (c) V, trigeminal. (d) Chewing, facial expression.

C4. (a) They are outside of the eyeballs, not intrinsic like the iris. (b) and (c) See Figure LG 11-3A. (d) III, lateral.

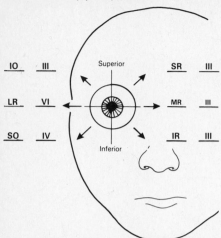

Figure LG 11-3A Right eyeball with arrows indicating directions of eye movements produced by the extrinsic eye muscles.

C5. No. The left eye contracts its medial rectus, while the right eye uses its lateral rectus and exerts some tension upon both oblique muscles.

C6. (a) Posterior and superior. (b) Inferior. (c) Anterior and inferior.

C7. Same answers as for C6.

C8. XII, hypoglossal.

C9. (a) D. (b) E. (c) C. (d) B. (e) A.

C10. Opposite, flexes.

C11. Extending and then hyperextending; posterior; longissimus capitis, semispinalis capitis and splenius capitis.

D1. Four. (a) Vertical; pubic crest and symphysis pubis; lower eight ribs; flexion. (b) External oblique, internal oblique, transversus abdominis. (c) No; strength is provided by the three different directions.

D2. (a) Dome, around the bottom of the rib cage and on lumbar vertebrae, central tendon. (b) Increase, inspiration. (c) Between ribs (costa), internal.

D5. Across, wall, thorax, pelvic.

E1. (a) LS, RMM, upper fibers of T. (b) PM, lower fibers of T. (c) RMM, T. (d) SA.

E2. Trapezius, serratus anterior.

E3. Superficial, humerus.

E4. (a) C, St, H; Ad, F. (b) C, H, Sc; Ab, EH (posterior fibers), F (anterior fibers). (c) H, VS; Ad, EH.

E5. (a) sternocleidomastoid; (b) three capitis muscles; (c) pectoralis major, coracobrachialis, deltoid (anterior fibers), biceps brachii; (d) latissimus dorsi, teres major, deltoid (posterior fibers), triceps brachii; (e) deltoid and supraspinatus; (f) pectoralis major, latissimus dorsi, teres major; (g) biceps brachii, brachialis, brachioradialis; (h) triceps brachii.

E6.

Muscle Name	Origin	Insertion	Action on Forearm
a. Biceps brachii	Scapula (2 sites)	Radial tuberosity (anterior)	Flexion, supination (*Note*: also flexes humerus)
b. Brachialis	Anterior humerus	Ulna (coronoid process)	Flexion
c. Triceps brachii	Scapula and 2 sites on posterior of humerus	Posterior of ulna (olecranon)	Extension (also extension of humerus)

E7. (a) Anterior, flex. (b) Adducts wrist; those lying over radius, such as flexor and extensor carpi radialis. (c) Superficialis is more superficial, and profundus lies deep. (d) Extensor indicis.

E8. (a) Triceps brachii. (b) Supinator and biceps brachii. (c) Serratus anterior. (d) Flexor digitorum superficialis and profundus. (e) Trapezius. (f) Pectoralis major. (g) Trapezius. (h) Triceps brachii. (i) Latissimus dorsi.

F1. (a) Erector spinae. (b) Iliocostalis, longissimus, spinalis. (c) Hipbone (ilium), ribs and vertebrae. (d) Extensors, antagonists.

F2. Extra weight of the abdomen demands extra support (contraction) by sacrospinalis muscles.

F3. C.

G2. (a) QF. (b) Ham. (c) Il. (d) GMax. (e) Gas. (f) Ad. (g) Sar. (h) Gmed.

G3.

	Movements of Thigh (hip joint)				Movement of Leg (knee)	
	Ab	Ad	EH	F	E	F
a. Iliopsoas				X		
b. Gluteus maximus	X		X			
c. Adductor mass		X		X		
d. Tensor fasciae latae	X					
e. Quadriceps femoris				X	X	
f. Hamstrings			X			X
g. Gracilis		X				X
h. Sartorius				X		X

G4. (a) Adductors: *G H I J;* Hamstrings: *KLM;* Quadriceps: *N O P Q.* (b) See key to Figure LG 11-4.

G5. (i) Iliacus + psoas (iliopsoas), rectus femoris, adductors, sartorius; (j) gluteus maximus, hamstrings, adductor magnus (posterior portion); (k) hamstrings, gracilis, sartorius, gastrocnemius; (l) quadriceps femoris; (m) same as i, but bilateral; (n) same as j but bilateral; (o) tensor fasciae latae, gluteus (medius and minimus), piriformis, and obturator internus; (p) adductors (longus, magnus and brevis), pectineus, and gracilis.

G6. (a) PF. (b) DF. (c) F. (d) Ex. (e) In. (f) Ev.

G7.

	DF	PF	In	Ev	F	Ex
Posterior compartment:						
a. Gastrocnemius and soleus		X				
b. Tibialis posterior		X	X			
c. Flexor digitorum longus		X	X		X	
d. Flexor hallucis longus*		X		X	X*	
Lateral compartment:						
e. Peroneus (longus and brevis)	X					
Anterior compartment:						
f. Tibialis anterior	X		X			
g. Extensor digitorum longus	X			X		X
h. Extensor hallicis longus*	X			X		X*

G8. (a) Intrinsic. (b) Plantar, 4, superficial. (c) Adductor hallucis.

G9. Figure LG 11-2a: flexors: G, J, K, S, T (anterior portion), U, V, W, AA, BB, EE; extensors: BB, CC, DD. Figure 11-2b: flexors: J, K, S; extensors: F, G, H, I, J, K, L, M (posterior fibers), N, O, R.

H2. The gluteus medius is superior and lateral to the sciatic nerve which runs deep to the gluteus maximus muscle.

H3. Vastus lateralis, deltoid.

H4. (a) Knee. (b) "Runner's knee," the patella tracks (glides) laterally, causing pain. (c) Tibia. (c) Replace worn shoes with new, supportive ones; do stretching and strengthening exercises; build up gradually; get proper rest; consider other exercise if prone to leg injuries since most (70 percent) runners do experience some injuries.

MASTERY TEST: Chapter 11

Questions 1–2: Arrange the answers in correct sequence.

_____ _____ _____ 1. Abdominal wall muscles, from superficial to deep:

    A. Transversus abdominis

    B. External oblique

    C. Internal oblique

_____ _____ _____ 2. From superior to inferior in location:

    A. Sternocleidomastoid

    B. Pelvic diaphragm

    C. Diaphragm and intercostal muscles

Questions 3–12: Circle T (true) or F (false). If the statement is false, change the underlined word or phrase so that the statement is correct.

T F 3. In extension of the thigh the hip joint serves as the fulcrum (F), while the hamstrings and gluteus maximus serve as the effort (E).

T F 4. The wheelbarrow and gastrocnemius are both examples of the action of second-class levers.

T F 5. The hamstrings are antagonists to the quadriceps femoris.

T F 6. The name deltoid is based on the action of that muscle.

T F 7. The most important muscle used for normal breathing is the diaphragm.

T F 8. The insertion end of a muscle is the attachment to the bone that does move.

T F 9. In general, adductors (of the arm and thigh) are located more on the medial than on the lateral surface of the body.

T F 10. Both the pectoralis major and latissimus dorsi muscles extend the humerus.

T F 11. The capitis muscles (such as splenius capitis) are extensors of the head and neck.

T F 12. The biceps brachii and biceps femoris are both muscles with two heads of origin located on the arm.

13. All of these muscles are located in the lower extremity EXCEPT:

    A. Hamstrings
    B. Gracilis
    C. Tensor fasciae latae
    D. Deltoid
    E. Peroneus longus

14. All of these muscles are located on the anterior of the body EXCEPT:

    A. Tibialis anterior
    B. Rectus femoris
    C. Sacrospinalis
    D. Pectoralis major
    E. Rectus abdominis

15. All of the following muscles are intrinsic muscles of the foot EXCEPT:

    A. Abductor digiti minimi
    B. Extensor digitorum brevis
    C. Extensor digitorum longus
    D. Dorsal interossei

16. All of these muscles are attached to ribs EXCEPT:

    A. Serratus anterior
    B. Intercostals
    C. Trapezius
    D. Iliocostalis
    E. External oblique

17. All of these muscles are directly involved with movement of the scapulae EXCEPT:

    A. Levator scapulae
    B. Pectoralis major
    C. Pectoralis minor
    D. Rhomboideus major
    E. Serratus anterior

18. All of these muscles have attachments to the coxal bones EXCEPT:

    A. Adductor muscles (longus, magnus, brevis)
    B. Biceps femoris
    C. Rectus femoris
    D. Vastus medialis
    E. Latissimus dorsi

19. The masseter and temporalis muscles are used for:

    A. Chewing
    B. Pouting
    C. Frowning
    D. Depressing tongue
    E. Elevating tongue

20. All of these muscles are used for facial expression EXCEPT:

    A. Zygomaticus major
    B. Orbicularis oculi
    C. Platysma
    D. Rectus abdominis
    E. Mentalis

Questions 21–25: fill-ins. Refer to Figures LG 11-2a and 11-2b. Write the word or phrase or key letters of muscles that best complete the statement or answer the questions.

_____ 21. Muscles *G, S, U, V,* and *W* on Figure LG 11-2a all have in common the fact that they carry out the action of ____.

_____ 22. Muscles *G, N, O,* and *R* on Figure LG 11-2b all have in common the fact that they carry out the action ____.

_____ 23. If you colored all flexors red and all extensors blue on these two figures, the view of the ____ surface of the body would appear more blue.

_____ 24. Choose the letters of all of the muscles listed below which would contract as you raise your left arm straight in front of you, as if to point toward a distant mountain: Figure 11-2a: *D T U V;* Figure LG 11-2b: *N O P*

_____ 25. Choose the letters of all of the muscles listed below that would contract as you raise your knee and extend your leg straight out in front of you, as if you are starting to march off to the distant mountain: Figure 11-2a: *AA BB CC DD;* Figure LG 11-2b: *H I J K R*

ANSWERS TO MASTERY TEST: ■ Chapter 11

Arrange

1. B C A
2. A C B

True-False

3. T
4. T
5. T
6. F. Shape
7. T
8. T
9. T
10. F. The latissimus dorsi extends, but the pectoralis major flexes
11. T
12. F. Two heads of origin; but origins of biceps brachii are on the scapula, and origins of biceps femoris are on ischium and femur

Multiple Choice

13. D   17. B
14. C   18. D
15. C   19. A
16. C   20. D

Fill-ins

21. Flexion
22. Extension
23. Posterior
24. Figure LG 11-2a: *T* (anterior fibers) *U V;* Figure LG 11-2b: none.
25. Figure LG 11-2a: *AA BB CC DD;* Figure LG 11-2b: *I.*

Control Systems
of the Human Body

FRAMEWORK 12

Nervous Tissue

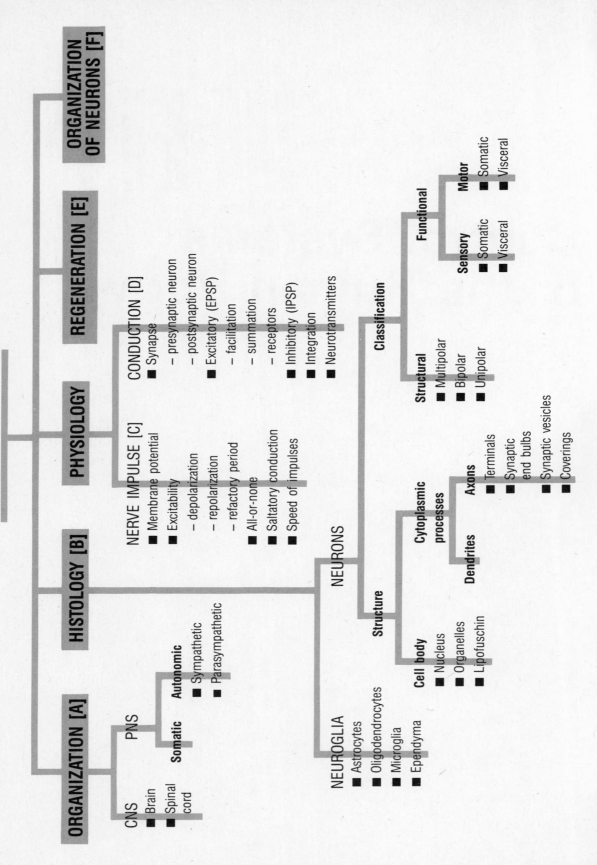

ORGANIZATION [A] HISTOLOGY [B] PHYSIOLOGY REGENERATION [E] ORGANIZATION OF NEURONS [F]

ORGANIZATION [A]

CNS
- ■ Brain
- ■ Spinal cord

PNS

Somatic

Autonomic
- ■ Sympathetic
- ■ Parasympathetic

HISTOLOGY [B]

NEUROGLIA
- ■ Astrocytes
- ■ Oligodendrocytes
- ■ Microglia
- ■ Ependyma

NEURONS

Structure

Cell body
- ■ Nucleus
- ■ Organelles
- ■ Lipofuschin

Cytoplasmic processes

Dendrites

Axons
- ■ Terminals
- ■ Synaptic end bulbs
- ■ Synaptic vesicles
- ■ Coverings

Classification

Structural
- ■ Multipolar
- ■ Bipolar
- ■ Unipolar

Functional

Sensory
- ■ Somatic
- ■ Visceral

Motor
- ■ Somatic
- ■ Visceral

PHYSIOLOGY

NERVE IMPULSE [C]
- ■ Membrane potential
- ■ Excitability
  - – depolarization
  - – repolarization
  - – refactory period
- ■ All-or-none
- ■ Saltatory conduction
- ■ Speed of impulses

CONDUCTION [D]
- ■ Synapse
  - – presynaptic neuron
  - – postsynaptic neuron
- ■ Excitatory (EPSP)
  - – facilitation
  - – summation
  - – receptors
- ■ Inhibitory (IPSP)
- ■ Integration
- ■ Neurotransmitters

Nervous Tissue

Two systems—nervous and endocrine—are responsible for regulating our diverse body functions. Each system exerts its control with the help of specific chemicals, namely neurotransmitters (nervous system) and hormones (endocrine system). In Unit III both systems of regulation will be considered, starting with the tissue of the nervous system.

Nervous tissue consists of two types of cells: neurons and neuroglia. The name neuroglia (*glia* = glue) offers a clue as to function of these cells: they bind, support, and protect neurons. Neurons perform the work of transmitting nerve impulses. These long and microscopically slender cells sometimes convey information several feet along a single neuron. Their function relies on an intricate balance between ions (Na^+ and K^+) found in and around nerve cells. Neurons release transmitters that bridge the gaps between adjacent neurons and at nerve–muscle or nerve–gland junctions. Analogous to a complex global telephone system, the nervous tissue of the human body boasts the design and organization that permits accurate communication, coordination, and integration of virtually all thoughts, sensations, and movements.

First introduce to your own nervous tissue the Chapter 12 Framework and note the organization of key terms in each section.

TOPIC OUTLINE AND OBJECTIVES

A. Organization

1. Identify the three basic functions of the nervous system in maintaining homeostasis.
2. Classify the organs of the nervous system into central and peripheral divisions.

B. Histology

3. Contrast the histological characteristics and functions of neuroglia and neurons.

C. Nerve impulses

4. List the sequence of events involved in the generation and conduction of a nerve impulse.
5. Define the all-or-none principle of nerve impulse transmission.
6. Discuss the factors that determine the speed of nerve impulse conductions.

D. Conduction across synapses; neurotransmitters

7. Define a synapse and list the factors involved in the conduction of a nerve impulse across a synapse.
8. Compare the functions of excitatory transmitter–receptor interactions and inhibitory transmitter–receptor interactions in helping to maintain homeostasis.

E. Regeneration of neurons

9. List the necessary conditions for the regeneration of nervous tissue.

F. Organization of neurons

10. Explain the organization of neurons in the nervous system.

Now study the following parts of words that may help you better understand terminology in this chapter.

Wordbyte	Meaning	Example	Wordbyte	Meaning	Example
astro-	star	*astro*cytes	neuro-	nerve	*neuro*n
dendr-	tree	*dendr*ite	olig-	few	*olig*odendroglia
-glia	glue	neuro*glia*	syn-	together	*syn*apse
lemm-	sheath	neuri*lemm*a			

CHECKPOINTS

A. Organization (page 332)

A1. List three principal functions of the nervous system.

■ **A2.** Check your understanding of organization of the nervous system by selecting answers that best fit descriptions below.

Aff. Afferent	Eff. Efferent
ANS. Autonomic nervous system	PNS. Peripheral nervous system
CNS. Central nervous system	SNS. Somatic nervous system

_____ a. Brain and spinal cord

_____ b. Sensory nerves

_____ c. Carry information from CNS to skeletal muscles

_____ d. Consists of sympathetic and parasympathetic divisions

B. Histology (pages 332–339)

■ **B1.** Write *neurons* or *neuroglia* after descriptions of these cells.

a. Conduct impulses from one part of the nervous system to another:

_____

b. Provide support and protection for the nervous system: _____
c. Bind nervous tissue to blood vessels, form myelin, and serve phagocytic functions:

_____

d. Smaller in size, but more abundant in number: _____

B2. Complete the table describing neuroglia cells.

Type	Description	Functions
a.	Star-shaped cells	
b. Oligoden-drocytes		
c.		Phagocytic
d.		Line ventricles of brain and central canal of spinal cord

■ **B3.** On Figure LG 12-1, label all structures with leader lines. Next draw arrows beside the figure to indicate direction of nerve impulses. Then color structures with color code ovals.

■ **B4.** Match the parts of a neuron listed in the box with the descriptions below.

> A. Axon
> CB. Cell body
> CS. Chromatophilic substance
>   (Nissl bodies)
> D. Dendrite
> L. Lipofuscin
> M. Mitochondria
> NF. Neurofibrils

_____ a. Contains nucleus; cannot regenerate since lacks mitotic apparatus

_____ b. Yellowish pigment that increases with age; appears to be byproduct of lysosomes

_____ c. Provide energy for neurons

_____ d. Long, thin fibrils composed of microtubules; may function in transport

_____ e. Orderly arrangement of rough ER; site of protein synthesis

_____ f. Conducts impulses toward cell body

_____ g. Conducts impulses away from cell body; has synaptic knobs that secrete chemical transmitter

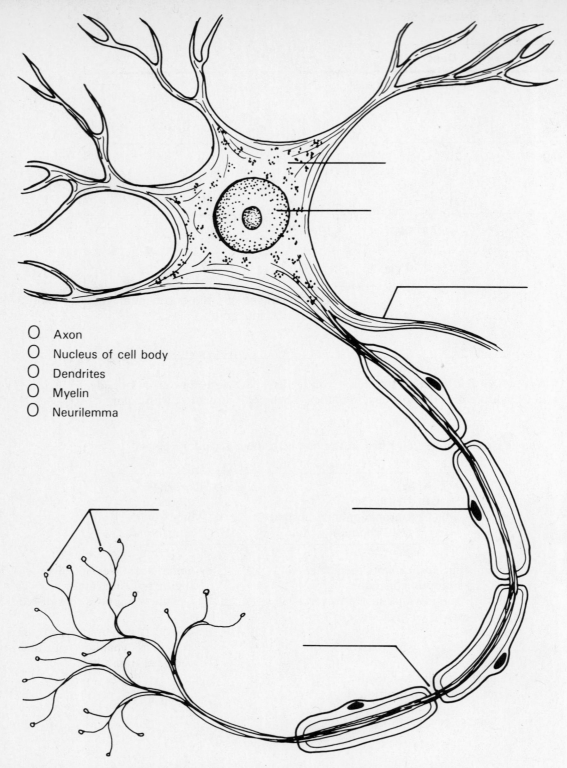

○ Axon
○ Nucleus of cell body
○ Dendrites
○ Myelin
○ Neurilemma

Figure LG 12-1 Structure of a typical neuron as exemplified by an efferent (motor) neuron. Complete the figure as directed in Checkpoint B3.

■ **B5.** Briefly explain how myelination relates to:

a. The color of white matter

b. The color of gray matter

c. Speed of nerve impulse transmission

■ **B6.** Describe how myelin is laid down on nerve fibers in the:

a. PNS

b. CNS

B7. Name and give a brief description of three types of neurons based on structural characteristics and three types of neurons classified according to functional differences.

Structural: *Functional:*

_____-polar   _____-fferent

_____-polar   _____-fferent

_____-polar   _____

■ **B8.** Identify phrases that describe a *neuron* and those that describe a *nerve*.

a. A single nerve cell; consists of a cell body with axon and dendrites:

_____

b. Bundle of axons and dendrites of many neurons, both afferent and efferent,

somatic and autonomic; contains no cell bodies: _____

c. Located entirely outside of the CNS and inside the PNS; macroscopic in diameter:

_____

■ **B9.** Using the list in the box, identify the type of nerve fiber that transmits each of the kinds of nerve impulses listed.

| GSA. General somatic afferent | GVA. General visceral afferent |
| GSE. General somatic efferent | GVE. General visceral efferent |

_____ a. All ANS fibers are of this type.

_____ b. Pain from a thorn in your skin is sensed via fibers of this types.

_____ c. Pain from a spasm of smooth muscle in the gallbladder is sensed by means of fibers of this type.

_____ d. With your eyes closed, you can tell the position of your skeletal muscles and joints due to nerve impulses that pass along this type of fiber.

_____ e. In order to increase your heart rate, impulses pass from your brain to your heart via this type of nerve fiber.

_____ f. These nerve fibers carry impulses from the CNS to muscles in your fingers used in writing.

C. Nerve impulses (pages 339–345)

■ **C1.** Refer to Figure LG 12-2a and complete this Checkpoint.

a. Note that one ion is 14× more concentrated outside the nerve cell; this is (K^+? Na^+?).

b. Potassium is _____ × more concentrated (*outside? inside?*) the neuron.

c. Now label the major cation in each site. Also label the major anion inside the neuron. (Values for all of these ions are shown graphically on Figure 27-2, page 864 in the text).

d. Color arrows across membranes using color code ovals to indicate types of transport processes for each ion. (Differences in thickness of arrows indicate relative rates of processes.)

e. The inside of a resting membrane is more (*positive? negative?*) than the outside. Explain why based on Figure LG 12-2a.

f. Since a difference in electrical charge exists between inside and outside of the neuron, the resting plasma membrane is said to be (*polarized? depolarized?*).

g. This difference in charge (between inside and outside of the membrane) is said to

be the membrane _____. This value is _____ millivolts (mV) in a resting membrane.

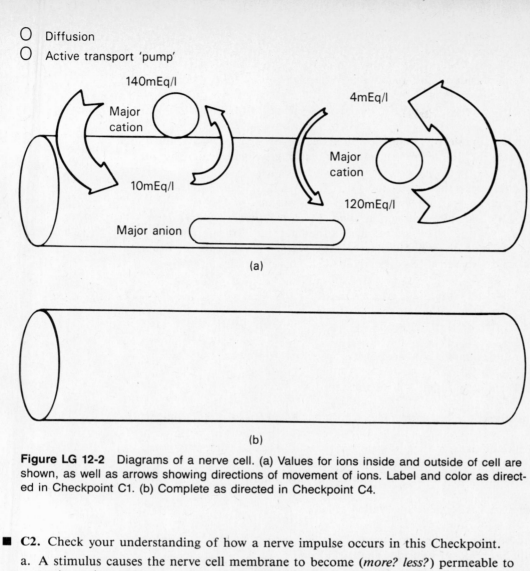

○ Diffusion
○ Active transport 'pump'

140mEq/l

Major cation

10mEq/l

Major anion

4mEq/l

Major cation

120mEq/l

(a)

(b)

Figure LG 12-2   Diagrams of a nerve cell. (a) Values for ions inside and outside of cell are shown, as well as arrows showing directions of movement of ions. Label and color as directed in Checkpoint C1. (b) Complete as directed in Checkpoint C4.

■ **C2.** Check your understanding of how a nerve impulse occurs in this Checkpoint.

a. A stimulus causes the nerve cell membrane to become (*more? less?*) permeable to Na$^+$. Na$^+$ can then enter the cell as voltage-sensitive Na$^+$_____ become activated and open.

b. At rest, the membrane had a potential of _____ mV. As Na$^+$ enters the cell, the inside of the membrane becomes more (*positive? negative?*). The potential will tend to go towards (*−80? −60?*) mV. The process of (*polarization? depolarization?*) is occurring. This process causes structural changes in more Na$^+$ channels so that even more Na$^+$ enters. This is an example of a (*positive? negative?*) feedback

mechanism. The result is a nerve impulse or nerve _____.

c. The membrane is completely depolarized at exactly (*−50? 0? +30?*) mV. Na$^+$ channels stay open until the inside of the membrane potential is (*reversed? repolarized?*) at +30 mV.

d. The nerve impulse is essentially a wave of negativity along the (*inside? outside?*) of

the nerve cell membrane. The impulse is propagated (or _____) along the nerve, as adjacent areas are depolarized, causing more channels to be activated and more (*Na$^+$? K$^+$?*) to enter.

e. After a fraction of a second, K⁺ voltage-sensitive channels at the site of the original stimulus become activated and open. K⁺ is more concentrated (*outside? inside?*) the cell (as you showed on Figure LG 12-2a); therefore K⁺ diffuses (*in? out?*). This causes the inside of the membrane to become more negative and return

to its resting potential of _____ mV. The process is known as (*de? re?*) -polarization.

f. During depolarization, the nerve cannot be stimulated at all. This period is known as the (*absolute? relative?*) refractory period. Only a stronger-than-normal stimulus

will result in an action potential during the _____ refractory period, which corresponds roughly with (*de? re?*)-polarization.

g. Depolarization and repolarization can occur (*only one time? thousands of times?*) before sufficient Na⁺ and K⁺ have shifted across the neuron cell membrane so as to interfere with continued nerve impulse formation. When that does finally occur, then Na⁺ and K⁺ are returned to their original sites. This occurs by an (*active?*

passive?) process known as the _____.

■ **C3.** Write the correct letter label of Figure LG 12-3a next to each description.

_____ a. The stimulus is applied at this point.

_____ b. Resting membrane potential is at this level.

_____ c. Membrane becomes so permeable to K⁺ that K⁺ diffuses rapidly out of the cell.

_____ d. The membrane is becoming more positive inside as Na⁺ enters; its potential is −30 mV. The process of depolarization is occurring.

_____ e. The membrane is completely depolarized at this point.

_____ f. The membrane is repolarizing at this point.

_____ g. Reversed polarization occurs; enough Na⁺ has entered so that this part of the cell is more positive inside than outside.

C4. *For extra review.* Show the events that occur during initiation and transmission of a nerve impulse on Figure LG 12-2b. Compare your diagram to Figure 12-8a to 12-8d (page 341) in the text.

C5. Define these terms.

a. Threshold stimulus

b. All-or-none principle

■ **C6.** Saltatory transmission occurs along (*myelinated? unmyelinated?*) nerve fibers. State two advantages of saltatory conduction.

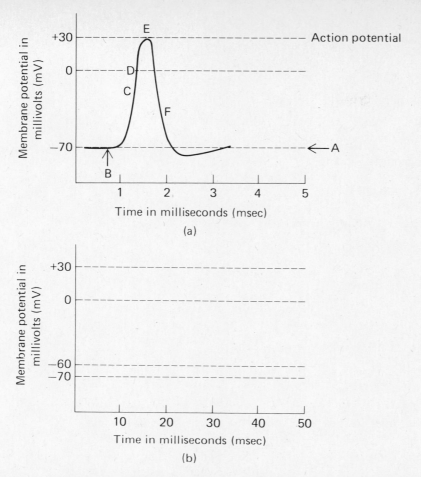

Figure LG 12-3 Diagrams for showing nerve action potentials. (a) Identify letter labels in Checkpoint C3. (b) Complete as directed in Checkpoints D5 and D7.

■ **C7.** The absolute refractory period of a large nerve fiber is about _____ msec; for a

small nerve fiber this period is about _____ msec. A nerve with a short refractory period can transmit (*more? fewer?*) impulses per second than a nerve with a long refractory period. So you can conclude that thick nerve fibers transmit more (*rapidly? slowly?*) than thin ones. Which type of nerve fiber is thickest? (*A? B? C?*)

D. Conduction across synapses; neurotransmitters (pages 345–350)

■ **D1.** In Chapter 10, you studied the point at which a neuron comes close to contacting

a muscle. This is known as a _____ junction, shown in Figure LG 10-2, page 165. The minute space between two neurons is known as a

_____, quite similar in structure. Does a nerve impulse actually "jump" across the space between two neurons? Explain briefly.

D2. Contrast terms in the following pairs:

a. Synapse/neuroglandular junction

b. Presynaptic neuron/postsynaptic neuron

c. Axosomatic synapse/axodendritic synapse

d. Divergence at synapse/convergence at synapse

e. Electrical synapse/chemical synapse

■ **D3.** Write these structures in order to show the pathway of transmission at a chemical synapse.

EB. End bulb of presynaptic neuron   SC. Synaptic cleft
 N. Neurotransmitter   SV. Synaptic vesicle
NR. Neurotransmitter receptor

_____ → _____ → _____ → _____ → _____
 First   Last

■ **D4.** Explain what accounts for one-way impulse conduction.

■ **D5.** Complete the following exercise.

a. A transmitter-receptor interaction refers to the effect that a certain chemical has

   upon the receptors of a _____-synaptic neuron.
b. An excitatory interaction (*raises? lowers?*) the −70 mV resting potential of the postsynaptic neuron, for example, to −65 mV. Such a potential is called an EPSP,

   which stands for _____.
c. If the excitation is not sufficient to cause a threshold potential, for example, if an

   EPSP of −65 is produced, then _____ is said to have occurred. Diagram a facilitation EPSP in green on Figure LG 12-3b.
d. If transmitter substance is released from several (or hundreds of) presynaptic knobs at one time, their combined effect may produce threshold EPSP. This phenomenon is known as (*spatial? temporal?*) summation. Temporal summation is that due to accumulation of transmitter from (*one? many?*) synaptic knob(s) over a period of time.

■ **D6.** In this Checkpoint, describe two theories explaining how neurotransmitters, binding to receptors on postsynaptic neurons, may cause Na^+ channels there to open so that a nerve impulse begins "on the other side of the synapse."

a. Binding of a transmitter such as _____ simply makes the channels more permeable to Na^+, which enters and begins (*de? re?*)-polarization of the postsynaptic neuron.

b. Binding of other transmitters activates the enzyme _____, which converts ATP to (*ADP? cyclic AMP?*). This chemical then activates enzymes that open Na^+ channels. _____ is the first messenger, whereas

_____ is the second messenger.

■ **D7.** Explain why an inhibitory postsynaptic potential (IPSP) is referred to as *hyperpolarization*. Diagram an IPSP in red on Figure LG 12-3b.

■ **D8.** In this Checkpoint, describe another mechanism besides IPSP that results in inhibition of nerve impulses.

a. An inhibitory neuron releases its inhibitory transmitter at a synapse with the synaptic end bulb of a(n) _____ neuron. The result is (*in? de?*)-crease in release of excitatory transmitter.

b. This type of inhibition is known as _____ inhibition. It may last for (*milliseconds? minutes or hours?*), for example, in sensory pathways.

■ **D9.** Seven statements about the action of the neurotransmitter acetylcholine (ACh) at a neuromuscular junction are listed below. Arrange these in correct sequence by writing letters on lines provided. ___ ___ ___ ___ ___ ___ ___

A. Quanta of ACh enter the synaptic cleft.
B. Alteration of the structure of receptors permits Na^+ to pass through channels into muscle cell. This depolarization leads to muscle contraction.
C. Calcium ions enter a presynaptic axon end bulb.
D. Acetylcholinesterase (AChase) which is present in subneural clefts of muscle cell membranes inactivates ACh.
E. As a result, ACh is released from synaptic vesicles or cytoplasm of the axon.
F. ACh binds to protein receptors in a sarcolemma (muscle cell membrane).
G. In the absence of ACh, the muscle fiber is repolarized so that it stops contracting until it receives another nerve impulse.

■ **D10.** What is the ultimate fate of the choline that results from the action of AChase?

■ **D11.** Name two transmitters that cause IPSP.

■ **D12.** Alkalosis tends to (*stimulate? depress?*) CNS activity, leading to convulsions, whereas acidosis tends to (*stimulate? depress?*) neuronal activity, leading to

_____.

■ **D13.** Check your understanding of chemicals that affect transmission at synapses and neuromuscular or neuroglandular junctions by completing this activity. Write E if the effect is excitatory or I if the effect is inhibitory. (Lines following descriptions are for the *clinical challenge* activity below.)

_____I_____ a. A chemical that inhibits release of ACh __3__

_____ b. A chemical that competes for the ACh receptor sites on muscle cells ____

_____ c. A chemical that inactivates acetylcholinesterase ____

_____ d. A chemical that increases threshold (for example, from −60 to −40) of a neuron ____

_____ e. A chemical that decreases threshold ____

A clinical challenge. Match the following chemicals with related mechanisms of actions. Write the number of the chemicals on lines to the right of the above descriptions. One is done for you.

1. Hypnotics, tranquilizers, anesthetics
2. Caffeine, benzedrine
3. Botulism toxin, inhibiting muscle contraction
4. Curare, a muscle relaxant
5. Nerve gas such as diisopropyl fluorophosphate

■ **D14.** *A clinical challenge.* Crack, a form of _____, (*causes? interferes with?*) inactivation of certain transmitters, such as

_____. This drug creates a short-term feeling of euphoria since the transmitter continues to function. Long-term effects include depression related to eventual (*excesses? depletion?*) of transmitters.

E. Regeneration of neurons (pages 350–351)

■ **E1.** Which of the following can regenerate if destroyed? Briefly explain why in each case.

a. Neuron cell body

b. CNS nerve fiber

c. PNS nerve fiber

F. Organization of neurons (pages 351–352)

F1. Define *neuronal pool, discharge zone,* and *facilitated zone.*

■ **F2.** Match the types of circuits in the box with related descriptions. Answers may be used more than once.

C. Converging	P. Parallel after-discharge
D. Diverging	R. Reverberating

_____ a. Impulse from a single presynaptic neuron causes stimulation of increasing numbers of cells along the circuit.

_____ b. An example is a single motor neuron in the brain that stimulates many motor neurons in the spinal cord, therefore activating many muscle fibers

_____ c. Branches from a second and third neuron in a pathway may send impulses back to the first, so the signal may last for hours, as in coordinated muscle activities

_____ d. One postsynaptic neuron receives impulses from several nerve fibers

_____ e. A single presynaptic neuron stimulates intermediate neurons which synapse with a common postsynaptic neuron, allowing this neuron to send out a stream of impulses, as in precise mathematical calculations

ANSWERS TO SELECTED CHECKPOINTS

A2. (a) CNS. (b) Aff. (c) SNS or Eff (*Hint:* remember S A M E: *Sensory = Afferent; Motor = Efferent.*) (d) ANS.

B1. (a) Neurons. (b) Neuroglia. (c) Neuroglia. (d) Neuroglia.

B3.

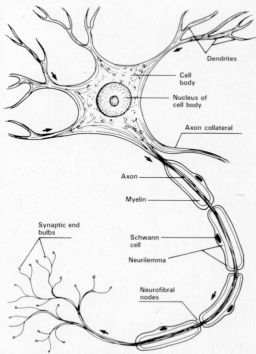

Figure LG 12-1A Structure of a typical neuron as exemplified by an efferent (motor) neuron.

B4. (a) CB. (b) L. (c) M. (d) NF. (e) CS. (f) D. (g) A.

B5. (a) Consists of myelinated fibers; myelin is white. (b) Consists of unmyelinated fibers, cell bodies, and neuroglia. (c) Speeds conduction by saltatory effect.

B6. (a) By neurolemmocytes wrapping their cell membranes around nerve fibers. (b) By a similar process involving oligodendroglial cells.

B8. (a) Neuron. (b) Nerve. (c) Nerve.

B9. (a) GVE. (b) GSA. (c) GVA. (d) GSA. (e) GVE. (f) GSE.

C1.

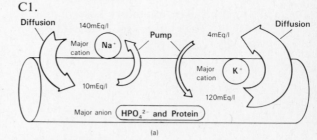

Figure LG 12-2A Diagrams of a nerve cell.

(a) Na^+. (b) 28–30, inside. (c, d) See Figure LG 12-2A. (e) Negative. Three reasons: more protein (negative ions) inside the cell, since protein is synthesized in the cell and is too large to cross the plasma membrane; $100\times$ more K^+ exits by diffusion than Na^+ enters by diffusion; 3 Na^+ are pumped out for every 2 K^+ pumped in. See Figure LG 12-2A. (f) Polarized. (g) Potential, -70.

215

C2. (a) More, channels. (b) −70, positive, −60, depolarization, positive, action potential. (c) 0, reversed. (d) Outside, transmitted, Na^+. (e) Inside, out, −70, re-. (f) Absolute, relative, re-. (g) Thousands of times, active, Na^+–K^+ pump.

C3. (a) B. (b) A. (c) F. (d) C. (e) D. (f) F. (g) E.

C6. Myelinated. Speed of saltatory impulse conduction is much greater than continuous (point-to-point) conduction. Saltatory conduction is more efficient: less energy is expended since fewer Na^+ and K^+ ions are displaced (that is, only those at neurofibral nodes of Ranvier), so less use of Na^+–K^+ pump is required.

C7. 0.4, 4.0, more, rapidly, A.

D1. Neuromuscular junction, synapse. No. The nerve impulse travels along the presynaptic axon. At the end of the the axon it stimulates synaptic vesicles there to release chemicals (transmitters) that enter the synaptic cleft and may then affect the postsynaptic neuron (either excite it or inhibit it).

D3. EB SV N SC NR

D4. Transmitters are released from axons (not dendrites or cell bodies).

D5.

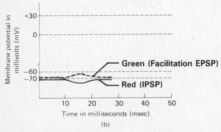

Figure LG 12-3bA   Diagram of a facilitation EPSP and an IPSP.

(a) Post. (b) Raises, excitatory postsynaptic potential. (c) Facilitation, see Figure LG 12-3bA. (d) Spatial, one.

D6. (a) Acetylcholine, de-. (b) Adenylate cyclase, cyclic AMP; the neurotransmitter, cyclic AMP.

D7. The inside of the postsynaptic membrane is even more negative (such as −90 mV) than the usual resting membrane due to its increased K^+ and/or Cl^- permeability. See Figure LG 12-3bA.

D8. Excitatory, de. (b) Presynaptic, minutes or hours.

D9. C, E, A, F, B, D, G

D10. It is recycled: it reenters the presynaptic axon and combines with acetate made in the end bulb to reform ACh with the help of the enzyme choline acetyltransferase.

D11. Gamma aminobutyric acid (GABA) and glycine.

D12. Stimulate; depress, decreased level of consciousness with possible coma and death.

D13. (b) I (4). (c) E (5). (d) I (1). (e) E (2).

D14. Cocaine, interferes with, dopamine, norepinephrine, serotonin; depletion.

E1. (a) No. About the time of birth, the mitotic apparatus is lost. (b) No. CNS nerve fibers lack the neurilemma necessary for regeneration, and scar tissue builds up by proliferation of astroglia cells. (c) Yes. PNS fibers do have a neurilemma. (See Chapter 13 for more about PNS nerve fiber regeneration.)

F2. (a) D. (b) D. (c) R. (d) C. (e) P.

Questions 1–2: Arrange the answers in correct sequence.

_____ _____ _____ 1. In order of transmission across synapse, from first structure to last:

   A. Presynaptic end bulb
   B. Postsynaptic neuron
   C. Synaptic cleft

_____ _____ _____ 2. Membrane potential values, from most negative to zero:

   A. Resting membrane potential
   B. Depolarized membrane potential
   C. Threshold potential

Questions 3–9: Circle the letter preceding the one best answer to each question.

3. Choose the one FALSE statement.

   A. Neurons in a discharge zone are more likely to receive enough transmitters to fire (start a nerve impulse) than neurons in a facilitated zone of a neuronal pool.
   B. In a resting membrane, permeability to K^+ ions is about 100 times less than permeability to Na^+ ions.
   C. C fibers are thin, unmyelinated fibers with relatively slow rate of nerve transmission.
   D. C fibers are more likely to innervate the heart and bladder than structures (such as skeletal muscles) that must make instantaneous responses.

4. Choose the one FALSE statement about the membrane of a resting neuron.

   A. It has a membrane potential of -70 mV.
   B. It is more negative inside than outside.
   C. It is depolarized.
   D. Its membrane potential is called resting potential.

5. Synaptic end bulbs are located:

   A. At ends of telodendria of axons
   B. On axon hillocks
   C. On neuron cell bodies
   D. At ends of dendrites
   E. At ends of both axons and dendrites

6. Which of these is equivalent to a nerve fiber?

   A. A neuron
   B. A neurofibril
   C. An axon or dendrite
   D. A nerve, such as sciatic nerve

7. *AChE* is an abbreviation for:

   A. Acetylcholine
   B. Norepinephrine
   C. Acetylcholinesterase
   D. Serotonin
   E. Inhibitory postsynaptic potential

8. A term that means the same thing as *afferent* is:

   A. Autonomic
   B. Somatic
   C. Peripheral
   D. Motor
   E. Sensory

9. An excitatory transmitter substance that changes the membrane potential from -70 to -65 mV causes:

   A. Impulse conduction
   B. Facilitation
   C. Inhibition
   D. Hyperpolarization

T F 10. The concentration of potassium ions (K^+) is about 28 to 30 times <u>greater</u> inside a resting cell than outside of it.

T F 11. Since CNS fibers contain no neurilemma, and the neurilemma produces myelin, CNS <u>fibers are all unmyelinated.</u>

T F 12. Transmitter substances are released at <u>synapses and also at neuromuscular junctions.</u>

T F 13. Generally, release of excitatory transmitter by <u>a single presynaptic knob</u> is sufficient to develop an action potential in the postsynaptic neuron.

T F 14. <u>Neuroglia</u> are a common source of tumors of the nervous system.

T F 15. In the <u>converging</u> circuit, a single presynaptic neuron influences several postsynaptic neurons (or muscle or gland cells) at the same time.

T F 16. Action potentials are measured in <u>milliseconds, which are thousandths of a second.</u>

T F 17. Nerve fibers with a short absolute refractory period can respond to <u>more rapid</u> stimuli than nerve fibers with a long absolute refractory period.

T F 18. Most neurons in the central nervous system (CNS) are classified as <u>unipolar.</u>

T F 19. A stimulus that is adequate will temporarily <u>increase permeability of the nerve membrane to Na^+.</u>

T F 20. The <u>brain and spinal nerves</u> are parts of the peripheral nervous system (PNS).

Questions 21–25: fill-ins. Complete each sentence with the word or phrase that best fits.

_____ 21. A postsynaptic neuron is called a(n) ____ since it receives both excitatory and inhibitory signals and determines its response according to the relative magnitude of these two types of signals.

_____ 22. Application of cold to a painful area can decrease pain in that area because ____.

_____ 23. One-way nerve impulse transmission can be explained on the basis of release of transmitters only from the ____ of neurons.

_____ 24. The neurilemma is found only around nerve fibers of the ____ nervous system, so that these fibers can regenerate.

_____ 25. In an axosomatic synapse, an axon's synaptic end bulb transmits nerve impulses to the ____ of a postsynaptic neuron.

ANSWERS TO MASTERY TEST: ■ Chapter 12

Arrange

1. A C B
2. A C B

Multiple Choice

3. B   7. C
4. C   8. E
5. A   9. B
6. C

True-False

10. T
11. F. May be myelinated since oligodendrocytes myelinate CNS fibers.
12. T
13. F. A number of presynaptic knobs
14. T
15. F. Diverging
16. T
17. T
18. F. Multipolar
19. T
20. F. Spinal nerves (as well as some other structures; but not the brain)

Fill-ins

21. Integrator
22. Cooling of neurons slows down the speed of nerve transmission, for example, of pain impulses
23. End bulbs of axons
24. Peripheral
25. Cell body (soma)

FRAMEWORK 13

The Spinal Cord & the Spinal Nerves

GROUPING OF NEURAL TISSUE [A]

SPINAL CORD

SPINAL NERVES [D]

DISORDERS [E]

PROTECTION, COVERINGS [B]
- Vertebrae
- Meninges
- Spaces
- Spinal tap

STRUCTURE, FUNCTION [C]

Features
- Length
- Enlargements
- Cross section

Horns (gray)

Columns (white)
- Ascending
- Descending

Functions

Impulse Conduction

Reflex Center

Reflex Arc
- Receptor
- Sensory neuron
- Synapse
- Motor neuron
- Effector

Types

Monosynaptic
- Stretch

Polysynaptic
- Tendon
- Flexor
- Crossed extensor

Diagnostic

Superficial

Deep Tendon
- Patellar
- Achilles
- Babinski
- Abdominal

NAMES (31 PAIRS)

COMPOSITION, COVERINGS
- Roots
  - – sensory
  - – motor
- Coverings

DISTRIBUTION

Ventral Rami
- Cervical plexus
- Brachial plexus
- Lumbar plexus
- Sacral plexus

Other Rami
- Dorsal
- Meningeal
- Communicantes

The Spinal Cord and the Spinal Nerves

CHAPTER
13

The spinal cord and spinal nerves serve as the major links in the communication pathways between the brain and all other parts of the body. Nerve impulses are conveyed along routes (or tracts) in the spinal cord, laid out much as train tracks: some head north to regions of the brain, and others carry nerve messages south from the brain toward specific body parts. Spinal nerves branch off from the spinal cord, perhaps like a series of bus lines that pick up passengers (nerve messages) to or from train depots (points along the spinal cord) en route to their final destinations. Organization is critical in this nerve impulse transportation network. Any structural breakdowns—by trauma, disease, or other disorders—can lead to interruption in service with resultant chaos (such as spasticity) or standstill (like sensory loss or paralysis).

Welcome aboard, and make your first stop the Chapter 13 Framework with key terms in each section.

TOPIC OUTLINE AND OBJECTIVES

A. Grouping of neural tissue

1. Describe how neural tissue is grouped.

B. Spinal cord: protection and coverings

C. Spinal cord: structure and function

2. Describe the protection, gross anatomical features, and cross sectional structure of the spinal cord.
3. List the location, origin, termination, and function of the principal ascending and descending tracts of the spinal cord.
4. Describe the components of a reflex arc and its relationship to homeostasis.
5. List and describe several clinically important reflexes.

D. Spinal nerves

6. Describe the composition and coverings of a spinal nerve.
7. Define a plexus and describe the composition and distribution of nerves of the cervical, brachial, lumbar, and sacral plexuses.

E. Disorders

8. Describe spinal cord injury and list the immediate and long-range effects.
9. Identify the effects of peripheral nerve damage and conditions necessary for its regeneration.
10. Explain the causes and symptoms of neuritis, sciatica, and shingles.

Now study the following parts of words that may help you better understand terminology in this chapter.

Wordbyte	Meaning	Example	Wordbyte	Meaning	Example
arachn-	spider	*arach*noid	pia	tender	*pia* mater
cauda-	tail	*cauda*te	rhiz-	root	*rhiz*otomy
dura	hard	*dura* mater	soma-	body	*soma*tic

CHECKPOINTS

A. Grouping of neural tissue (page 356)

■ **A1.** Locations that contain concentrations of cell bodies appear (*gray? white?*). Nerves that contain mostly unmyelinated nerve fibers look (*gray? white?*), whereas those that contain myelinated fibers appear (*gray? white?*).

■ **A2.** Complete the table about structures in the nervous system.

Structure	Color (Gray or White)	Composition (Cell Bodies or Nerve Fibers)	Location (CNS or PNS)
a. Nerve	White		
b. Tract			CNS
c. Ganglia		Cell bodies	
d. Nucleus	Gray		

B. Spinal cord: protection and coverings (pages 356–357)

■ **B1.** Refer to Figure LG 13-1 and do the following exercise.

a. Color the meninges to match color code ovals.
b. Label these spaces: *epidural space, subarachnoid space,* and *subdural space.*
c. Write next to each label of a space the contents of that space. Choose these answers: *cerebrospinal (CSF), serous fluid, fat and blood vessels.*

■ **B2.** *A clinical challenge.* Answer these questions related to meninges.

a. At what level of the vertebral column (not spinal cord) is the needle for a spinal tap (or lumbar puncture) inserted? Explain why it is done at this level. Into which space does it enter?

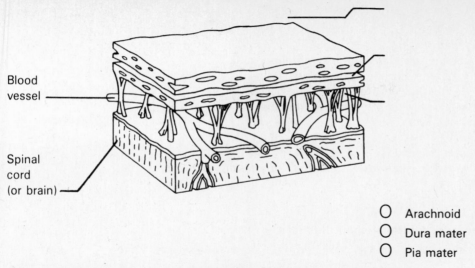

Blood vessel

Spinal cord (or brain)

○ Arachnoid
○ Dura mater
○ Pia mater

Figure LG 13-1 Meninges. Color and label as indicated in Checkpoint B1.

b. Explain why an epidural anesthetic exerts its effects on nerves with slower and more prolonged effects than a lumbar puncture. (*Hint:* Work this out logically based on the site at which anesthetic is deposited.) Also state one example of a procedure for which an "epidural" may be used.

c. *Pachymeningitis* is inflammation of the _____, whereas

leptomeningitis is inflammation of the _____.

C. Spinal cord: structure and function (pages 357–369)

■ **C1.** Match the names of the structures listed in the box with descriptions given.

| Ca. Cauda equina | F. Filum terminale |
| Co. Conus medullaris | S. Spinal segment |

_____ a. Tapering inferior end of spinal cord

_____ b. Any region of spinal cord from which one pair of spinal nerves arises

_____ c. Nonnervous extension of pia mater; anchors cord in place

_____ d. "Horse's tail"; extension of spinal nerves in lumbar and sacral regions within subarachnoid space

■ **C2.** Identify the following structures on Figure LG 13-2.

  a. On the right side of the diagram, write labels next to leader lines for the following structures: *anterior median fissure, anterior gray horn, lateral gray horn, posterior gray horn, gray commissure,* and *lateral white column.*

  b. Color the five tracts, using color code ovals. (*Note:* Tracts are actually present on both sides of the cord, but shown on only one side here.)

  c. Write next to the name of each tract its correct functions. Use these answers:
  Movements of axial skeleton
  Pain, temperature
  Precise movements (as of fingers)
  Proprioception, stereognosis, weight discrimination, vibration, and two-point discriminatory touch
  Touch and pressure

■ **C3.** Circle or fill in the correct answers about the tracts.

  a. Tracts conduct nerve impulses in the (*central? peripheral?*) nervous system. Functionally they are comparable to _____ in the peripheral nervous system.

  b. Tracts are located in (*gray horns? white columns?*). They appear white since they consist of bundles of (*myelinated? unmyelinated?*) nerve fibers.

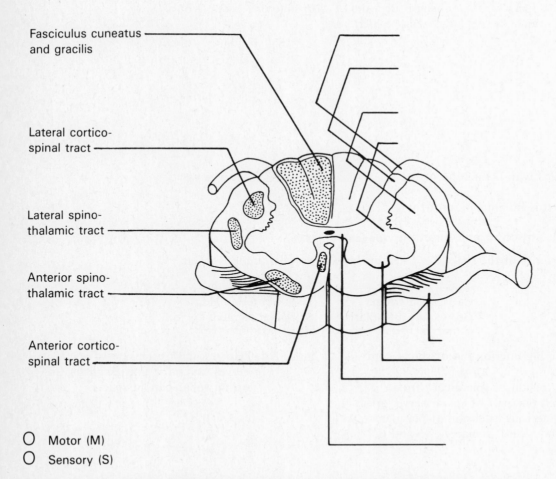

Fasciculus cuneatus and gracilis

Lateral cortico-spinal tract

Lateral spino-thalamic tract

Anterior spino-thalamic tract

Anterior cortico-spinal tract

O Motor (M)
O Sensory (S)

Figure LG 13-2 Outline of the spinal cord, roots, and nerves. Color and label according to Checkpoints C2, C4, and C5.

c. Ascending tracts are all (*sensory? motor?*), conveying impulses between the spinal

cord and the _____.

d. All motor tracts in the cord are (*ascending? descending?*).

e. The name *lateral corticospinal* indicates that the tract is located in the

_____ white column, that it originates in the (*cerebral cortex?*

thalamus? spinal cord?), and that it ends in the _____.

f. The lateral corticospinal tract is (*ascending, sensory? descending, motor?*).

C4. *For extra review.* Draw and label the following tracts on Figure LG 13-2. Then color them according to the color code ovals on the figure: *anterior spinocerebellar, posterior cerebellar, rubrospinal,* and *vestibulospinal.*

■ **C5.** Describe the two major functions of the spinal cord in this exercise.

a. The first main function of the spinal cord discussed above is to permit

_____ between the periphery, such as arms and legs, and the brain by means of the (*tracts? nerves? ganglia?*) located in the white columns of the cord.

b. The second main function of the cord is to serve as a _____ center by means of

spinal nerves. These are attached by _____ roots. The (*anterior? posterior?*) root

contains sensory nerve fibers, and the _____ root contains motor fibers. Color the sensory and motor roots on Figure LG 13-2. Select colors according to the color code ovals there.

■ **C6.** Label structures *A–H* on Figure LG 13-3, using the following terms: *effector, motor neuron axon, motor neuron cell body, receptor, sensory neuron axon, sensory neuron cell body, sensory neuron dendrite,* and *synapse.* Note that these structures are lettered in alphabetical order along the conduction pathway of a reflex arc. Add arrows showing the direction of nerve transmission in the arc.

■ **C7.** Answer these questions about the conduction pathway in Figure LG 13-3.

a. How many neurons does this reflex contain? _____ The neuron that conveys the

impulse toward the spinal cord is a _____ neuron; the one

that carries the impulses toward the effector is a _____ neuron.

b. This is a (*monosynaptic? polysynaptic?*) reflex arc. The synapse, like all somatic synapses, is located in the (*CNS? PNS?*).

c. Receptors, in this case located in skeletal muscle, are called

_____. They are sensitive to changes in

_____. This type of reflex might therefore be called a

_____ reflex.

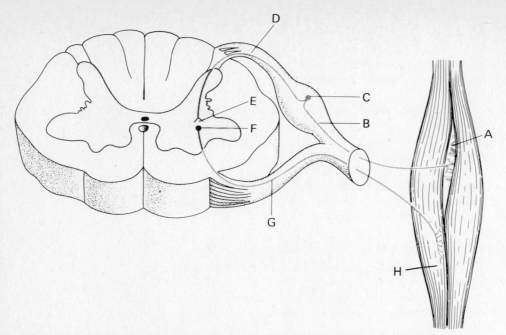

Figure LG 13-3 Reflex arc: stretch reflex. Label as directed in Checkpoints C6 and C7.

d. Since sensory impulses enter the cord on the same side as motor impulses leave, the reflex is called (*ipsilateral? contralateral? intersegmental?*).

e. What structure is the effector? _____

f. One example of a stretch reflex is the _____, in which stretching of the patellar tendon initiates the reflex.

■ **C8.** Explain how the brain may get the message that a stretch reflex (or other type of reflex) has occurred.

■ **C9.** Contrast stretch, flexor, and crossed extensor reflexes in this learning activity.

    a. A flexor reflex (*does? does not?*) involve association neurons, and so it is (*more? less?*) complex than a stretch reflex.

    b. A flexor reflex sends impulses to (*one? several?*) muscle(s), whereas a stretch reflex, such as the knee jerk, activates (*one? several?*) muscle(s), such as the quadriceps.

    A flexor reflex is also known as a _____ reflex.

    c. If you simultaneously contract the flexor and extensor (biceps and triceps) muscles of your forearm with equal effort, what action occurs? (Try it.)

    _____ In order for movement to occur, it is necessary for extensors (triceps) to be inhibited while flexors (biceps) are stimulated. The nervous system exhibits such control by a phenomenon known as

    _____ innervation.

    d. Reciprocal innervation also occurs in the following instance. Suppose you step on a tack under your right foot. You quickly withdraw that foot by (*flexing? extending?*) your right leg using your hamstring muscles. (Stand up and try it.) What happens to your left leg? You (*flex? extend?*) it. This is an example of reciprocal

    innervation involving a _____ reflex.

    e. A crossed extensor reflex is (*ipsilateral? contralateral?*). It (*may? may not?*) be intersegmental. Therefore many muscles may be contracted to extend your left thigh and leg to shift weight to your left side and provide balance during the tack episode.

■ **C10.** Do this exercise describing how tendon reflexes protect tendons.

    a. Receptors located in tendons are named (*muscle spindles? Golgi tendon organs?*). They are sensitive to changes in muscle (*length? tension?*), as when the hamstring muscles are contracted excessively, pulling on tendons.

    b. When this occurs, association neurons cause (*excitation? inhibition?*) of this same muscle, so that the hamstring fibers (*contract further? relax?*).

    c. Simultaneously, other association neurons fire impulses which stimulate (*synergistic? antagonistic?*) muscles (such as the quadriceps in this case). These muscles then (*contract? relax?*).

    d. The net effect of such a (*mono? poly?*)-synaptic tendon reflex is that the tendons are (*protected? injured?*).

■ **C11.** Why are deep tendon reflexes (that involve stretching a tendon such as that of the quadriceps femoris muscle in the patellar reflex) particularly helpful diagnostically?

C12. Describe how reinforcement can be used to increase reflex activity.

■ **C13.** Complete the table about reflexes that are of clinical significance.

	Patellar Reflex	**Achilles Reflex**	**Babinski Sign**
a. Procedure used to demonstrate reflex	Tap patellar ligament		
b. Nature of positive response		Plantar flexion	
c. Spinal nerves and muscles evaluated by procedure in reflex			Determines if corticospinal tract is myelinated yet
d. Cause of negative response			Plantar reflex: all toes curl under; normal after age 1½
e. Cause of exaggerated response		Damage to tracts in S1 or S2 segments of cord	

D. Spinal nerves (pages 369–378)

■ **D1.** Complete this exercise about spinal nerves.

a. There are ____ pairs of spinal nerves. Write the number of pairs in each region.

_____ Cervical _____ Thoracic _____ Lumbar _____ Sacral _____ Coccygeal

b. Which of these spinal nerves form the cauda equina?

c. Spinal nerves are attached by two roots. The posterior root is (*sensory? motor? mixed?*), the anterior root is _____, whereas the spinal nerve

is _____.

d. Individual nerve fibers are wrapped in a connective tissue covering known as (*endo-? epi-? peri-?*) neurium. Groups of nerve fibers are held in bundles (fascicles) by

_____-neurium. The entire nerve is wrapped with _____-neurium.

e. Spinal nerves branch when they leave the intervertebral foramen. These branches

are called _____. Since they are extensions of spinal nerves, rami are (*sensory? motor? mixed?*).

f. Which ramus is larger? (*ventral? dorsal?*) What areas does it supply?

g. What area does the dorsal ramus innervate?

h. Name two other branches (rami) and state their functions.

■ **D2.** Match the plexus names in the box with descriptions. Refer to Figure 13-1a (page 358 in your text) for help.

B. Brachial	L. Lumbar
C. Cervical	S. Sacral
I. Intercostal	

_____ a. Provides the entire nerve supply for the arm

_____ b. Contains origin of phrenic nerve (nerve that supplies diaphragm)

_____ c. Forms median, radial, and axillary nerves

_____ d. Not a plexus at all, but rather segmentally arranged nerves

_____ e. Supplies nerves to scalp, neck, and part of shoulder and chest

_____ f. Supplies fibers to the femoral nerve which innervates the quadriceps, so injury to this plexus would interfere with actions such as touching the toes

_____ g. Forms the largest nerve in the body (the sciatic) which supplies posterior of thigh and the leg

■ **D3.** *For extra review.* Match names of spinal nerves in the box with their descriptions. On the line following the description, write the site of origin of the nerve. The first one is done for you.

Axillary	Pudendal
Inferior gluteal	Radial
Musculocutaneous	Sciatic

_____**Sciatic**_____ a. Consists of two nerves, the tibial and common peroneal; supplies the

hamstrings, adductors, and all muscles distal to the knee. __**L4–S3**__

_____ b. Supplies the deltoid muscle. _____

_____ c. Innervates the major flexors of the arm. _____

_____ d. Supplies most extensor muscles of the forearm, wrist, and fingers; may be

damaged by extensive use of crutches. _____

_____ e. Supplies the gluteus maximus muscle. _____

_____ f. May be anesthetized in childbirth since it innervates external genitalia and

lower part of the vagina. _____

■ **D4.** *A clinical challenge.* If the cord were completely transected (severed) just below the C7 spinal nerves, how would the functions listed below be affected? (Remember that nerves that originate below this point would not communicate with the brain, so would lose much of their function.) Explain your reasons in each case. (*Hint:* Refer to Figure 13-1a, page 358 in the text.)

a. Breathing via diaphragm

b. Movement and sensation of thigh and leg

c. Movement and sensation of the arm

d. Use of muscles of facial expression and muscles that move jaw, tongue, eyeballs

D5. Describe the general pattern of dermatomes:

a. In the trunk

b. In the extremities

E. Disorders (pages 378–380)

E1. List several causes of spinal cord injury.

■ **E2.** Match the terms in the box with descriptions below.

H. Hemiplegia	P. Paraplegia
M. Monoplegia	Q. Quadriplegia

_____ a. Paralysis of one extremity only

_____ b. Paralysis of both legs

_____ c. Paralysis of both arms and both legs

_____ d. Paralysis of the arm, leg, and trunk on one side of the body

E3. Briefly describe changes that occur following cord transaction and discuss surgical intervention. Include these: *spinal shock, areflexia, delayed nerve grafting.*

■ **E4.** Answer these questions about nerve regeneration.

a. In order for damaged neurons to be repaired, they must have an intact cell body

and also a _____ .

b. Can axons in the CNS regenerate? Explain.

c. When a nerve fiber (axon or dendrite) is injured, the changes that follow in the cell body are called (*Chromatolysis? Wallerian degeneration?*). Those that occur in the

portion of the fiber distal to the injury are known as _____ .

E5. Explain how peripheral nerves regenerate by describing each of these events.

a. Chromatolysis

b. Wallerian degeneration

c. Retrograde degeneration

d. Accelerated protein synthesis

231

■ **E6.** Shingles is an infection of the (*central? peripheral?*) nervous system. The causative virus is also the agent of (*chickenpox? measles? cold sores?*). Following recovery from chickenpox, the virus remains in the body in the (*spinal cord? dorsal root ganglia?*). At times it is activated and travels along (*sensory? motor?*) neurons, causing (*pain? paralysis?*).

■ **E7.** Match names of disorders in the box with definitions below.

A. Areflexia	Sc. Sciatica
N. Neuritis	Sh. Shingles

_____ a. Inflammation of a single nerve

_____ b. Lack of reflex activity

_____ c. Acute inflammation of the nervous system by *Herpes zoster* virus

_____ d. Neuritis of a nerve in the posterior of hip and thigh; often due to a slipped disc in the lower lumbar region

ANSWERS TO SELECTED CHECKPOINTS

A1. Gray, gray, white.

A2.

Structure	Color (Gray or White)	Composition (Cell Bodies or Nerve Fibers)	Location (CNS or PNS)
a. Nerve	White	Nerve fibers	PNS
b. Tract	White	Nerve fibers	CNS
c. Ganglia	Gray	Cell bodies	PNS
d. Nucleus	Gray	Cell bodies	CNS

B1.

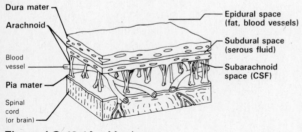

Figure LG 13-1A Meninges.

B2. (a) Between L3 and L4 (or L4 and L5) vertebrae; the spinal cord ends at about L1, so the cord cannot be injured at this lower level; subarachnoid. (b) The anesthetic must penetrate epidural tissues and then all three layers of meninges before reaching nerve fibers; childbirth. (c) Dura mater; arachnoid and pia mater.

C1. (a) Co. (b) S. (c) F. (d) Ca.

C2.

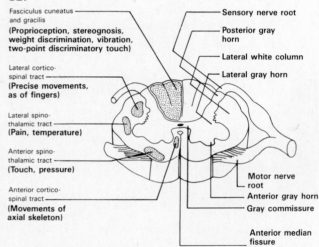

Figure LG 13-2A Outline of the spinal cord, roots, and nerves.

C3. (a) Central, nerves. (b) White columns, myelinated. (c) Sensory, brain. (d) Descending. (e) Lateral, cerebral cortex, spinal cord (in anterior gray horn). (f) Descending, motor.

C5. (a) Conduction, tracts. (b) Reflex, 2, posterior (dorsal), anterior. See Figure LG 13-2A.

C6.

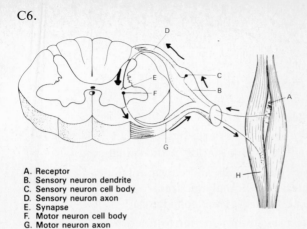

A. Receptor
B. Sensory neuron dendrite
C. Sensory neuron cell body
D. Sensory neuron axon
E. Synapse
F. Motor neuron cell body
G. Motor neuron axon
H. Effector

Figure LG 13-3A   Reflex arc: stretch reflex.

C7. (a) 2, sensory, motor. (b) Monosynaptic, CNS. (c) Neuromuscular spindles, length (or stretch), stretch. (d) Ipsilateral. (e) Skeletal muscle. (f) Knee jerk (patellar reflex).

C8. Branches from axons of sensory or association neurons can travel through tracts to the brain.

C9. (a) Does, more. (b) Several, one, withdrawal. (c) No action, reciprocal. (d) Flexing, extend, crossed extensor. (e) Contralateral, may.

C10. (a) Golgi tendon organ, tension. (b) Inhibition, relax. (c) Antagonistic, contract. (d) Poly, protected

C11. They can readily pinpoint a disorder of a specific spinal nerve or plexus, or portion of the cord, since they do not involve the brain.

C13.

	Patellar Reflex	Achilles Reflex	Babinski Sign
a. Procedure used to demonstrate reflex	Tap patellar ligament	**Tap on calcaneal (Achilles) tendon**	**Light stimulation of outer margin of sole of foot**
b. Nature of positive response	**Extension of leg**	Plantar flexion	**Great toe extends, with or without fanning of other toes. Abnormal after age 1½; shows incomplete myelination.**
c. Spinal nerves and muscles evaluated by procedure in reflex	**L2–4 (quadriceps muscles)**	**L4–5, S1–3 (gastrocnemius muscles)**	Determines if corticospinal tract is myelinated yet
d. Cause of negative response	**Chronic diabetes mellitus, neurosyphilis**	**Diabetes, neurosyphilis, alcoholism**	Plantar reflex: all toes curl under; normal after age 1½
e. Cause of exaggerated response	**Injury to corticospinal tracts**	Damage to motor tracts in S1 or S2 segments of cord	**Positive Babinski after age 1½ indicates interruption of corticospinal tracts**

D1. (a) 31; 8, 12, 5, 5, 1. (b) Lumbar, sacral and coccygeal. (c) Sensory, motor, mixed. (d) Endo-, peri-, epi-. (e) Rami, mixed. (f) Ventral: all of the extremities and the ventral and lateral portions of the trunk. (g) Muscles and skin of the back. (h) Meningeal branch supplies primarily vertebrae and meninges; rami communicantes have autonomic functions.

D2. (a) B. (b) C. (c) B. (d) I. (e) C. (f) L. (g) S.

D3. (b) Axillary, C5–C6. (c) Musculocutaneous, C5–C7. (d) Radial, C5–C8, T-1. (e) Inferior gluteal, L5–S2. (f) Pudendal, S2–S4.

D4. (a) Not affected since (phrenic) nerve to the diaphragm orginates from the cervical plexus (at C3–C5), higher than the transection. So this nerve continues to receive nerve impulses from the brain. (b) Complete loss of sensation and paralysis since lumbar and sacral plexuses originate below the injury, and therefore no longer communicate with the brain. (c) Most arm functions are not affected. As shown on Figure 13-12, page 372 of the text, the brachial plexus originates from C-5 through T-1, so most nerves to the arm (those from C-5 through C-7) still communicate with the brain. (d) Not affected since all are supplied by cranial nerves which originate from the brain.

E2. (a) M. (b) P. (c) Q. (d) H.

E4. (a) Neurilemma. (b) No. They lack a neurilemma. (c) Chromatolysis, Wallerian degeneration.

E6. Peripheral, chickenpox, dorsal root ganglia, sensory, pain.

E7. (a) N. (b) A. (c) Sh. (d) Sc.

MASTERY TEST: Chapter 13

Questions 1–4: Arrange the answers in correct sequence.

_____ _____ _____ 1. From superficial to deep:

    A. Subarachnoid space
    B. Epidural space
    C. Dura mater

_____ _____ _____ 2. From anterior to posterior in the spinal cord:

    A. Fasciculus gracilis and cuneatus
    B. Anterior spinothalamic tract
    C. Central canal of the spinal cord

_____ _____ _____ _____ 3. The plexuses, from superior to inferior:

    A. Lumbar
    B. Brachial
    C. Cervical
    D. Sacral

_____ _____ _____ _____ _____ 4. Order of structures in a conduction pathway, from origin to termination:

    A. Motor neuron
    B. Sensory neuron
    C. Center
    D. Receptor
    E. Effector

Questions 5–10: Circle the letter preceding the one best answer to each question.

5. During the chromatolysis phase of nerve regeneration, all of the following occur EXCEPT:

    A. Chromatophilic substance (Nissl bodies) breaks down into finely granular masses
    B. The cell body shrinks in size
    C. The number of free ribosomes increases as the number of them on endoplasmic reticulum decreases
    D. The nucleus is notably off-center in position

6. All of these tracts are sensory EXCEPT:

    A. Anterior spinothalamic
    B. Lateral spinothalamic
    C. Fasciculus cuneatus
    D. Lateral corticospinal
    E. Posterior spinocerebellar

7. Choose the FALSE statement about the spinal cord.

    A. It has enlargements in the cervical and lumbar areas.
    B. It lies in the vertebral foramen.
    C. It extends from the medulla to the sacrum.
    D. It is surrounded by meninges.
    E. In cross section an H-shaped area of gray matter can be found.

8. All of these structures are composed of white matter EXCEPT:

    A. Posterior root (spinal) ganglia
    B. Tracts
    C. Lumbar plexus
    D. Sciatic nerve
    E. Ventral ramus of a spinal nerve

9. Which is a FALSE statement about the patellar reflex?

    A. It is also called the knee jerk.
    B. It involves a two-neuron, monosynaptic reflex arc.
    C. It results in extension of the leg by contraction of the quadriceps femoris.
    D. It is contralateral.

10. The cauda equina is:

    A. Another name for the cervical plexus
    B. The lumbar and sacral nerves extending below the end of the cord and resembling a horse's tail.
    C. The inferior extension of the pia mater
    D. A denticulate ligament
    E. A canal running through the center of the spinal cord

T   F   11. In order for regeneration of a nerve to occur, the injured nerve must have an <u>intact cell body and a myelin sheath</u>.

T   F   12. The two main functions of the spinal cord are that it serves as a <u>reflex center</u> and it is <u>the site where sensations are felt</u>.

T   F   13. Dorsal roots of spinal nerves are <u>sensory, ventral roots are motor, and spinal nerves are mixed</u>.

T   F   14. A tract is a bundle of nerve fibers <u>inside the central nervous system (CNS)</u>.

T   F   15. Synapses <u>are</u> present in posterior (dorsal) root ganglia.

T   F   16. After a person reaches 18 months, the Babinski sign should be <u>negative, as indicated by plantar flexion (curling under of toes and foot)</u>.

T   F   17. Visceral reflexes are used diagnostically <u>more often than somatic ones since it is easy to stimulate most visceral receptors</u>.

T   F   18. Transection of the spinal cord at level C-6 will result in <u>greater</u> loss of function than transection at level T-6.

T   F   19. The ventral root of a spinal nerve contains <u>axons and dendrites</u> of <u>both motor and sensory</u> neurons.

T   F   20. A lumbar puncture (spinal tap) is usually performed at about the level of vertebrae <u>L1 to L2</u> since the cord ends between about <u>L3 and L4</u>.

_____ 21. Clusters of cell bodies located outside of the central nervous system (CNS) are known as ____.

_____ 22. An inflammation of the dura mater, arachnoid, and/or pia mater is known as ____.

_____ 23. Tendon reflexes are ____-synaptic and ____-lateral.

_____ 24. Lateral gray horns are found only in ____ regions of the spinal cord.

_____ 25. The filum terminale and denticulate ligaments are both composed of ____ mater.

ANSWERS TO MASTERY TEST: ■ Chapter 13

Arrange

1. B C A
2. B C A
3. C B A D
4. D B C A E

Multiple Choice

5. B   8. A
6. D   9. D
7. C   10. B

True-False

11. F. Intact cell body and neurilemma
12. F. Reflex center, conduction site
13. T
14. T
15. F. Are not
16. T
17. F. Less often than somatic ones since it is difficult to stimulate most visceral receptors
18. T
19. F. Axons of motor
20. F. L3–L4, L1–L2.

Fill-ins

21. Ganglia
22. Meningitis
23. Poly, ipsi
24. Thoracic, upper lumbar, sacral (further explanation in Chapter 16)
25. Pia

FRAMEWORK 14

Brain & Cranial Nerves

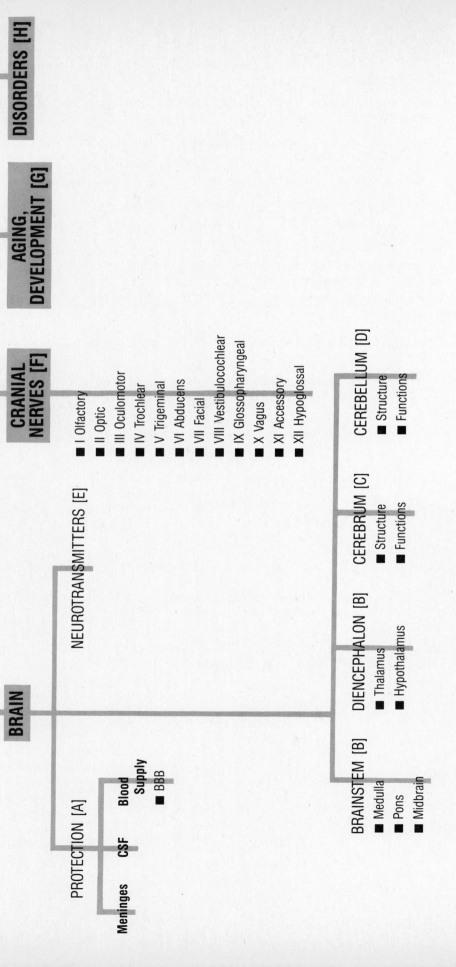

DISORDERS [H]

AGING, DEVELOPMENT [G]

CRANIAL NERVES [F]

- I Olfactory
- II Optic
- III Oculomotor
- IV Trochlear
- V Trigeminal
- VI Abducens
- VII Facial
- VIII Vestibulocochlear
- IX Glossopharyngeal
- X Vagus
- XI Accessory
- XII Hypoglossal

NEUROTRANSMITTERS [E]

BRAIN

PROTECTION [A]

Meninges   CSF   Blood Supply
- BBB

BRAINSTEM [B]
- Medulla
- Pons
- Midbrain

DIENCEPHALON [B]
- Thalamus
- Hypothalamus

CEREBRUM [C]
- Structure
- Functions

CEREBELLUM [D]
- Structure
- Functions

The Brain and Cranial Nerves

The brain is the major control center for the global communication network of the body: the nervous system. The brain requires round-the-clock protection and maintenance afforded by bones, meninges, cerebrospinal fluid, a special blood-brain barrier, along with a fail-safe blood supply. This vital control center consists of four major substructures: the brain stem, diencephalon, cerebrum, and cerebellum. Each brain part carries out specific functions and each releases specific chemical neurotransmitters. Twelve pairs of cranial nerves convey information to and from the brain.

Structural defects may occur in construction (brain development) and also with the normal wear-and-tear that accompanies aging. Just as a giant computer network may experience minor disruptions in service or major shutdowns, disorders within the brain or cranial nerves may lead to minor, temporary changes in nerve functions or profound and fatal outcomes.

First study the Chapter 14 Framework and the key terms associated with each section.

TOPIC OUTLINE AND OBJECTIVES

A. Brain: introduction

1. Identify the principal parts of the brain, and describe how the brain is protected.
2. Explain the formation and circulation of cerebrospinal fluid (CSF).
3. Describe the blood supply to the brain and the concept of the blood-brain barrier (BBB).

B. Brain: brain stem and diencephalon

C. Brain: cerebrum

D. Brain: cerebellum

4. Compare the structure and functions of the brain stem, diencephalon, cerebrum, and cerebellum.
5. Describe the structure and functions of the limbic system.

E. Neurotransmitters in the brain

6. Discuss the various neurotransmitters found in the brain, as well as the different types of neuropeptides and their functions.

F. Cranial nerves

7. Define a cranial nerve and identify the 12 pairs of cranial nerves by name, number, type, location, and function.

G. Aging and developmental anatomy of the nervous system

8. Describe the effects of aging on the nervous system.
9. Describe the development of the nervous system.

H. Disorders, medical terminology

10. List the clinical symptoms of these disorders of the nervous system: cerebrovascular accidents (CVAs), epilepsy, transient ischemic attacks (TIAs), brain tumors, poliomyelitis, cerebral palsy (CP), Parkinson's disease (PD), multiple sclerosis (MS), dyslexia, Tay-Sachs disease, headache, trigeminal neuralgia (tic douloureux), Reye's syndrome (RS), Alzheimer's disease (AD), and delirium.

11. Define medical terminology associated with the central nervous system.

WORDBYTES

Now study the following parts of words that may help you better understand terminology in this chapter.

Wordbyte	Meaning	Example	Wordbyte	Meaning	Example
cephalo-	head	hydro*cephal*ic	ophthalm-	eye	*ophthalm*ic
cortico-	bark	cerebral *cort*ex	pons	bridge	*pons*
enceph-	brain	di*enceph*alon	quadri-	four	corpora
falx	sickle	*falx* cerebri			*quadri*gemina
glossi-	tongue	hypo*gloss*al	vita	life	arbor *vita*e
hemi-	half	*hemi*sphere			

CHECKPOINTS

A. Brain: introduction (pages 385–390)

■ **A1.** Identify numbered parts of the brain on Figure LG 14-1. Then complete this exercise.

a. Structures 1–3 are parts of the _____.

b. Structures 4 and 5 together form the _____.

c. Structure 6 is the largest part of the brain, the _____.

d. The second largest part is structure 7, the _____.

■ **A2.** List three ways in which the brain is protected.

A3. Review the layers of the meninges covering the brain and spinal cord by listing them here. *For extra review.* Label the layers on Figure LG 14-1 and review Chapter 13 Checkpoint B1 (page LG 222).

■ **A4.** To check your understanding of the pathway of cerebrospinal fluid (CSF), list in order the structures through which it passes. Use key letters on Figure LG 14-1. Start at the site of formaton of CSF.

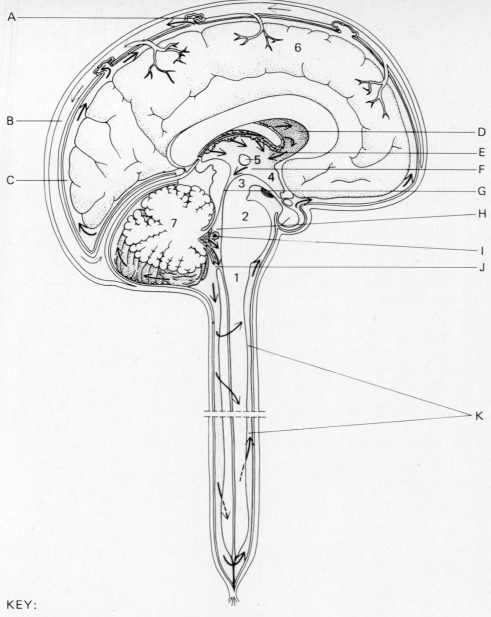

KEY:

1. Medulla oblongata
2. Pons varolii
3. Midbrain

4. Hypothalamus
5. Thalamus
6. Cerebrum
7. Cerebellum

A. Arachnoid villus
B. Cranial venous sinus
C. Subarachnoid space of brain
D. Lateral venticle
E. Interventricular foramen
F. Third ventricle

G. Cerebral aqueduct
H. Fourth ventricle
I. Lateral aperture
J. Median aperture
K. Subarachnoid space of spinal cord

Figure LG 14-1 Brain and meninges seen in sagittal section. Parts of the brain are numbered; refer to Checkpoint A1. Letters indicate pathway of cerebrospinal fluid (CSF); refer to Checkpoint A4.

■ **A5.** Circle all correct answers about CSF.

a. The entire nervous system contains 80–150 ml of CSF. This amount is equal to approximately:

A. 1 to 2 tablespoons   C. 1 to 2 cups
B. ⅓ to ⅔ cup   D. 1 quart

b. The color of CSF is:

A. Yellow   C. Red
B. Clear, colorless   D. Green

c. Choose the function(s) of CSF.

A. Serves as a shock absorber for brain and cord
B. Contains red blood cells
C. Contains white blood cells called lymphocytes
D. Contains nutrients

d. Which statement(s) describe its formation?

A. It is formed by diffusion of substances from blood.
B. It is formed by filtration and secretion.
C. It is formed from blood in capillaries called choroid plexuses.
D. It is formed in all four ventricles.

e. Which statement(s) describe its pathway?

A. It circulates around the brain, but not the cord.
B. It flow below the end of the spinal cord.
C. It bathes the brain by flowing through the epidural space.
D. It passes via projections (villi) of the arachnoid into blood vessels (venous sinuses) surrounding the brain.
E. It is formed initially from blood, and finally flows back to blood.

A6. Briefly state results of oxygen starvation of the brain for about 4 minutes.

Explain the role of lysosomes in this process.

Now list effects of glucose deprivation of the brain.

■ **A7.** Increase in the carbon dioxide level or decrease in oxygen level of blood circulating through the brain has the effect of (*dilating? constricting?*) cerebral blood vessels, and therefore helping to supply the brain with a better blood flow. The amount of

dilation of these vessels is directly related to the increase in the ____ ion in the cerebrum. Show the chemical reaction which explains how increased CO_2 level leads to increased H^+ level (or lower pH).

CO_2 + ⬚ ⟶ ⬚ ⟶ ⬚ + H^+

■ **A8.** Describe the blood–brain barrier (BBB) in this exercise.

a. Blood capillaries supplying the brain have (*more? less?*) densely packed cells plus a

layer of _____ cells and a continuous

_____ membrane.

b. What advantages are provided by this membrane?

c. List two types of substances that do normally cross the blood–brain barrier.

d. What problems may result from the fact that some chemicals cannot cross
this barrier?

B. Brain: brain stem, diencephalon (pages 390–396)

■ **B1.** Describe the principal functions of the medulla in this exercise.

a. The medulla serves as a _____ pathway for all ascending and
descending tracts. Its white matter therefore transmits (*sensory? motor? both sensory and motor?*) impulses.

b. Included among these tracts are the triangular _____ tracts
which are the principal (*sensory? motor?*) pathways. The main fibers that pass
through the pyramids are the (*spinothalamic? corticospinal?*) tracts.

c. Crossing (or _____) of fibers occurs in the medulla. This
explains why movements of your right hand are initiated by motor neurons that
originate in the (*right? left?*) side of your cerebrum. The (*axons? dendrites?*) of
these motor neurons decussate in the medulla to proceed down the right lateral

_____-spinal tract.

d. The medulla contains gray areas as well as white matter. Several important nuclei
lie in the medulla. Two of these are synapse points for sensory axons that have
ascended in the posterior columns of the cord. These two nuclei, named

_____ and _____, transmit impulses for

sensations of _____, _____, and

_____.

e. You maintain a conscious state as a result of all kinds of sensory input. Sensory

impulses pass into the _____ formation, which consists of a
dispersed area in the brain stem, diencephalon, and spinal cord.

f. Some input to the medulla arrives by means of cranial nerves; these nerves may serve motor functions also. Which cranial nerves are attached to the medulla?

_____

(Functions of these nerves will be discussed later in this chapter.)

g. A hard blow to the base of the skull can be fatal since the medulla is the site of

three vital centers: the _____ center, regulating the heart; the

_____ center, adjusting the rhythm of breathing; and the

_____ center, regulating blood pressure by altering diameter of blood vessels.

■ **B2.** Summarize important aspects of the pons in this learning activity.

a. The name *pons* means _____. It serves as a bridge in two ways. It contains longitudinally arranged fibers that connect the

_____ and _____ with the upper parts of the brain. It has transverse fibers that connect the two sides of the

_____.

b. Cell bodies associated with fibers in cranial nerves numbered

_____ lie in nuclei in the pons.

c. (*Respiration? Heartbeat? Blood pressure?*) is controlled by the pneumotaxic and apneustic areas of the pons.

■ **B3.** Describe the thalamus in this exercise.

a. If you look for the thalamus in most figures of the brain, you will not find it. It is located (*deep within? on the surface of?*) the cerebrum.

b. The thalamus is H-shaped. The cross bar of the H is known as the

_____ mass. It passes through the center of the slitlike

_____ ventricle. (See Figure LG 14-1.) The two side bars of the H form the lateral walls of the third ventricle.

c. The thalamus is the principal relay station for (*motor? sensory?*) impulses. For example, spinothalamic and lemniscal tracts convey general sensations such as pain,

_____, _____,

_____, and temperature to the thalamus where they are relayed to the cerebral cortex. Special sense impulses (for vision and hearing) are relayed through the (*geniculate? reticular? ventral posterior?*) nuclei of the thalamus.

d. The thalamus (*does? does not?*) contain nuclei controlling motor functions. However, its principal role is conveying (*motor? sensory?*) impulses.

■ **B4.** The name hypothalamus indicates that this structure lies (*above? below?*) the thalamus, forming the floor and part of the lateral walls of the

_____ ventricle.

B5. Expanding on the key words listed below, write a sentence describing major hypothalamic functions.

a. Regulator of visceral activities

b. Psychosomatic

c. Regulating factors to anterior pituitary

d. ADH and oxytocin to posterior pituitary

e. Thermostat

f. Thirst

g. Feeding and satiety center

h. Reticular formation: arousal and consciousness

■ **B6.** *For extra review.* Check your understanding of these parts of the brain stem and diencephalon by matching them with the descriptions given below.

H. Hypothalamus	P. Pons
Med. Medulla	T. Thalamus
Mid. Midbrain	

_____ a. It is the main regulator of visceral activities since it acts as a liaison between cerebral cortex and autonomic nerves that control viscera.

_____ b. It is the site of the red nucleus, the origin of rubrospinal tracts concerned with muscle tone and posture.

_____ c. Cranial nerves III–IV attach to this brain part.

_____ d. Cranial nerves V–VIII attach to this brain part.

_____ e. Cranial nerves VIII–XII attach to this brain part.

_____ f. Feelings of hunger, fullness, and thirst stimulate centers here so that you can respond accordingly.

_____ g. All sensations except smell are relayed through here.

_____ h. Regulation of heart, blood pressure, and respiration occurs by centers located here.

_____ i. It constitutes four-fifths of the diencephalon.

_____ j. It lies under the third ventricle, forming its floor.

_____ k. It forms most of side walls of the third ventricle.

_____ l. Tumor in this region could compress cerebral aqueduct and cause internal hydrocephalus.

C. Brain: cerebrum (pages 396–404)

■ **C1.** Complete this exercise about cerebral structure.

a. The outer layer of the cerebrum is called _____. It is composed of (*white? gray?*) matter. This means that it contains mainly (*cell bodies? neurons?*).

b. In the margin, draw a line the same length as the thickness of the cerebral cortex. Use a metric ruler. Note how thin the cortex is.

c. The surface of the cerebrum looks much like a view of tightly packed mountains

or ridges, called _____. The parts where the cerebral cortex

dips down into valleys are called _____ (deep valleys) or

_____ (shallow valleys).

d. The cerebrum is divided into halves called _____. Connecting

them is a band of (*white? gray?*) matter called the _____.
Notice this structure in Figure LG 14-1 and in Figure 14-5a, page 393 of your text.

e. The falx cerebri is composed of (*nerve fibers? dura mater?*). Where is it located?

_____ At its superior and inferior margins, the falx is dilated
to form channels for venous blood flowing from the brain; these enclosures are

called _____.

■ **C2.** Label the following structures using leader lines on Figure LG 14-2: *frontal lobe, occipital lobe, parietal lobe, temporal lobe, central sulcus, lateral sulcus, precentral gyrus, postcentral gyrus.*

■ **C3.** Match the three types of white matter fibers with these descriptions.

A. Association	C. Commissural	P. Projection

_____ a. The corpus callosum contains these fibers and connects the two cerebral hemispheres.

_____ b. Sensory and motor fibers passing between cerebrum and other parts of the CNS are this type of fiber; the internal capsule is an example.

_____ c. These fibers transmit impulses among different areas of the same hemisphere.

C4. List names of structures that are considered parts of the basal ganglia.

Now list two functions of the basal ganglia.

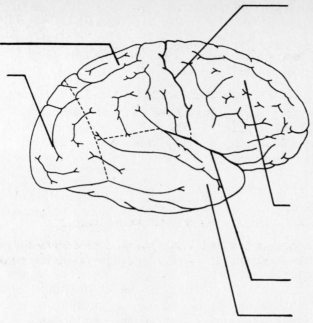

○ Broca's (motor speech) area   ○ Primary somesthetic (general sensory) area

○ Premotor area   ○ Primary visual area

○ Primary auditory area   ○ Auditory association area

○ Primary motor area

Figure LG 14-2 Right lateral view of lobes and fissures of the cerebrum. Label and color as directed in Checkpoint C2 and C6.

C5. Describe the limbic system in this exercise.

a. This system consists of portions of the cerebrum, thalamus, and hypothalamus. List several component structures.

b. The limbic system is shaped much like a _____ surrounding the brain stem.

c. Explain why the limbic system is sometimes called the "visceral" or "emotional" brain.

d. One other function of the limbic system is _____. Forgetfulness, such as inability to recall recent events, results partly from impairment of this system.

■ **C6.** Color functional areas of the cerebral cortex listed with color code ovals on Figure LG 14-2.

■ **C7.** Check your understanding of functional areas of the cerebral cortex by doing this matching exercise. Answers may be used more than once.

> B. Broca's area
> G. Gnostic area
> PA. Primary auditory area
> PM. Primary motor area
>
> PO. Primary olfactory area
> PS. Primary somesthetic area
> PV. Primary visual area

_____ a. In the occipital lobe

_____ b. In the postcentral gyrus

_____ c. Receives sensations of pain, touch, pressure, and temperature

_____ d. In the parietal lobe

_____ e. In the temporal lobe; interprets sounds

_____ f. Integrates general and special sensations to form a common thought about them

_____ g. In precentral gyrus of the frontal lobe; controls specific muscles or groups of muscles

_____ h. In the frontal lobe; translates thoughts into speech

_____ i. Controls smell

C8. Contrast: aphasia/agraphia.

■ **C9.** Identify the type of brain waves associated with each of the following situations.

> A. Alpha   T. Theta
> B. Beta   D. Delta

_____ a. Occur in persons experiencing stress and in certain brain disorders

_____ b. Lowest frequency brain waves, normally occurring when adult is in deep sleep; presence in awake adult indicates brain damage

_____ c. Highest frequency waves, noted during periods of mental activity

_____ d. Present when awake but resting, these waves are intermediate in frequency between alpha and delta waves

■ **C10.** *A clinical challenge.* You are helping to care for Laura, a hospital patient with a contusion of the right side of the brain. Which of the following would you be most likely to observe in Laura? Circle letters of correct answers.

A. Her right leg is paralyzed.
B. Her left am is paralyzed.
C. She cannot speak out loud to you nor write a note to you.
D. She used to sing well, but cannot seem to stay on tune now.
E. She has difficulty with concepts involving numbers.

D. Brain: cerebellum (pages 404–406)

D1. Describe the cerebellum.

a. Where is it located?

b. Describe its functions. Use these key terms: *coordinated movements, posture, equilibrium,* and *emotions.*

■ **D2.** *For extra review.* Arrange the following sentences in correct order to form a paragraph describing how coordinated movement occurs. Write letters in order on these

lines: ____ ____ ____ ____ ____

A. Cerebellum sends subconscious motor impulses along inferior cerebellar peduncles to medulla and spinal cord.

B. Once movement has started, cerebral cortex continually receives proprioceptive feedback by means of dorsal columns, nuclei cuneatus and gracilis, medial lemniscal tracts, and thalamus.

C. Cerebral cortex simultaneously activates the brain stem, which relays impulses over superior cerebellar peduncles to cerebellum.

D. Impulses from medulla and spinal cord pass downward to motor neurons in anterior gray horn of cord; these regulate coordinated muscle action.

E. Motor areas of cerebral cortex voluntarily initiate muscle contraction.

■ **D3.** How would you assess a study partner for cerebellar damage? *Note:* Symptoms of such damage can be expected to be (*ipsi? contra?*)-lateral.

E. Neurotransmitters in the brain (pages 406–408)

■ **E1.** Match the neurotransmitter or neuropeptide with the description that best fits.

A. Angiotensin	GABA. Gamma-aminobutyric acid
ACh. Acetylcholine	5-HT. Serotonin
DA. Dopamine	NE. Norepinephrine
EED. Enkephalins, endorphins, and dynorphin	RF. Regulating factors

_____ a. Usually excitatory; inactivated by COMT or MAO.

_____ b. The most common inhibitory transmitter in the brain

_____ c. Affects release of pituitary hormones

_____ d. Mostly excitatory; inactivated by acetylcholinesterase (AChE)

_____ e. Inhibitory; released by neurons and basal ganglia involved with emotions and gross subconscious movements

_____ f. Neuropeptides with morphine-like pain-suppressing qualities

_____ g. Produced by renin released by kidneys

_____ h. Concentrated in brain stem neurons, this neurotransmitter helps control sleep, temperature, and mood.

F. Cranial nerves (pages 408–413)

■ **F1.** Answer these questions about cranial nerves.

a. There are ____ pairs of cranial nerves. They are all attached to the

_____; they leave the _____ via
foramina.

b. They are numbered by Roman numerals in the order that they leave the cranium. Which is most anterior? (*I? XII?*) Which is most posterior (*I? XII?*)

c. All spinal nerves are (*purely sensory? purely motor? mixed?*). Are all cranial nerves mixed (*Yes? No?*)

F2. Complete the table about cranial nerves. (For extra review on locations of cranial nerves, refer back to Chapter 7, Checkpoint B7, page LG 118).

Number	Name	Functions
a.	Olfactory	
b.		Vision (not pain or temperature of the eye)
c. III		
d.	Trochlear	
e. V		
f.		Stimulates lateral rectus muscle to abduct eye; proprioception of the muscle
g.	Facial	
h.		Hearing; equilibrium
i. IX		
j.	Vagus	
k. XI		
l.		Supplies muscles of the tongue with motor and sensory fibers

■ **F3.** Check your understanding of cranial nerves by completing this exercise. Write the name of the correct cranial nerve following the related description.

a. Differs from all other cranial nerves in that it originates from the brain stem and

from the spinal cord: _____

b. Eighth cranial nerve (VIII): _____

c. Is widely distributed into neck, thorax, and abdomen: _____

d. Senses toothache, pain under a contact lens, wind on the face:

_____

e. The largest cranial nerve; has three parts (ophthalmic, maxillary, and mandibular):

_____

f. Controls contraction of muscle of the iris, causing constriction of pupil:

_____

g. Innervates muscles of facial expression: _____

h. Two nerves that contain taste fibers and autonomic fibers to salivary glands:

_____ , _____

i. Three purely sensory cranial nerves: _____ ,

_____ , _____

■ **F4.** Write the number of the cranial nerve related to each of the following disorders.

_____ a. Bell's palsy   _____ e. Blindness

_____ b. Inability to shrug shoulders or turn   _____ f. Trigeminal neuralgia (tic douloureux)
head

_____ c. Anosmia   _____ g. Paralysis of vocal cords; loss of sensation of many organs

_____ d. Strabismus and diplopia   _____ h. Vertigo and nystagmus

G. Aging and developmental anatomy of the nervous system   (pages 409–414)

■ **G1.** Do this exercise describing effects of aging on the nervous system.

a. The total number of nerve cells ____-creases with age. Since conduction velocity

(*increases? slows down?*), the time required for a typical reflex is ____-creased.

b. Parkinson's disease is the most common (*motor? sensory?*) disorder that is likely to occur in the elderly.

c. Problems with sense organs include presbyopia which is an inability to focus on objects that are (*near? far away?*) and presbycussis which is impaired (*hearing? taste?*). The sense of touch is likely to be (*impaired? heightened?*) in old age, placing the older person at (*higher? lower?*) risk for burns or other skin injury.

■ **G2.** Check your understanding of the early development of the nervous system by completing this exercise.

a. The nervous system begins to develop during the third week of gestation when the

_____-derm forms a thickening called the

_____ plate. Soon a longitudinal depression is found in the

plate; this is the neural _____. As neural folds on the sides of

the groove grow and meet, they form the neural _____.

b. Three types of cells form the walls of this tube. Two of these are the marginal layer which forms (*gray? white?*) matter and the mantle layer which becomes (*gray? white?*) matter.

c. The neural crest forms (*peripheral? central?*) nervous system structures such as

ganglia as well as spinal and cranial _____.

d. The anterior portion of the neural plate and tube develops into three enlarged

_____ by the fourth week. These fluid-filled cavities even-

tually become the _____ which are filled with

_____ fluid.

e. By week five the primary vesicles have flexed (bent) to form a total of (*five? ten?*) secondary vesicles. (More about these in the next exercise.)

■ **G3.** Fill in the blanks in this table outlining brain development.

a. Primary brain vesicles (3)	b. Secondary brain vesicles (5)	c. Parts of brain formed (7)
Prosencephalon (_____ -brain) →	1. Diencephalon →	1A. _____ 1B. _____
	2. _____ →	2A. Cerebrum
_____ -encephalon (Midbrain) →	3. Mesencephalon →	3A. _____
_____ -encephalon (_____) →	4. _____ -encephalon →	4A. _____
	5. Met- _____ →	5A. Pons 5B. _____

H. Disorders, medical terminology (pages 414–419)

H1. CVA refers to _____.
CVAs are (*rare? common?*) brain disorders. Describe three causes of CVAs.

H2. Contrast *CVA* with *TIA*.

H3. Write three signs or symptoms of brain tumors.

■ **H4.** Complete this Checkpoint about Parkinson's disease.

a. This condition involves a decrease in the neurotransmitter

_____. It is normally produced by cell bodies located in the

substantia nigra which is part of the _____. Axons lead from

here to the _____ where dopamine (DA) is released. DA is an
(*excitatory? inhibitory?*) transmitter.

b. The most common sign of Parkinsonism is _____. Explain
how tremor as well as muscle rigidity are related to low DA level.

c. Voluntary movements may be slower than normal; this is the condition of (*brady?
tachy?*)-kinesia.

■ **H5.** Do this exercise describing some main points about multiple sclerosis (MS).

a. MS affects (*cell bodies? myelin sheaths?*) within the (*central? peripheral?*) nervous
system. At sites along the nerves where myelin is destroyed, many scars (or

_____) form; this is the basis for the name MS.

b. Initial symptoms are most likely to occur at ages (*10–15? 20–40? 60–70?*). MS
affects (*motor? sensory? both motor and sensory?*) neurons. List signs or symptoms that may suggest MS.

c. How do immunosuppressants such as ACTH or prednisone help MS?

H6. Contrast headaches of intracranial and extracranial origins.

H7. Describe the following aspects of Alzheimer's disease (AD):

a. Progressive changes in behavior

b. Pathological findings of brain tissue

c. The possible roles of amyloid and A68 in the causation of AD and DS (Down's syndrome).

■ **H8.** Arrange in order from highest to lowest level of consciousness. Patients who are

experiencing: ____ ____ ____

A. Stupor   B. Lethargy and torpor   C. Coma

■ **H9.** *For extra review.* Match the name of the disorder with the related description.

Cerebral palsy	Parkinsonism
Cerebrovascular accident (CVA)	Poliomyelitis
Dyslexia	Reye's syndrome
Epilepsy	Tay-Sachs disease
Multiple sclerosis	Transient ischemic attack (TIA)
Neuralgia	

a. Degeneration of myelin sheath to form hard plaques in many regions causing loss of motor and sensory function:

_____

b. Degeneration of dopamine-releasing neurons in basal ganglia; characterized by tremor or rigidity of muscles: _____

c. Temporary cerebral dysfunction caused by interference of blood supply to the brain.

_____

d. Most common brain disorder; also called stroke: _____

e. Characterized by difficulty in handling words and symbols, for example, reversal of letters (*b* for *d*); cause unknown:

_____

f. Attack of pain along an entire nerve, for example, tic douloureux:

_____

g. A disease of children causing swelling of brain cells and fatty infiltration of liver; usually follows viral infection, especially if aspirin is taken: _____

h. Inherited disease especially among Jewish populations due to excessive lipids in brain cells; affects infants: _____

i. Characterized by seizures due to abnormal discharges of electricity from neurons of the brain: _____

j. Of viral origin; often affects only the respiratory system; when nervous system involved, affects movement, but not sensation:

_____

k. Motor disorder caused by damage during fetal life, birth, or infancy; not progressive; apparent mental retardation often actually only speaking or hearing disability:

_____

252

ANSWERS TO SELECTED CHECKPOINTS

A1. (a) Brain stem. (b) Diencephalon.
(c) Cerebrum. (d) Cerebellum.

A2. Skull bones, meninges, cerebrospinal fluid (CSF).

A4. D E F G H I J K C A B.

A5. (a) B. (b) B. (c) A, C, D. (d) B, C. D.
(e) B, D, E.

A7. Dilating, H^+; complete reaction:
$CO_2 + H_2O \rightarrow H_2CO_3 \rightarrow HCO_3^- + H^+$.

A8. (a) More, neuroglia (astroglia), basement.
(b) The brain is protected from many substances which could harm the brain, but are kept from the brain by this barrier. (c) Glucose, oxygen, and some ions. (d) Certain helpful medications, such as most antibiotics, cannot cross this barrier. (Fortunately, at the time they are most needed — as in a brain infection — the BBB is more permeable, and so may permit passage of these drugs.)

B1. (a) Conduction, both sensory and motor.
(b) Pyramidal, motor, corticospinal. (c) Decussation, left, axons, cortico-. (d) Cuneatus, gracilis; fine touch (two-point discrimination), proprioception, vibrations. (See Exhibit 13-1, pages 361–362 of text.) (e) Reticular. (f) Part of VIII, as well as IX–XII. (g) Cardiac, medullary rhythmicity, vasomotor.

B2. (a) Bridge; spinal cord, medulla; cerebellum. (b) V–VII and part of VIII. (c) Respiration.

B3. (a) Deep within. (b) Intermediate, third. (c) Sensory; touch, pressure, proprioception; geniculate. (d) Does, sensory.

B4. Below, third.

B6. (a) H. (b) Mid. (c) Mid. (d) P. (e) Med. (f) H. (g) T. (h) Med. (i) T. (j) H. (k) T. (l) Mid.

C1. (a) Cerebral cortex, gray, cell bodies. (b) 2 to 4 mm. (c) Gyri, fissures, sulci. (d) Hemispheres, white, corpus callosum. (e) Dura mater, between cerebral hemispheres, superior and inferior sagittal sinuses (See Figure 14-2, pages 387–388 in the text.)

C2.

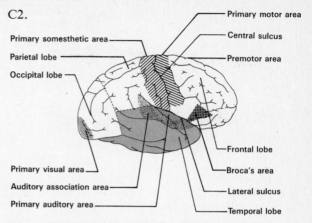

Figure LG 14-2A Right lateral view of lobes and fissures of the cerebrum.

C3. (a) C. (b) P. (c) A.

C6. See Figure LG 14-2A above.

C7. (a) PV. (b) PS. (c) PS. (d) PS. (e) PA. (f) G. (g) PM. (h) B. (i) PO.

C9. (a) T. (b) D. (c) B. (d) A.

C10. B D

D2. E C A D B

D3. Ask partner to walk (check gait), touch finger to nose while eyes closed. Effects are ipsilateral.

E1. (a) NE. (b) GABA. (c) RF. (d) ACh. (e) DA. (f) EED. (g) A. (h) 5-HT.

F1. (a) 12, brain, cranium. (b) I, XII. (c) Mixed; no (three are purely sensory).

F3. (a) Accessory. (b) Vestibulocochlear. (c) Vagus. (d) Trigeminal. (e) Trigeminal. (f) Oculomotor. (g) Facial. (h) Facial, glossopharyngeal. (i) Olfactory, optic, vestibulocochlear.

F4. (a) VII. (b) XI. (c) I. (d) III. (e) II. (f) V. (g) X. (h) VIII (vestibular branch).

G1. (a) De, slows down, in. (b) Motor. (c) Near, hearing, impaired, higher.

G2. (a) Ecto, neural, groove, tube. (b) White, gray. (c) Peripheral, nerves. (d) Primary vesicles, ventricles of brain, cerebrospinal (CSF). (e) Five.

G3.

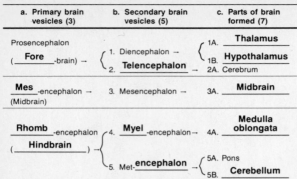

a. Primary brain vesicles (3)	b. Secondary brain vesicles (5)	c. Parts of brain formed (7)
Prosencephalon (__Fore__-brain) →	1. Diencephalon → 2. __Telencephalon__ →	1A. __Thalamus__ 1B. __Hypothalamus__ 2A. Cerebrum
__Mes__-encephalon → (Midbrain)	3. Mesencephalon →	3A. __Midbrain__
__Rhomb__-encephalon (__Hindbrain__) →	4. __Myel__-encephalon → 5. Met-__encephalon__ →	4A. __Medulla oblongata__ 5A. Pons 5B. __Cerebellum__

H4. (a) Dopamine, midbrain, basal ganglia, inhibitory. (b) Tremor; lack of inhibitory DA causes a double-negative effect: extra unnecessary movements occur (tremor) which may then interfere with normal movements (rigidity), (c) Brady.

H5. (a) Myelin sheaths, central, multiple scleroses. (b) 20–40; both motor and sensory; lack of control in writing or walking, double vision, bladder infections. (c) Suppress autoimmune response in which the body is making antibodies which destroy the oligodendroglia necessary for CNS myelin formation.

H8. B A C.

H9. (a) Multiple sclerosis. (b) Parkinsonism. (c) TIA. (d) CVA. (e) Dyslexia. (f) Neuralgia (g) Reye's syndrome. (h) Tay-Sachs disease. (i) Epilepsy. (j) Poliomyelitis. (k) Cerebral palsy.

MASTERY TEST: Chapter 14

Questions 1–8: Circle the letter preceding the one best answer to each question.

1. All of these are located in the medulla EXCEPT:

   A. Nucleus cuneatus and nucleus gracilis
   B. Cardiac center, which regulates heart
   C. Site of decussation (crossing) of pyramidal tracts
   D. The olive
   E. Origin of cranial nerves V to VIII

2. Which of these is a function of the postcentral gyrus?

   A. Controls specific groups of muscles, causing their contraction
   B. Receives general sensations from skin, muscles, and viscera
   C. Receives olfactory impulses
   D. Primary visual reception area
   E. Somesthetic association area

3. All of these structures contain cerebrospinal fluid EXCEPT:

   A. Subdural space
   B. Ventricles of the brain
   C. Central canal of the spinal cord
   D. Subarachnoid space

4. Damage to the accessory nerve would result in:

   A. Inability to turn the head or shrug shoulders
   B. Loss of normal speech function
   C. Hearing loss
   D. Changes in heart rate
   E. Anosmia

5. Which statement about the trigeminal nerve is FALSE?

   A. It sends motor fibers to the muscles used for chewing.
   B. Pain in this nerve is called tic douloureux.
   C. It carries sensory fibers for pain, temperature, and touch from the face, including eyes, lips, and teeth area.
   D. It is the smallest cranial nerve.

6. All of these are functions of the hypothalamus EXCEPT:

   A. Control of body temperature
   B. Release of chemicals (regulating factors) that affect release or inhibition of hormones
   C. Principal relay station for sensory impulses
   D. Center for mind-over-body (psychosomatic) phenomena
   E. Involved in maintaining sleeping or waking state

7. Damage to the occipital lobe of the cerebrum would most likely cause:

   A. Loss of hearing
   B. Loss of vision
   C. Loss of ability to smell
   D. Paralysis
   E. Loss of feeling in muscles (proprioception)

8. Choose the one FALSE statement.

   A. Multiple sclerosis is a progressive disorder that worsens as time lapses
   B. Cerebral palsy (CP) is a progressive disorder that gets worse with time.
   C. About 70 percent of CP victims appear mentally retarded, but, in fact, this may be a reflection of inability to speak or walk well.
   D. Poliomyelitis may affect motor nerves, but does not affect sensations.

Questions 9–12: Arrange the answers in correct sequence.

_____ _____ _____ 9. From anterior to posterior:

   A. Fourth ventricle
   B. Pons and medulla
   C. Cerebellum

_____ _____ _____ 10. From superior to inferior:

   A. Thalamus
   B. Hypothalamus
   C. Corpus callosum

_____ _____ _____ 11. Order in which impulses are relayed in conduction pathway for vision:

   A. Optic nerve
   B. Optic tract
   C. Optic chiasma

_____ _____ _____ _____ 12. Pathway of cerebrospinal fluid, from formation to final destination:

   A. Choroid plexus in ventricle
   B. Subarachnoid space
   C. Cranial venous sinus
   D. Arachnoid villi

Questions 13–20: Circle T (true) or F (false). If the statement is false, change the underlined word or phrase so that the statement is correct.

T   F   13. GABA, serotonin, and acetylcholine all function as excitatory transmitter substances.

T   F   14. The thalamus, hypothalamus, and cerebrum are all developed from the forebrain (prosencephalon).

T   F   15. Endorphins, enkephalins, and dopamine are all chemicals which are considered the "body's own pain killers."

T   F   16. The language areas are located in the cerebellar cortex.

T   F   17. Cranial nerves I, III, and VIII are all purely sensory nerves.

T   F   18. The limbic system functions in control of emotional aspects of behavior.

T   F   19. The reticular formation controls arousal and consciousness.

T   F   20. Delta brain waves have the lowest frequency of the four kinds of waves produced by normal individuals.

Questions 21–25: fill-ins. Complete each sentence with the word or phrase that best fits.

_____ 21. Portions of the cerebrum that connect sensory and motor areas, and are concerned with memory, personality, judgment, and intelligence are called ____ areas.

_____ 22. The superior and inferior colliculi, associated with movements of eyeballs and head in response to visual stimuli, are located in the ____ .

_____ 23. If the brain is deprived of oxygen for more than 4 minutes, ____ of brain cells break open and release enzymes that can self-destruct these cells.

_____ 24. If blood flowing through the hypothalamus is warmer than normal, the hypothalamus directs heat loss activities such as ____ .

_____ 25. The corpus striatum, including the caudate and lentiform nuclei, is part of the ____ .

Multiple Choice

1. E   5. D
2. B   6. C
3. A   7. B
4. A   8. B

Arrange

9. B A C
10. C A B
11. A C B
12. A B D C

True-False

13. F. Serotonin and acetylcholine
14. T
15. F. Endorphins, enkephalins and dynorphin
16. F. Cerebral cortex
17. F. I, II and VIII
18. T
19. T
20. T

Fill-ins

21. Association
22. Midbrain
23. Lysosomes
24. Sweating, dilation of blood vessels of skin
25. Basal ganglia

FRAMEWORK 15

Sensory, Motor & Integrative Systems

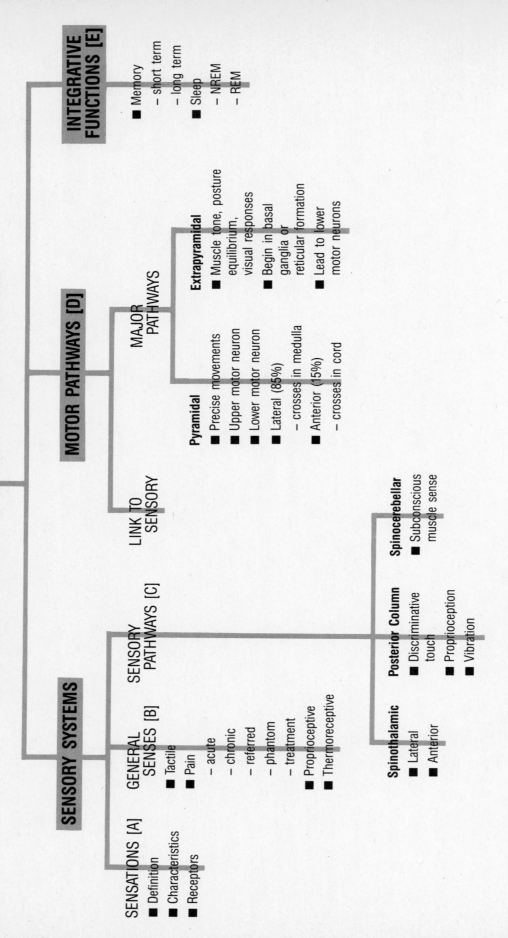

SENSORY SYSTEMS

SENSATIONS [A]
- Definition
- Characteristics
- Receptors

GENERAL SENSES [B]
- Tactile
- Pain
  - acute
  - chronic
  - referred
  - phantom
  - treatment
- Proprioceptive
- Thermoreceptive

SENSORY PATHWAYS [C]

Spinothalamic
- Lateral
- Anterior

Posterior Column
- Discriminative touch
- Proprioception
- Vibration

Spinocerebellar
- Subconscious muscle sense

MOTOR PATHWAYS [D]

LINK TO SENSORY

MAJOR PATHWAYS

Pyramidal
- Precise movements
- Upper motor neuron
- Lower motor neuron
- Lateral (85%)
  - crosses in medulla
- Anterior (15%)
  - crosses in cord

Extrapyramidal
- Muscle tone, posture equilibrium, visual responses
- Begin in basal ganglia or reticular formation
- Lead to lower motor neurons

INTEGRATIVE FUNCTIONS [E]
- Memory
  - short term
  - long term
- Sleep
  - NREM
  - REM

The Sensory, Motor, and Integrative Systems

The past three chapters have introduced the components of the nervous system and demonstrated the arrangement of neurons in the spinal cord, brain, spinal nerves, and cranial nerves. Chapter 15 integrates this information in the study of major nerve pathways including those for (a) general senses (such as touch, pain and temperature), (b) motor control routes for precise, conscious movements and for unconscious movements, and (c) complex integrative functions such as memory and sleep.

To begin this chapter, consider the organization of concepts in the Chapter 15 Framework and note key terms associated with each section.

TOPIC OUTLINE AND OBJECTIVES

A. Sensations

1. Define a sensation and list the characteristics of sensations.
2. Classify receptors on the basis of location, stimulus detected, and simplicity or complexity.

B. General senses

3. List the location and function of the receptors for tactile sensations (touch, pressure, vibration), thermoreceptive sensations (heat and cold), and pain.
4. Distinguish somatic, visceral, referred, and phantom pain.
5. Identify the proprioceptive receptors and indicate their functions.

C. Sensory pathways

6. Discuss the origin, neuronal components, and destination of the posterior column and spinothalamic pathways.

7. Explain the neural pathways for pain and temperature; light touch and pressure; and discriminative touch, proprioception, and vibration.

D. Motor pathways

8. Describe how sensory input and motor responses are linked in the central nervous system.
9. Compare the location and functions of the pyramidal and extrapyramidal motor pathways.

E. Integrative functions

10. Compare integrative functions such as memory, wakefulness, and sleep.

Now study the following parts of words that may help you better understand terminology in this chapter.

Wordbyte	Meaning	Example	Wordbyte	Meaning	Example
-ceptor	receiver	proprio*ceptor*	ipsi-	same	*ipsi*lateral
contr-	against, opposed	*contr*alateral	proprio-	one's own	*proprio*ceptor

CHECKPOINTS

A. Sensations (pages 425–426)

■ **A1.** List the four prerequisites for a sensation.

■ **A2.** Contrast *generator potential (GP)* with *nerve action potential (NAP)* in this Checkpoint. Write *GP* or *NAP* next to the characteristics that describe each of these potentials.

_____ a. Results from stimulation of a receptor.

_____ b. Decreases in intensity as it travels along a nerve fiber.

_____ c. Obeys the all-or-none principle, so the frequency and magnitude of the stimulus does not alter the potential.

_____ d. Does not have a refractory period, so summation of potentials is possible.

■ **A3.** What do all receptors have in common regardless of complexity?

A4. Defend or dispute this statement: "You *seem* to see with your eyes, and hear with your ears, but this is not actually so."

■ **A5.** Match the characteristics of sensations in the box with the related descriptions.

Ad. Adaptation	M. Modality
Af. Afterimage	P. Projection

_____ a. The statement you just discussed (Checkpoint A4) is explained by this characteristic.

_____ b. After a while in the presence of a strong-smelling aroma, you do not seem to notice it as much.

_____ c. With your eyes closed you can distinguish sensations: whether you are hearing a voice, feeling a breeze, experiencing pain.

_____ d. You look into a flash camera while your picture is taken, and even moments later you see a light spot in the field of view where the flash was. Since the sensation persists even after the stimulus is removed, this phenomenon is the reverse of adaptation.

■ **A6.** Identify the class of receptor that fits each description.

E. Exteroceptor	P. Proprioceptor	V. Visceroceptor

_____ a. These receptors inform you that you are hungry or thirsty; these also inform your brain of need for adjustment of blood pressure

_____ b. Your ears, eyes, and receptors in your skin for pain, touch, hot, and cold are of this type.

_____ c. With your eyes closed you can tell your exact body position, thanks to these receptors.

■ **A7.** Complete these sentences describing types of receptors based on nature of stimulus detected.

a. You can maintain balance by means of inner ear

_____-receptors sensitive to change in position.

b. Different tastes and smells are distinguished with the help of

_____-receptors in nose and mouth.

c. _____-receptors inform you of the temperature around you.

■ **A8.** Choose the descriptions associated with general (G) or special (S) receptors.

_____ a. These are numerous and widespread in the body.

_____ b. Relatively complex receptors are of this type.

_____ c. Sight, hearing, smell, and taste involve receptors of this type.

_____ d. Sensations of the skin (cutaneous) are of this type.

B. General senses (pages 426–432)

B1. Name the cutaneous senses.

■ **B2.** Do the following exercise about tactile sensations.
a. Name three categories of tactile sensations.

b. (*Discriminative? Light?*) touch refers to ability to determine that something has touched the skin, but its precise location, shape, size, or texture cannot be distinguished.
c. In general, pressure receptors are more (*superficial? deep?*) in location than touch receptors.

d. Sensations of _____ result from rapid repetition of sensory impulses from tactile receptors.

■ **B3.** Match receptors with descriptions.

FNE. Free nerve endings	MD. Merkel's discs
MC. Meissner's corpuscles	PC. Pacinian corpuscles

_____ a. Egg-shaped masses located in dermal papillae, especially in fingertips, palms, and soles

_____ b. Onion-shaped structures sensitive to pressure and high frequency vibration

_____ c. Touch receptors that may be located in epidermis (2 answers)

B4. *For extra review.* Refer to Figure LG 5-1 (page 85). Identify and label types of receptors on that figure. Add your own drawings of three other types of receptors in appropriate layers of skin. Label those receptors also.

■ **B5.** The body quickly adapts to certain stimuli, such as clothing on the body or

odors in the environment. Does the body adapt rapidly to pain? _____
How is this advantageous?

■ **B6.** *A clinical challenge.* Explain why patients experiencing visceral pain (as during a heart attack or gallbladder attack) may feel pain in locations quite distant from those two organs.

a. Pain impulses that originate in the heart, for example, during a "heart attack," enter the spinal cord at the same level as do sensory fibers from skin covering the

_____. This level of the cord is about _____ to _____.

b. Refer to Figure 15-2 (page 429) in your text. Notice that some liver and gallbladder pain refers to a region quite distant from these organs, that is, to the

_____ region. These organs lie just inferior to the

dome-shaped _____ which is supplied by the phrenic nerve from the (*cervical? brachial? lumbar?*) plexus. A painful gallbladder can send impulses to the cord via the phrenic nerve which happens to enter the cord at the

same level (about C _____ or C _____) as nerves from the neck and shoulder.

So gallbladder pain is said to be _____ to the neck and shoulder area.

■ **B7.** A friend asks you how acupuncture is used to control pain. Explain how acupuncture may exert its effect: by enhancing the release of (*enkephalin? substance*

P?) which then acts to inhibit release of transmitter _____ from sensory neurons in pain pathways.

B8. Contrast general and spinal anesthesia.

B9. Define *proprioception* and briefly describe two types of proprioceptors: *muscle spindles* and *tendon organs.*

■ **B10.** Arrange these levels of sensation from simplest to most complex.

Write letters on these lines: _____ _____ _____

A. Sensory impulse reaching the spinal cord
B. Sensation reaching the cerebral cortex
C. Sensory pathway reaching the brain stem or thalamus

C. Sensory pathways (pages 432–437)

■ **C1.** Do this learning activity about sensory reception in the cerebral cortex.

a. Picture the brain as roughly similar in shape to a small, rounded loaf of bread.

The crust would be comparable to the cerebral _____.

b. Suppose you cut a slice of bread in midloaf that represents the postcentral gyrus. It looks somewhat like the "slice" or section of brain shown at the top of Figure LG 15-1. Locate the *Z* on that figure. This portion (next to the

_____ fissure) receives sensations from the (*face? hand? hip? foot?*) (Refer to Figure 15-4, page 433 of your text for help.)

c. Suppose a person suffered a loss of blood supply (CVA or "stroke") to the area of the brain marked *X* in Figure LG 15-1. Loss of sensation in the (*face? hand? leg?*) area would result. Damage to the area marked *W* would be likely to lead to loss of the sense of (*vision? hearing? taste?*) since the W is located on the temporal lobe.

d. The letters *W, X, Y,* and *Z* are marked on the (*right? left?*) side of the brain. (Note that this is an anterior view of this brain section. If a friend stands and faces you, the *right* side of your friend's brain will be in your *left* field of view.) What effects would brain damage to area *Y* of the postcentral gyrus have? Loss of (*sensation? movement?*) to the (*face? hand? foot?*) on the (*right? left?*) side of the body.

e. Certain regions of the body are represented by large areas of somesthetic or

_____ cortex, indicating that these areas (*are? are not?*) very sensitive. Name two of these areas.

■ **C2.** Check your understanding of pathways by drawing them according to the directions below. Try to draw them from memory; do not refer to the text figures or the table until you finish. Label the *first-, second-,* and *third-order neurons* as I, II and III.

a. Use Figure LG 15-1 to show the pathway for pain and temperature in the left hand. Draw the pathway in red.

b. Also draw on Figure LG 15-1 the pathway for two-point discrimination, vibration, and proprioception in the left hand. Use blue for this pathway.

■ **C3.** Answer these questions about the spinocerebellar tracts.

a. They are concerned with (*conscious? subconscious?*) muscle sense.

b. The spinocerebellar tracts permit reflex adjustments for _____

and _____.

c. The left posterior spinocerebellar tract conveys impulses from muscles on (*the left side? the right side? both sides?*) of the body.

d. The left anterior spinocerebellar tract conveys impulses from muscles on (*the left side? the right side? both sides?*) of the body.

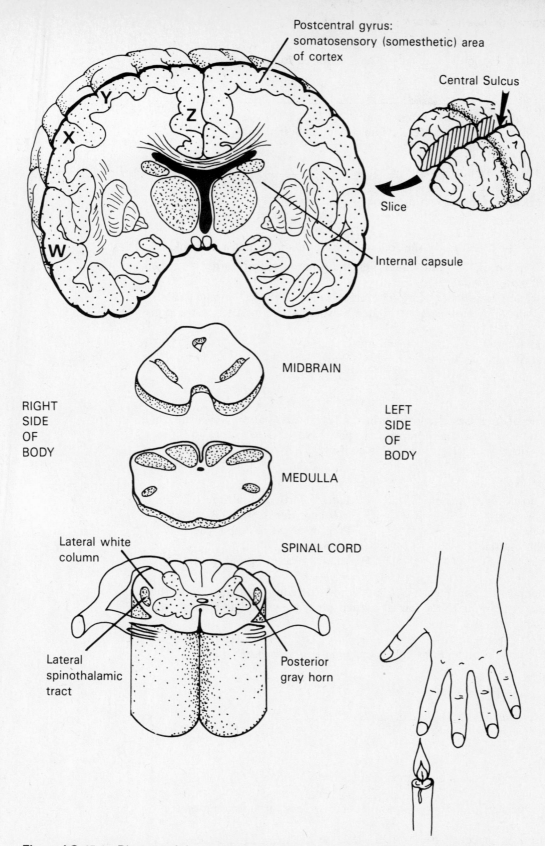

Postcentral gyrus:
somatosensory (somesthetic) area
of cortex

Central Sulcus

Slice

Internal capsule

X   Y   Z

W

MIDBRAIN

RIGHT
SIDE
OF
BODY

LEFT
SIDE
OF
BODY

MEDULLA

Lateral white
column

SPINAL CORD

Lateral
spinothalamic
tract

Posterior
gray horn

Figure LG 15-1 Diagram of the central nervous system. Refer to Checkpoints C1 and C2.

■ **D1.** The linking of sensory to motor response occurs at different levels in the central nervous system. Indicate which is the highest level required for each of these responses.

B. Brain stem	CC. Cerebral cortex
C. Cerebellum	S. Spinal cord

_____ a. Quick withdrawal of a hand when it touches a hot object

_____ b. Unconscious responses to proprioceptive impulses

_____ c. Playing piano or writing a letter

■ **D2.** Refer to Figure LG 15-2 and do the following exercise about motor control.

a. This "slice" or section of the brain containing the primary motor cortex is located in the (*frontal? parietal?*) lobe.

b. Area *P* in this figure controls (*sensation? movement?*) to the (*left? right?*) (*arm? leg? side of the face?*). Motor neurons in area *P* would send impulses to the pons

to activate cranial nerve _____ to facial muscles.

c. Describe effects of damage to area *R*.

■ **D3.** Show the route of impulses along the principal pyramidal pathway by listing in correct sequence the structures that comprise the pathway. Refer to Figure 15-9 (page 439) in the text if you have difficulty with this activity.

___ ___ ___ ___ ___ ___ ___ ___

A. Anterior gray horn (lower motor neuron)
B. Midbrain and pons
C. Effector (skeletal muscle)
D. Internal capsule

E. Lateral corticospinal tract
F. Medulla, decussation site
G. Precentral gyrus (upper motor neuron)
H. Ventral root of spinal nerve

■ **D4.** Now draw this pathway on Figure LG 15-2. Show the control of muscles in the left hand by means of neurons in the lateral corticospinal pathway. Label the *upper* and *lower motor neurons*.

D5. Why are the lower motor neurons called the *final common pathway*?

Distinguish between *spasticity* and *flaccid paralysis*.

■ **D6.** Complete this exercise about the anterior corticospinal tract.

a. Only about ____ percent of the upper motor neurons pass through anterior corticospinal tracts; ____ percent pass through the large _____ corticospinal (pyramidal) tracts.

b. Left anterior corticospinal tracts consist of axons that originated in upper motor neurons on the (*right? left?*) side of the motor cortex.

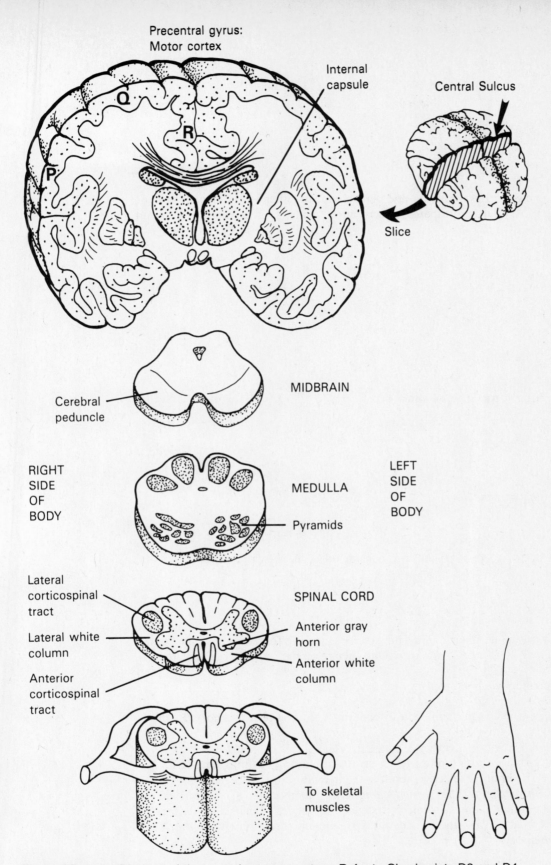

Precentral gyrus:
Motor cortex

Internal
capsule

Central Sulcus

Q

R

P

Slice

Cerebral
peduncle

MIDBRAIN

RIGHT
SIDE
OF
BODY

MEDULLA

LEFT
SIDE
OF
BODY

Pyramids

Lateral
corticospinal
tract

SPINAL CORD

Lateral white
column

Anterior gray
horn

Anterior white
column

Anterior
corticospinal
tract

To skeletal
muscles

Figure LG 15-2 Diagram of the central nervous system. Refer to Checkpoints D2 and D4.

■ **D7.** In general, the extrapyramidal tracts begin in the _____ and

end in the _____. List three extrapyramidal tracts and their
functions.

D8. Summarize how the basal ganglia and cerebellum, together with association neurons, help you to carry out coordinated, precise movements, such as those involved in a game of tennis.

E. Integrative functions   (pages 440–444)

■ **E1.** List three examples of activities that require complex integration processes by the brain.

E2. Define *learning* and *memory*.

E3. List several parts of the brain that are considered to assist with memory.

■ **E4.** Do this exercise on memory.
   a. Looking up a phone number, dialing it, and then quickly forgetting it is an example of (*activated? long-term?*) memory. One theory of such memory is that the

   phone number is remembered only as long as a _____ neuronal circuit is active.
   b. Repeated use of a telephone number can commit it to long-term memory. Such

   reinforcement is known as memory _____. (*Most? Very little?*) of the information that comes to conscious attention goes into long-term memory, since the brain is selective about what it retains.
   c. Some evidence indicates that (*activated? long-term?*) memory involves electrical and chemical events rather than anatomical changes. Supporting this idea is the fact that chemicals used for anesthesia and shock treatments interfere with (*recent? long-term?*) memory.

E5. Discuss the following anatomical or biochemical changes that may be associated with enhanced long-term memory:

a. Presynaptic neurons and neurotransmitter

b. RNA

c. NMDA receptors

■ **E6.** Complete this exercise about sleep and wakefulness.

a. _____ *rhythm* is a term given to the usual daily pattern of sleep and wakefulness in humans.

b. Whether you are awake or asleep depends upon whether the reticular formation,

   also known as the _____ system (RAS) is active. The RAS has two principal parts; the (*thalamic? pons and midbrain?*) portion is responsible

   for your waking from a deep sleep, and the _____ portion helps you to maintain a conscious state.

c. List three types of sensory input that can stimulate the RAS.

d. Stimulation of the reticular formation leads to (*increased? decreased?*) activity of

   the cerebral cortex, as indicated by _____ recordings of brain waves.

e. Continued feedback between the RAS, the _____, and the

   _____ maintains the state called *consciousness*. Inactivation of

   the RAS produces a state known as _____.

■ **E7.** Do the following learning activity about sleep.

a. Name two kinds of sleep.

b. (*REM? NREM?*) sleep is required first in order for (*REM? NREM?*) sleep to occur.

c. In NREM sleep, a person gradually progresses from stage (*1 to 4? 4 to 1?*) into deep sleep. Alpha waves are present in stage(s) (*1? 3 and 4?*), whereas slow delta waves characterize stage(s) (*1? 3 and 4?*) sleep.

d. REM sleep is also called _____ sleep since it follows, but is much more active than, deep stage of sleep. Respirations and pulse are

much ____-creased over their levels in stage 4; alpha waves are present as in stage

____ sleep. The name REM indicates that _____ movements can be observed. Dreaming occurs during (*stages 1–4 of NREM? REM?*) sleep.

e. Periods of REM and NREM sleep alternate throughout the night in about

____-minute cycles. During the early part of an 8-hour sleep period REM periods last about (*5–10? 30? 50?*) minutes; they gradually increase in length until the final

REM period lasts about ____ minutes.

ANSWERS TO SELECTED CHECKPOINTS

A1. Stimulus, receptor, conduction, translation.
A2. (a) GP. (b) GP. (c) NAP. (d) GP.
A3. They all contain dendrites of sensory neurons.
A5. (a) P. (b) Ad. (c) M. (d) Af.
A6. (a) V. (b) E. (c) P.
A7. (a) Mechano-. (b) Chemo-. (c) Thermo-.
A8. (a) G. (b) S. (c) S. (d) G.
B2. (a) Touch, pressure, vibration. (b) Light. (c) Deep. (d) Vibration.
B3. (a) MC. (b) PC. (c) FNE and MD.

B5. No; it is protective.
B6. (a) Medial aspects of left arm, T1, T4. (b) Shoulder and neck (right side), diaphragm, cervical, 3, 4, referred.
B7. (b) Enkephalin, substance P.
B10. A C B
C1. (a) Cortex. (b) Longitudinal, foot. (c) Face, hearing. (d) Right; sensation, hand, left. (e) General sensory or somatosensory, are; lips, face, thumb and other fingers (see Figure 15-4, page 433 in the text).

C2.

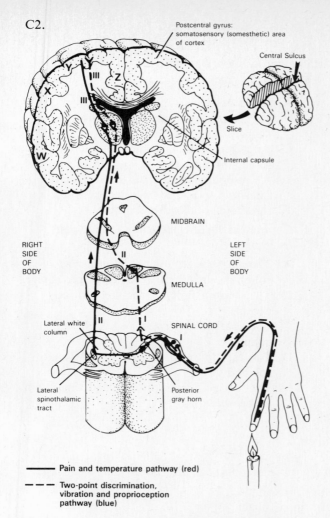

Pain and temperature pathway (red)

- - - Two-point discrimination, vibration and proprioception pathway (blue)

Figure LG 15-1A Diagram of the central nervous system.

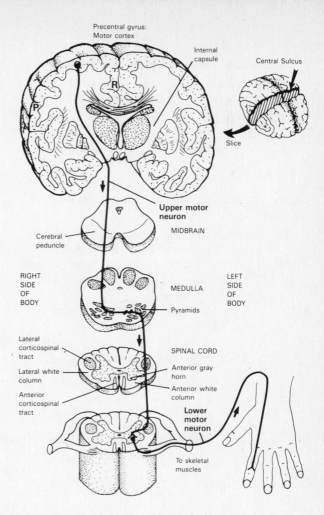

Figure LG 15-2A Diagram of the central nervous system.

C3. (a) Subconscious. (b) Posture, muscle tone. (c) The left side. (d) Both sides.
D1. (a) S. (b) C. (c) CC.
D2. (a) Frontal. (b) Movement, left, side of the face; VII. (c) Movement of left foot affected (spasticity).
D3. G D B F E A H C.
D4. See Figure LG 15-2A above.
D6. (a) 15, 85, lateral. (b) Left.
D7. Basal ganglia and reticular formation; spinal cord. Rubrospinal, tectospinal, vestibulospinal, and reticulospinal.

E1. Memory, sleep/wakefulness, emotions.
E4. (a) Activated, reverberating. (b) Consolidation, very little. (c) Activated, recent.
E6. (a) Circadian. (b) Reticular activating; thalamic; pons and midbrain (mesencephalic). (c) Examples; light, sounds, touch, impulses from cerebral cortex. (d) Increased, electroencephalograph (EEG). (e) Cerebral cortex, spinal cord, sleep.
E7. (a) Rapid eye movement (REM) and non-rapid eye movement (NREM). (b) NREM, REM. (c) 1 to 4; 1; 3 and 4. (d) Paradoxical, in, 1, rapid eye, REM. (e) 90, 5–10, 50.

Questions 1–4: Circle the letter preceding the one best answer to each question.

1. All of these sensations are conveyed by the posterior column pathway EXCEPT:

   A. Pain and temperature
   B. Proprioception
   C. Fine touch, two-point discrimination
   D. Vibration
   E. Stereognosis

2. Choose the FALSE statement about REM sleep.

   A. Infant sleep consists of a higher percentage of REM sleep than does adult sleep.
   B. Dreaming occurs during this type of sleep.
   C. The eyes move rapidly behind closed lids during REM sleep.
   D. EEG readings are similar to those of stage 4 of NREM sleep.
   E. REM sleep occurs periodically throughout a typical 8-hour sleep period.

3. Choose the FALSE statement about receptors.

   A. Receptors are very excitable.
   B. All receptors contain dendrites of sensory neurons.
   C. Some receptors have simple structure, some complex.
   D. All receptors except pain have a high threshold to a specific stimulus and a low threshold to all other stimuli.

4. Choose the FALSE statement about pain receptors.

   A. They may be stimulated by any type of stimulus, such as heat, cold, or pressure.
   B. They have a simple structure with no capsule.
   C. They are characterized by a high level of adaptation.
   D. They are found in almost every tissue of the body.
   E. They are important in helping to maintain homeostasis.

Questions 5–8: Arrange the answers in correct sequence.

_____ _____ _____ 5. Density of sense receptors, from most dense to least:

   A. In tip of tongue
   B. In tip of finger
   C. In back of neck

_____ _____ _____ _____ 6. In order of occurrence in a sensation:

   A. Reception by a sense organ or other receptor
   B. Translation of the impulse into a sensation
   C. Conduction along a nervous pathway
   D. A stimulus

_____ _____ _____ _____ 7. Levels of sensation, from those causing simplest, least precise reflexes to those causing most complex and precise responses:

   A. Thalamus
   B. Brain stem
   C. Cerebral cortex
   D. Spinal cord

_____ _____ _____ _____ _____ 8. Pathway for conduction of most of the impulses for voluntary movement of muscles:

   A. Anterior gray horn of the spinal cord
   B. Precentral gyrus
   C. Internal capsule
   D. Location where decussation occurs
   E. Lateral corticospinal tract

T F   9. In general, the <u>left</u> side of the the brain controls the right side of the body.

T F 10. Generator (receptor) potentials <u>obey</u> the all-or-none principle.

T F 11. The final common pathway consists of <u>upper motor</u> neurons.

T F 12. Circadian rhythm pertains to the <u>complex feedback circuits involved in producing coordinated momements.</u>

T F 13. <u>Sight, hearing, smell, and pressure</u> are all special senses.

T F 14. The diaphragm seems to refer pain to the shoulder and neck region since the <u>phrenic nerve enters the cord at the same levels (C3 to C5) as cutaneous nerves of the shoulder and neck.</u>

T F 15. The neuron that crosses to the opposite side in sensory pathways is usually the <u>second-order</u> neuron.

T F 16. Muscle spindles, tendon organs, joint kinesthetic receptors are all examples of <u>proprioceptive</u> receptors.

T F 17. Damage to the left lateral spinothalamic tract would be most likely to result in loss of awareness of <u>pain and vibration sensations in the left</u> side of the body.

T F 18. Conscious sensations, such as those of sight and touch, can occur only in the <u>cerebral cortex.</u>

T F 19. Pain experienced by an amputee as if the amputated limb were still there is an example of <u>phantom pain.</u>

T F 20. Damage to the final common pathway will result in <u>flaccid paralysis.</u>

_____ 21. ____ is a neurotransmitter released from axons of posterior root ganglia in pain pathways.

_____ 22. Inactivation of the ____ results in the state of sleep.

_____ 23. Rubrospinal, tectospinal, and vestibulospinal tracts are all classified as ____ tracts.

_____ 24. A CVA affecting the medial portion of the postcentral gyrus of the right hemisphere (area *Z* in Figure LG 15-1, page LG 265) is most likely to result in symptoms such as ____ .

_____ 25. Short-term memory is also referred to as ____ memory.

ANSWERS TO MASTERY TEST: ■ Chapter 15

Multiple Choice

1. A   3. D
2. D   4. C

Arrange

5. A B C
6. D A C B
7. D B A C
8. B C D E A

True-False

9. T
10. F. Do not obey
11. F. Lower motor
12. F. Events that occur at approximately 24-hour intervals, such as sleeping and waking
13. F. Sight, hearing, smell and taste (not pressure)
14. T
15. T
16. T
17. F. Pain and temperature sensations in the right
18. T
19. T
20. T

Fill-ins

21. Substance P
22. Reticular activating system
23. Extrapyramidal
24. Loss of sensation of left foot
25. Activated

FRAMEWORK 16

The Autonomic Nervous System (ANS)

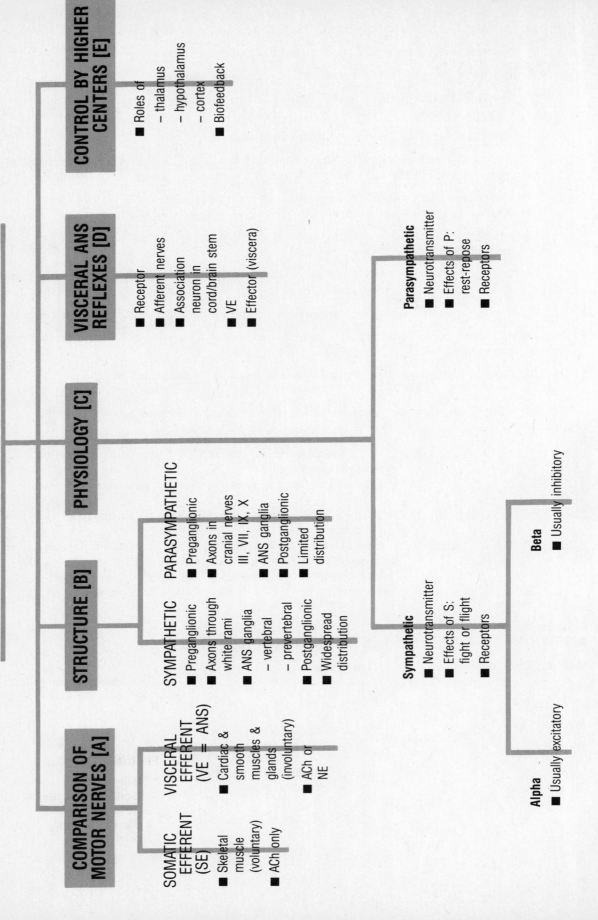

COMPARISON OF MOTOR NERVES [A]

SOMATIC EFFERENT (SE)
- Skeletal muscle (voluntary)
- ACh only

VISCERAL EFFERENT (VE = ANS)
- Cardiac & smooth muscles & glands (involuntary)
- ACh or NE

STRUCTURE [B]

SYMPATHETIC
- Preganglionic
- Axons through white rami
- ANS ganglia
  – vertebral
  – prevertebral
- Postganglionic
- Widespread distribution

PARASYMPATHETIC
- Preganglionic
- Axons in cranial nerves III, VII, IX, X
- ANS ganglia
- Postganglionic
- Limited distribution

PHYSIOLOGY [C]

Sympathetic
- Neurotransmitter
- Effects of S: fight or flight
- Receptors

Alpha
- Usually excitatory

Beta
- Usually inhibitory

Parasympathetic
- Neurotransmitter
- Effects of P: rest-repose
- Receptors

VISCERAL ANS REFLEXES [D]

- Receptor
- Afferent nerves
- Association neuron in cord/brain stem
- VE
- Effector (viscera)

CONTROL BY HIGHER CENTERS [E]

- Roles of
  – thalamus
  – hypothalamus
  – cortex
- Biofeedback

The Autonomic Nervous System (ANS)

The autonomic nervous system (ANS) consists of the special branch of the nervous system that exerts unconscious control over viscera. The ANS regulates activities such as heart rate and blood pressure, glandular secretion and digestion. The two divisions of the ANS—the sympathetic and parasympathetic—carry out a continual balancing act, readying the body for response to stress or facilitating rest and relaxation, according to momentary demands. ANS neutrotransmitters hold particular significance clinically since they are often mimicked or inhibited by medications.

First study the Chapter 16 Framework and the key terms associated with each section.

TOPIC OUTLINE AND OBJECTIVES

A. Comparison of efferent (motor) nerves

1. Compare the structural and functional differences between the somatic efferent and autonomic portions of the nervous system.

B. Structure

2. Identify the principal structural features of the autonomic nervous system.
3. Compare the sympathetic and parasympathetic divisions of the autonomic nervous system in terms of structure, physiology, and neurotransmitters released.

C. Physiology

4. Describe the various postsynaptic receptors involved in automatic responses.

D. Visceral autonomic reflexes

5. Describe a visceral autonomic reflex and its components.

E. Control by higher centers

6. Explain the role of the hypothalamus and its relationship to the sympathetic and parasympathetic divisions.
7. Explain the relationship between biofeedback and meditation, and the autonomic nervous system.

Now study the following parts of words that may help you better understand terminology in this chapter.

Wordbyte	Meaning	Example	Wordbyte	Meaning	Example
auto-	self	*auto*nomic	pre-	before	*pre*ganglionic
nomos	law, governing	auto*nomic*	ram(i)-	branch	white *ram*us
post-	after	*post*ganglionic	splanchn-	viscera	*splanchn*ic nerve

Checkpoints

A. Comparison of efferent (motor) nerves (page 448)

A1. Why is the autonomic nervous system so named? Is the autonomic nervous system (ANS) entirely independent of higher control centers? Explain.

■ **A2.** Give examples of structures innervated by each type of nerve.

a. Somatic efferent

b. Visceral efferent

For extra review. Study Checkpoint B9 in Chapter 12 (page 208).

■ **A3.** Name the two divisions of the ANS.

_____   _____

A4. Many visceral organs have dual innervation by the autonomic system.

a. What does dual innervation mean?

b. How do the sympathetic and parasympathetic divisions work in harmony to control viscera?

■ **A5.** Contrast the somatic efferent and autonomic (visceral efferent) nervous systems by identifying characteristics of each in this Checkpoint.

SE. Somatic efferent	VE. Visceral efferent (ANS)

_____ a. All neurons produce the transmitter acetylcholine (ACh).

_____ b. Some neurons release ACh, whereas others produce norepinephrine (NE).

_____ c. Involuntary control (effectors are not under conscious control).

_____ d. Effectors are all skeletal muscles.

B. Structure (pages 448-454)

■ **B1.** Complete this exercise describing structural differences between visceral efferent and somatic efferent pathways.

a. Between the spinal cord and effector, somatic pathways include _____ neuron(s), whereas visceral pathways require _____ neuron(s), known as the

pre-_____ and _____ neurons.

b. Somatic pathways begin at (*all? only certain?*) levels of the cord, whereas visceral

routes begin at _____ levels of the cord.

■ **B2.** Contrast preganglionic with postganglionic fibers in this exercise.

a. Consider the sympathetic division first. Its preganglionic cell bodies lie in the

lateral gray of segments _____ to _____ of the cord. For this reason, the

sympathetic division is also known as the _____ outflow.

b. Preganglionic axons (*are? are not?*) myelinated, and therefore they appear white.

These axons may branch to reach a number of different _____ where they synapse. These ganglia contain (*pre-? post-?*) ganglionic neuron cell bodies whose axons proceed to the appropriate organ. Postganglionic axons are (*gray? white?*) since they (*do? do not?*) have a myelin covering.

c. Parasympathetic preganglionic cell bodies are located in two areas. One is the

_____ from which axons pass out in cranial nerves _____,

_____, _____, and _____. A second location is the (*anterior? lateral?*) gray

horns of segments _____ through _____ of the cord.

d. Based on location of (*pre-? post-?*) ganglionic neuron cell bodies, the parasym-

pathetic division is also known as the _____ outflow. The axons of these neurons extend to (*the same? different?*) autonomic ganglia from those in sympathetic pathways (see Checkpoint B4.). There postganglionic cell bodies send out short (*gray? white?*) axons to innervate viscera.

e. Most viscera (*do? do not?*) receive fibers from both the sympathetic and the parasympathetic divisions. However, the origin of these divisions in the CNS and the pathways taken to reach viscera (*are the same? differ?*).

■ **B3.** Contrast these two types of ganglia.

	Posterior Root Ganglia	**Autonomic Ganglia**
a. Afferent or efferent neurons		
b. Neurons in somatic or visceral pathway or both.		
c. Synapsing does or does not occur		

■ **B4.** Complete the table contrasting types of autonomic ganglia.

	Sympathetic Trunk	Prevertebral	Terminal
a. Sympathetic or para-sympathetic		Sympathetic	
b. Alternate name	Paravertebral or lateral or sympathetic trunk or vertebral chain		
c. General location			Close to or in walls of effectors

■ **B5.** Refer to Figure 16-3 (page 451) in your text. Trace with your finger the route of a preganglionic neuron. Now describe the pathway common to **all** sympathetic pregan-

glionic neurons by listing the structures in correct sequence. ____ ____ ____ ____
   A. Ventral root of spinal nerve
   B. Sympathetic trunk ganglion
   C. Lateral gray of spinal cord (between T1 and L2)
   D. White ramus communicants

■ **B6.** Now consider the possible routes sympathetic preganglionic neurons may take once they are in the sympathetic trunk ganglion. (It may be helpful to refer again to Figure 16-3 in your text and again trace pathways with your finger.)

   a. What is the shortest possible path they can take to reach a postganglionic neuron cell body?

   b. They may ascend and/or descend the sympathetic chain to reach trunk ganglia at other levels. This is important since sympathetic preganglionic cell bodies are

   limited in location to _____ to _____ levels of the cord, yet the sympathetic

   trunk extends from _____ to _____ levels of the vertebral column. Sympathetic fibers must be able to reach these distant areas to provide the entire body with sympathetic innervation.

c. Viscera in skin and extremities (such as sweat glands, blood vessels, hair muscles) receive sympathetic innervation through the following route. Preganglionic neurons synapse with postganglionic cell bodies in sympathetic trunk ganglia (at the level of entry or after ascending or descending). Postganglionic fibers then pass through

_____ which connect to _____ .

d. Some preganglionic fibers do not synapse as described in (b) or (c), but pass on through trunk ganglia without synapsing there. They course through

_____ nerves to _____ ganglia. Here they

synapse with postganglionic neurons whose axons form _____ en route to viscera.

e. Prevertebral ganglia are located only in the _____ . In the neck, thorax, and pelvis the only sympathetic ganglia are those of the trunk (see Figure 16-2, page 450 in the text). As a result, cardiac nerves, for example, contain only (*preganglionic? postganglionic?*) fibers, as indicated by broken lines in the figure.

f. A given sympathetic preganglionic neuron is likely to have (*few? many?*) branches; these may take any of the paths described. Once a branch synapses, it (*can? cannot?*) synapse again, since any autonomic pathway consists of just

_____ neurons (preganglionic and postganglionic).

■ **B7.** Test your understanding of these routes by completing the sympathetic pathway shown in Figure LG 16-1. A preganglionic neuron located at level T5 of the cord has its axon drawn as far as the white ramus. Finish this pathway by showing how the axon may branch and synapse to eventually innervate these three organs: the heart, a sweat gland in skin of the thoracic region, and the stomach. Draw the preganglionic fibers in solid lines and the postganglionic fibers in broken lines.

■ **B8.** In the past several Checkpoints, you have been focusing on the (*sympathetic? parasympathetic?*) division of the ANS. Now that you are somewhat familiar with the rather complex sympathetic pathways, you will notice that parasympathetic pathways are much simpler. This is due to the fact that there is no parasympathetic chain of

ganglia; the only parasympathetic ganglia are _____ , which are in or close to the organs innervated. Also parasympathetic preganglionic fibers synapse with (*more? fewer?*) postganglionic neurons. What is the significance of the structural simplicity of the parasympathetic system?

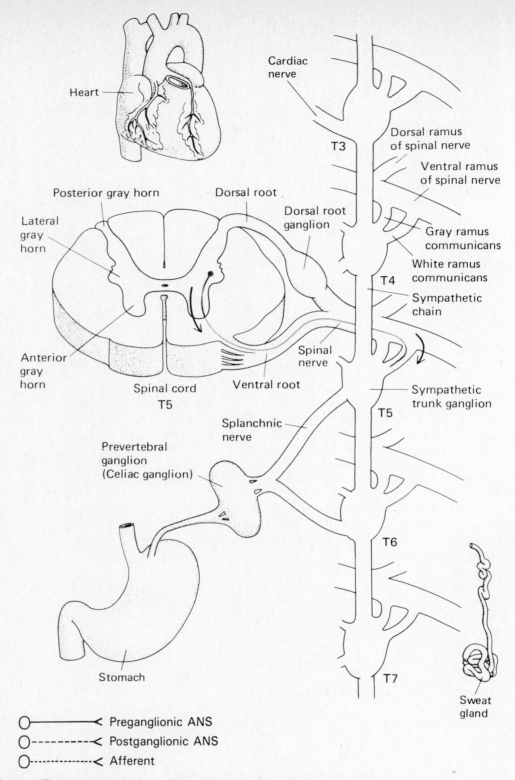

Heart

Cardiac
nerve

Dorsal ramus
of spinal nerve

Ventral ramus
of spinal nerve

T3

Posterior gray horn

Dorsal root

Dorsal root
ganglion

Lateral
gray
horn

Gray ramus
communicans

White ramus
communicans

T4

Sympathetic
chain

Anterior
gray
horn

Spinal
nerve

Sympathetic
trunk ganglion

Spinal cord
T5

Ventral root

T5

Splanchnic
nerve

Prevertebral
ganglion
(Celiac ganglion)

T6

Stomach

T7

Sweat
gland

O———< Preganglionic ANS

O----------< Postganglionic ANS

O-------------< Afferent

Figure LG 16-1 Typical sympathetic pathway beginning at level T5 of the cord. Complete pathways according to directions in Checkpoints B7 and D2.

■ **B9.** Complete the table about the cranial portion of the parasympathetic system.

Cranial Nerve		Name of Terminal Ganglion	Structures Innervated
Number	Name		
a.			Iris and ciliary muscle of eye
b.		1. Pterygopalatine 2. Submandibular	1. 2.
c.	Glossopharyngeal		
d.		Ganglia in cardiac and pulmonary plexuses and in abdominal plexuses	

■ **B10.** *For extra review.* Write S if the description applies to the sympathetic division of the ANS, P if it applies to the parasympathetic division, and P, S if it applies to both

_____ a. Also called thoracolumbar outflow

_____ b. Has long preganglionic fibers leading to terminal ganglia and very short postganglionic fibers

_____ c. Celiac and superior mesenteric ganglia are sites of postganglionic neuron cell bodies

_____ d. Sends some preganglionic fibers through cranial nerves

_____ e. Has some preganglionic fibers synapsing in vertebral chain (trunk)

_____ f. Has more widespread effect in the body, affecting more organs

_____ g. Has some fibers running in gray rami to supply sweat glands, hair muscles, and blood vessels

_____ h. Has fibers in white rami (connecting spinal nerve with vertebral chain)

_____ i. Contains fibers that supply viscera with motor impulses

C. Physiology (pages 454–457)

■ **C1.** Do this exercise about neurotransmitters.

a. Axons that release the transmitter acetylcholine (ACh) are known as (*adrenergic? cholinergic?*). Those that release norepinephrine (NE, also called noradrenalin) are

called _____ . Most sympathetic nerves are said to be (*cholinergic? adrenergic?*), while all parasympathetic nerves are

_____ .

b. During stress, the (*sympathetic? parasympathetic?*) division of the ANS prevails, so stress responses are primarily (*adrenergic? cholinergic?*) responses.

c. Another source of NE as well as epinephrine is the gland known as the

_____ . Therefore chemicals released by this gland (as in stress) will mimic the action of the (*sympathetic? parasympathetic?*) division of the ANS.

■ **C2.** *A clinical challenge.* Certain medications can mimic the effects of the body's own transmitters. Refer to Figure LG 16-2 and do this exercise.

a. A beta$_1$-stimulator (or beta-exciter) mimics (*ACh? NE?*), causing the heart rate and strength of contraction to ____-crease. Name one such drug. (For help, refer to a

pharamacology text.) _____ Such a drug has a structure that is close enough to the shape of NE that it can "sit on" the beta$_1$ receptor as NE would and activate it.

b. A beta-blocker (or beta-inhibitor) also has a shape similar to that of NE and so

can sit on _____ receptors. Name one such drug.

_____ . However this drug cannot activate the receptor, but simply prevents NE from doing so. So a beta-blocker ____-creases heart rate and force of contraction.

c. Other drugs known as anticholinergic (or ACh-blockers) can take up residence on ACh receptors so that ACh being released from vagal nerve stimulation cannot exert its normal effects. A person taking such a medication may exhibit a(n) (*increased? decreased?*) heart rate.

■ **C3.** *A clinical challenge.* Do this exercise on other types of ANS receptors and related medications.

a. Some slightly different types of beta receptors (known as beta$_2$) are located in smooth muscle of air passageways of lungs. Interaction of NE with these receptors causes the opposite response: relaxation of the muscle. So the effect of the sympathetic stimulation of lungs is to cause (*dilation? constriction?*) of airways, making breathing (*easier? more difficult?*) during stressful times.

b. The effect of NE on the smooth muscle of the stomach, intestines, bladder, and

uterus is also (*contraction? relaxation?*), so that during stress, the body ____-creases activity of these organs and can focus on more vital activities such as heart contractions.

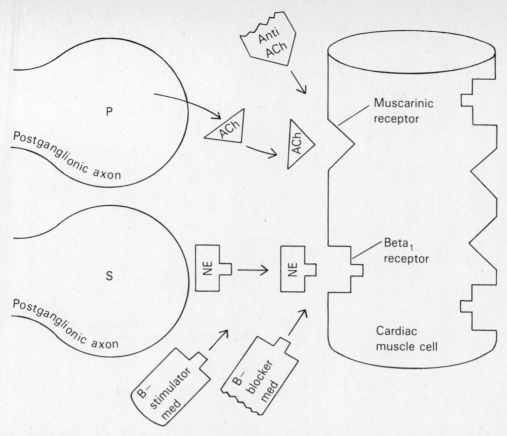

Figure LG 16-2 Diagram of ANS neurotransmitters at neuromuscular junction at the heart. Note that neurotransmitters and receptors are greatly enlarged. ACh, acetylcholine; NE, norepinephrine; P, parasympathetic; S, sympathetic. Refer to Checkpoint C2.

c. Alpha receptors in smooth muscle of blood vessels of the skin respond to NE by contracting. The result is (*dilation? constriction?*) of these vessels due to sympathetic nerve stimulation. Since blood is squeezed out of these skin vessels and

forced to enter major blood pathways, blood pressure ____-creases.

d. Alpha-blocker drugs have just the opposite effect, so tend to (*constrict? dilate?*)

blood vessels. Name one of these dilator drugs _____. These

can help a person to ____-crease blood pressure.

■ **C4.** Use arrows to show whether parasympathetic (P) or sympathetic (S) fibers stimulate (↑) or inhibit (↓) each of the following activities. Use a dash (−) to indicate that there is no parasympathetic innervation. The first one is done for you.

a. P __↓__   S __↑__   Dilation of pupil

b. P _____   S _____   Heart rate and blood flow to coronary (heart muscle) blood vessels

c. P _____   S _____   Constriction of skin blood vessels

d. P _____   S _____   Salivation and digestive organ contractions

e. P _____   S _____   Erection of genitalia

f. P _____   S _____   Dilation of bronchioles for easier breathing

g. P _____   S _____   Contraction of bladder and relaxation of internal urethral sphincter causing urination

h. P _____   S _____   Contraction of pili of hair follicles causing "goose bumps"

i. P _____   S _____   Contraction of spleen which transfers some of its blood to general circulation, causing increase in blood pressure

j. P _____   S _____   Release of epinephrine and norepinephrine from adrenal medulla

k. P _____   S _____   Coping with stress, fight-or-flight response

■ **C5.** Write an explanation based on anatomy of the ANS that tells why skin structures such as sweat glands, hair muscles, and blood vessels lack parasympathetic innervation.

C6. Write a paragraph outlining the changes effected by the ANS during a fight-or-flight response, such as running in fear.

D. Visceral autonomic reflexes   (page 458)

■ **D1.** Autonomic neurons can be stimulated by (*somatic afferents? visceral afferents? either somatic or visceral afferents?*).

■ **D2.** On Figure LG 16-1 draw an afferent neuron (in contrasting color) to show how the sympathetic neuron could be stimulated by stomach pain.

■ **D3.** Arrange in correct sequence structures in the pathway for a painful stimulus at your fingertip to cause a visceral (autonomic) reflex, such as sweating. Write the letters of the structures in order on the lines provided.

___ ___ ___ ___ ___ ___ ___ ___ ___ ___

A. Association neuron in spinal cord
B. Nerve fiber in gray ramus
C. Pain receptor in skin
D. Cell body of postganglionic neuron in trunk ganglion
E. Cell body of preganglionic neuron in lateral gray of cord
F. Nerve fiber in anterior root of spinal nerve
G. Nerve fiber in white ramus
H. Fiber in spinal nerve in brachial plexus
I. Sweat gland
J. Sensory neuron

■ **D4.** Most visceral sensations (*do? do not?*) reach the cerebral cortex, so most visceral sensations are at (*conscious? subconscious?*) levels. Hunger and nausea are exceptions.

E. Control by higher centers (pages 458–459)

E1. Explain the role of each of these structures in control of the autonomic system.

a. Thalamus

b. Hypothalamus (Which part controls sympathetic nerves? Which controls parasympathetic?)

c. Cerebral cortex (When is it most involved in ANS control? In stress or nonstress situations?)

E2. Explain how biofeedback is used to help in each of the following cases.

a. Migraine headache

b. Childbirth

E3. Explain how yoga and transcendental meditation (*TM*) may demonstrate whether or not the autonomic nervous system is truly autonomous.

ANSWERS TO SELECTED CHECKPOINTS

A2. (a) Skeletal muscles, such as gastrocnemius.
(b) Smooth muscle (digestive organs, urinary bladder, walls of blood vessels, hair muscles), cardiac muscle (heart), and glands (sweat, salivary, other digestive glands).

A3. Sympathetic and parasympathetic.

A5. (a) SE. (b) VE. (c) VE. (d) SE.

B1. (a) 1, 2, ganglionic, postganglionic. (b) All, only certain (See Checkpoint B2).

B2. (a) T1, L2, thoracolumbar. (b) Are, ganglia, post-, gray, do not. (c) Brain stem; III, VII, IX, X; lateral; S2, S4. (d) Pre-, craniosacral, different, gray. (e) Do, differ.

B3. (a) Afferent, efferent. (b) Both, visceral. (c) Does not, does.

B4.

	Sympathetic Trunk	Prevertebral	Terminal
a. Sympathetic or parasympathetic	**Sympathetic**	Sympathetic	**Para-sympathetic**
b. Alternate name	Paravertebral or lateral or sympathetic trunk or vertebral chain	**Collateral or prevertebral (celiac, superior and inferior mesenteric)**	**Intramural (in walls of viscera)**
c. General location	**In vertical chain along both sides of vertebral bodies from base of skull to coccyx**	**In three sites anterior to spinal cord and close to major abdominal arteries**	Close to or in walls of effectors

B5. C A D B.

B6. (a) Immediately synapse in trunk ganglion. (b) T1, L2; C3, sacral. (c) Gray rami, spinal nerves. (d) Splanchnic, prevertebral, plexuses. (e) Abdomen, postganglionic. (f) Many, cannot, 2.

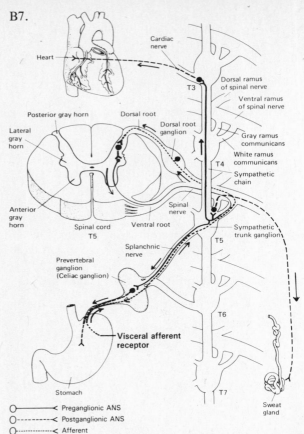

B7.

Figure LG 16-1A Typical sympathetic pathway beginning at level T5 of the cord.

B8. Sympathetic, terminal ganglia, fewer. Parasympathetic effects are much less widespread than sympathetic, distributed to fewer areas of the body, and characterized by a response of perhaps just one organ, rather than an integrated response in which many organs respond *in sympathy* with one another.

B9.

Cranial Nerve		Name of Terminal Ganglion	Structures Innervated
Number	Name		
a. III	**Oculomotor**	**Ciliary**	Iris and ciliary muscle of eye
b. VII	**Facial**	1. Pterygopalatine 2. Submandibular	1. **Nasal mucosa, palate, pharynx, lacrimal gland** 2. **Submandibular and submaxillary glands**
c. IX	Glossopharyngeal	**Otic**	**Parotid salivary gland**
d. X	**Vagus**	Ganglia in cardiac and pulmonary plexuses and in abdominal plexuses	**Thoracic and abdominal viscera**

B10. (a) S. (b) P. (c) S. (d) P. (e) S. (f) S. (g) S. (h) S. (i) P, S.

C1. (a) Cholinergic, adrenergic, adrenergic, cholinergic. (b) Sympathetic, adrenergic. (c) Adrenal medulla, sympathetic.

C2. (a) NE, in, isoproterinol (Isoprel) or epinephrine. (b) Beta, propranolol (Inderal), de. (c) Increased (a double negative: the inhibiting vagus is itself inhibited).

C3. (a) Dilation, easier, (b) Relaxation, de. (c) Constriction, in. (d) Dilate, phentolamine (Regitine), de.

C4. (b) P ↓, S ↑. (c) P−, S ↑. (d) P ↑, S ↓. (e) P ↑, S ↓. (f) P ↓, S ↑. (g) P ↑, S ↓. (h) P−, S ↑. (i) P−, S ↑. (j)P−, S ↑. (k) P ↓, S ↑.

C5. Parasympathetic nerves do not enter the sympathetic chain or gray rami to reach individual spinal nerves, so they have no access to skin.

D1. Either somatic or visceral afferents.

D2. See Figure LG 16-1A above.

D3. C J A E F G D B H I.

D4. Do not, subconscious.

MASTERY TEST: Chapter 16

Questions 1–6: Circle letters preceding ALL correct answers to each question.

1. Choose all TRUE statements about the vagus nerve.

   A. Its autonomic fibers are sympathetic.
   B. It supplies ANS fibers to viscera in the thorax and abdomen, but not in the pelvis.
   C. It causes the salivary glands and other digestive glands to increase secretions.
   D. Its ANS fibers are mainly preganglionic.
   E. It is a cranial nerve originating from the medulla.

2. Which activities are characteristic of the stress response, or fight-or-flight reaction?

   A. The liver breaks down glycogen to glucose.
   B. The heart rate decreases.
   C. Kidneys increase urine production since blood is shunted to kidneys.
   D. There is increased blood flow to genitalia, causing erect state.
   E. Hairs stand on end ("goose pimples") due to contraction of arrector pili muscles.
   F. In general, the sympathetic system is active.

3. Which fibers are classified as autonomic?

   A. Any visceral efferent nerve fiber
   B. Any visceral afferent nerve fiber
   C. Nerves to salivary glands and sweat glands
   D. Pain fibers from ulcer in stomach wall
   E. Sympathetic fibers carrying impulses to blood vessels
   F. All nerve fibers within cranial nerves
   G. All parasympathetic nerve fibers within cranial nerves

4. Choose all TRUE statements about gray rami.

   A. They contain only sympathetic nerve fibers.
   B. They contain only postganglionic nerve fibers.
   C. They carry impulses from trunk ganglia to spinal nerves.
   D. They are located at all levels of the vertebral column (from C1 to coccyx).
   E. They carry impulses between lateral ganglia and collateral ganglia.
   F. They carry preganglionic neurons from anterior ramus of spinal nerve to trunk ganglion.

5. Which of the following structures contain some sympathetic preganglionic nerve fibers?

   A. Splanchnic nerves
   B. White rami
   C. Sciatic nerve
   D. Cardiac nerves
   E. Ventral roots of spinal nerves
   F. The sympathetic chains

6. Which are structural features of the parasympathetic system?

   A. Ganglia close to the CNS and distant from the effector
   B. Forms the craniosacral outflow
   C. Distributed throughout the body, including extremities
   D. Supplies nerves to blood vessels, sweat glands, and adrenal gland
   E. Has some of its nerve fibers passing through lateral (paravertebral) ganglia

7. All of the following axons are cholinergic EXCEPT:

    A. Parasympathetic preganglionic
    B. Parasympathetic postganglionic
    C. Sympathetic preganglionic
    D. Sympathetic postganglionic to sweat glands
    E. Sympathetic postganglionic to heart muscle

8. All of the following are collateral ganglia EXCEPT:

    A. Superior cervical
    B. Prevertebral ganglia
    C. Celiac ganglion
    D. Superior mesenteric ganglion
    E. Inferior mesenteric ganglion

9. Which region of the cord contains no preganglionic cell bodies at all?

    A. Sacral
    B. Lumbar
    C. Cervical
    D. Thoracic

10. Which statement about postganglionic neurons is FALSE?

    A. They all lie entirely outside of the CNS.
    B. Their axons are nonmyelinated.
    C. They terminate in visceral effectors.
    D. Their cell bodies lie in the lateral gray matter of the cord.
    E. They are very short in the parasympathetic system.

Questions 11–20: Circle T (true) or F (false). If the statement is false, change the underlined word or phrase so that the statement is correct.

T   F   11. Control of the ANS by the cerebral cortex occurs primarily during times when a person is relaxed (nonstressed).

T   F   12. Synapsing occurs in both sympathetic and parasympathetic ganglia.

T   F   13. The sciatic, brachial, and femoral nerves all contain sympathetic postganglionic nerve fibers.

T   F   14. Biofeedback and meditation confirm that the ANS is independent of higher control centers.

T   F   15. Under stress conditions the sympathetic system dominates over the parasympathetic.

T   F   16. When one side of the body is deprived of its sympathetic nerve supply, as in Horner's syndrome, the following symptoms can be expected: constricted pupil (miosis) and lack of sweating (anhidrosis).

T   F   17. Sympathetic cardiac nerves stimulate heart rate, and the vagus slows down heart rate.

T   F   18. Viscera do have sensory nerve fibers, but they are not included in the autonomic nervous system.

T   F   19. All viscera have dual innervation by sympathetic and parasympathetic divisions of the ANS.

T   F   20. The sympathetic system has a more widespread effect in the body than the parasympathetic does.

Questions 21–25: fill-ins. Complete each sentence with the word or phrase that best fits.

_____ 21. The three types of tissue (effectors) innervated by the ANS nerves are ___.

_____ 22. ___ nerves convey impulses from the sympathetic trunk ganglia to collateral ganglia.

_____ 23. About 80 percent of the craniosacral outflow (parasympathetic nerves) is located in the ___ nerves.

_____ 24. Alpha (α) and beta (β) receptors are stimulated by the transmitter ___.

_____ 25. Drugs that block (inhibit) beta receptors in the heart will cause ___-crease in heart rate and blood pressure.

ANSWERS TO MASTERY TEST: ■ Chapter 16

Multiple Answers

1. B, C, D, E
2. A, E, F
3. A, C, E, G
4. A, B, C, D
5. A, B, E, F
6. B

Multiple Choice

7. E   9. C
8. A   10. D

True-False

11. F. Stressed
12. T
13. T
14. F. Dependent on
15. T
16. T
17. T
18. T
19. F. Some (or most)
20. T

Fill-ins

21. Cardiac muscle, smooth muscle, and glandular epithelium
22. Splanchnic
23. Vagus
24. NE (and also epinephrine)
25. De

FRAMEWORK 17

The Special Senses

OLFACTION (SMELL) [A]

GUSTATION TASTE [B]

- Structure of receptors
- Stimulation of receptors
- Adaptation, thresholds
- Pathways

EYE: VISION

STRUCTURES [C]
- Accessory
  - eyebrows
  - palpebrae
  - caruncle
  - tarsal plate & glands
  - conjunctiva
  - eyelashes
  - lacrimal apparatus
- Eyeball
  - layers (fibrous, vascular, retina)
  - other structures (lens, aqueous humor, vitreous humor)

PHYSIOLOGY [D]
- Formation of image
  - refraction
  - accommodation
  - constriction of pupil
  - convergence
- Stimulation of photoreceptors
  - rods
  - cones
- Visual pathway
  - rods, cones
  - bipolar cells
  - ganglion cells
  - thalamus
  - cerebral cortex

EAR [E]

STRUCTURE
- External
  - pinna
  - external auditory meatus
  - cerumen
  - eardrum
- Middle
  - ossicles (malleus, incus, stapes)
- Inner
  - cochlea
  - vestibule
  - semi-circular canals

PHYSIOLOGY
- Hearing
  - sound wave conduction
  - bone conduction
  - fluid conduction
  - oval & round windows
  - hair cells
  - medulla, thalamus, cerebral cortex
- Equilibrium
  - static
  - dynamic

DISORDERS [F]

VISION
- Cataract
- Glaucoma
- Conjunctivitis
- Trachoma

HEARING, EQUILIBRIUM
- Deafness
- Vertigo
- Otitis media
- Motion sickness

The Special Senses

All sensory organs contain receptors for increasing sensitivity to the environment. In Chapter 15 you considered the relatively simple receptors and pathways for the general senses of touch, pressure, temperature, pain, and vibration. The special senses include smell, taste, vision, hearing, and equilibrium. Special afferent pathways in many ways resemble general afferent pathways. However, a major point of differentiation is the arrangement of special sense receptors in complex sensory organs, specifically the nose, tongue, eyes and ears.

First study carefully the Chapter 17 Framework and note the key terms associated with each section.

TOPIC OUTLINE AND OBJECTIVES

A. Olfactory sensations

1. Locate the receptors for olfaction and describe the neural pathway for smell.

B. Gustatory sensations

2. Identify the gustatory receptors and describe the neural pathway for taste.

C. Visual sensations: anatomy

3. List and describe the structural divisions of the eye.

D. Visual sensations: physiology

4. Discuss retinal image formation by describing refraction, accommodation, constriction of the pupil, convergence, and inverted image formation.
5. Identify the afferent pathway of light impulses to the brain.

E. Auditory sensations and equilibrium

6. Describe the anatomical subdivisions of the ear.
7. List the principal events in the physiology of hearing.
8. Identify the receptor organs for static and dynamic equilibrium.

F. Disorders, medical terminology

9. Contrast the causes and symptoms of cataracts, glaucoma, conjunctivitis, trachoma, deafness, labyrinthine disease, Ménière's syndrome, vertigo, otitis media, and motion sickness.
10. Define medical terminology associated with the sense organs.

WORDBYTES

Now study the following parts of words that may help you better understand terminology in this chapter.

Wordbyte	Meaning	Example	Wordbyte	Meaning	Example
aqua-	water	*aqu*eous humor	sclera	hard	oto*scler*osis
macula	spot	*macula* lutea	tympano-	drum	*tympan*ic membrane
ossi-	bone	*ossi*cle			
ot(o)-	ear	*oto*laryngologist	vitr-	glassy	*vitr*eous humor

CHECKPOINTS

A. Olfactory sensations (pages 463–464)

■ **A1.** Describe olfactory receptors in this exercise.

a. Receptors for smell are located in the (*superior? inferior?*) portion of the nasal

cavity. Receptors consist of ____-polar neurons which have cell bodies located

between _____ cells. The distal end of each olfactory receptor

cell contains dendrites, known as olfactory _____.
b. What is the function of olfactory (Bowman's) glands in the nose?

c. The _____ theory of olfaction suggests that different chemical receptors in olfactory hair membranes respond to different substances, leading to

_____.
d. Adaptation to smell occurs (*slowly? rapidly?*) at first and then happens at a much slower rate.

■ **A2.** Complete this exercise about the nerves associated with the nose.

a. Upon stimulation of olfactory hairs impulses pass to cell bodies and axons; the axons of these olfactory cells form cranial nerves (*I? II? III?*), the olfactory nerves. These pass from the nasal cavity to the cranium to terminate in the olfactory

_____ located just inferior to the _____ lobes of the cerebrum.
b. Neurons in the olfactory bulb then convey impulses along the olfactory

_____ directly to the _____. Note that olfaction is one sense that (*does? does not?*) involve the thalamus.
c. The olfactory pathway just described transmits impulses for the (*general? special?*) sense of smell. General sensations of the nose, such as pain, cold, or tickling, are

carried primarily by cranial nerve ____, the trigeminal.

B. Gustatory sensations (pages 464-466)

■ **B1.** Describe receptors for taste in this exercise.

a. Taste, or _____ sensation, receptors are located in taste buds.

These consist of supporting cells that form a _____; inside

are 4 to 20 gustatory cells with _____ projecting from a taste
pore. The lifespan of each of these gustatory cells is about (*10 years? 10 days?*)

b. Taste buds are located on elevated projections of the tongue called

_____. The largest of these are located in a V-formation at

the back of the tongue. These are _____ papillae.

_____ papillae are mushroom-shaped and located on sides

and tip of the tongue. Pointed _____ papillae cover the

_____ of the tongue, and these (*do? do not?*) contain many
taste buds.

■ **B2.** Do the following exercise about taste.

a. There are only four primary taste sensations: _____,

_____, _____, and

_____. The taste buds at the tip of the tongue are most sensi-

tive to _____ and _____ tastes, whereas

the posterior portion is most sensitive to _____ taste.

b. The four primary tastes are modified by _____ sensations to
produce the wide variety of "tastes" experienced. If you have a cold (with a "stuffy
nose"), you may not seem to taste them, largely because of loss of the sense of

_____.

c. The threshold for smell is quite (*high? low?*), meaning that (*a large? only a small?*)
amount of odor must be present in order for smell to occur. The threshold varies
for the four primary tastes; the threshold for (*sour? sweet* and *salt? bitter?*) is

lowest, while those for _____ are highest.

d. Taste impulses are conveyed to the brain stem by cranial nerves _____, _____,

and _____; the _____ finally relays impulses to the cerebral
cortex.

C1. Examine your own eye structure in a mirror. With the help of Figure 17-3 (page 467) in your text, identify each of the accessory structures of the eye listed below.

Conjunctiva	Medial commissure
Eyebrow	Tarsal glands
Eyelashes	Palpebrae
Lacrimal glands: identify site	Tarsal plate

■ **C2.** *For extra review.* Match the structure listed above with the following descriptions. Write the name of the structure on the line provided.

a. Eyelid: _____

b. Thick fold of connective tissue that forms much of the eyelid: _____

c. Covers the orbicularis oculis muscle; provides protection: _____

d. Short hairs; infection of glands at the base of these hairs is *sty:* _____

e. Angle where eyes meet medially; site of caruncle: _____

f. Located in superolateral region of orbit; secrete tears: _____

g. Secrete oil; infection is *chalazion:* _____
h. Mucous membrane lining the eyelids and covering anterior surface of the eye:

_____

C3. The lacrimal glands produce _____ which contain the bacteriocidal enzyme named _____. State three functions of lacrimal fluid (tears).

■ **C4.** On Figure LG 17-1, color the three layers (or tunics) of the eye using color code ovals. Next to the ovals write letters of labeled structures that form each layer. One is done for you. Then label all structures in the eye and check your answers against the key.

■ **C5.** Contrast the two types of photoreceptor cells by writing R for rods or C for cones before related descriptions.

_____ a. Most concentrated in the central fovea of the macula lutea.

_____ b. Over 100 million in each eye, these are located mainly in peripheral regions of the eye.
_____ c. Sense color and acute (sharp) vision.

_____ d. Used for night vision.

KEY

A. Pupil
B. Scleral venous sinus
   (Canal of Schlemm)
C. Suspensory ligament
D. Ciliary body
E. Retina
F. Choroid layer
G. Sclera
H. Optic disk (blind spot)
I. Optic nerve

J. Conjunctiva
K. Cornea
L. Anterior chamber of anterior cavity
M. Iris
N. Posterior chamber of anterior cavity
O. Lens
P. Ora serrata
Q. Extrinsic eye muscle
R. Posterior cavity
S. Central fovea (in macula lutea)

O Fibrous tunic   __G__
O Vascular tunic   ____ ____ ____
O Nervous tunic   ____ ____ ____

Figure LG 17-1 Structure of the eyeball in horizontal section. Color as directed in Check-point C4.

■ **C6.** Complete this exercise about blood vessels of the eye.

a. Retinal blood vessels (*can? cannot?*) be viewed through an ophthalmoscope. Of what advantage is this?

b. The central retinal artery enters the eye at the (*optic disc? central fovea?*).

■ **C7.** Do this exercise that describes the retina.

a. After light strikes the back of the retina, nerve impulses pass anteriorly through three zones of neurons. First is the region of rods and cones, or

_____ zone; next is the _____ layer;

most anterior is the _____ layer.

b. Axons of the ganglion layer converge to form the optic

_____. This nerve exits from the eye at the point known as

the _____ disc, or _____. The name indicates that no image formation can occur here since no rods or cones are present.

C8. *A clinical challenge.* Briefly explain these clinical terms.

a. Detachment of the retina (Where does it occur? How can it be reattached?)

b. Senile macular degeneration (Which age group does it usually affect? How can you test for it?)

■ **C9.** Describe the lens and the cavities of the eye in this exercise.

a. The lens is composed of (*lipid? protein?*) arranged in layers, much like an onion. Normally the lens is (*clear? cloudy?*). A loss of transparency of the lens occurs in

the condition called _____.

b. The lens divides the eye into anterior and posterior (*chambers? cavities?*). The

posterior cavity is filled with _____ humor which has a (*watery? jellylike?*) consistency. This humor helps to hold the

_____ in place. Vitreous humor is formed during embryonic life and (*is? is not?*) replaced in later life.

c. The anterior cavity is subdivided by the _____ into two chambers. Which is larger? (*Anterior chamber? Posterior chamber?*) A watery fluid

called _____ humor is present in this chamber. It is formed by

the _____ bodies.

d. How is the formation and final destination of aqueous fluid similar to that of cerebrospinal fluid (CSF)?

e. State two functions of aqueous humor.

f. Aqueous humor creates pressure within the eye, normally about _____ mm Hg. In glaucoma this pressure (*increases? decreases?*), leading to destruction of the optic nerve and subsequent blindness.

■ **C10.** *For extra review.* Check your understanding by matching eye structures in the box with descriptions below. Use each answer once.

CF. Central fovea	OD. Optic disc
Cho. Choroid	P. Pupil
Cor. Cornea	R. Retina
CM. Ciliary muscle	S. Sclera
I. Iris	SVS. Scleral venous sinus
L. Lens	(Canal of Schlemm)

_____ a. "White of the eye"

_____ b. Normally clear; if cloudy, it is a cataract

_____ c. Blind spot; area in which there are no cones or rods

_____ d. Area of sharpest vision; area of densest concentration of cones

_____ e. Nonvascular, transparent, fibrous coat; most anterior eye structure

_____ f. Layer containing neurons; if detached, causes blindness

_____ g. Dark brown later; prevents reflection of light rays; also nourishes eyeball since it is vascular

_____ h. A hole; appears black, like a circular doorway leading into a dark room

_____ i. Regulates the amount of light entering the eye; colored part of the eye

_____ j. Attaches to the lens by means of radially arranged fibers called the suspensory ligaments

_____ k. Located at the junction of iris and cornea; drains aqueous humor

D. Visual sensations: physiology (pages 474–480)

■ **D1.** List the three principal processes necessary in order for vision to occur.

a.

b.

c.

D2. Define these four processes necessary for formation of an image on the retina.

a. Refraction of light

b. Accommodation of the lens

c. Constriction of the pupil

d. Convergence of the eyes

■ **D3.** Explain how accommodation of the lens enables your eyes to focus clearly.

a. The lens must become (*more? less?*) convex for near vision. This occurs by a thickening and bulging forward of the lens caused by the (*contraction? relaxation?*) of

the _____ muscle. In fact, after long periods of close work (such as reading), you may experience eye strain.

b. The normal (or _____ eye) can refract rays from a distance of

_____ meters and form a clear image on the retina.

c. In the nearsighted (or _____) eye, images focus (*in front of? behind?*) the retina. This may be due to an eyeball that is too (*long? short?*). Corrective lenses should be slightly (*concave? convex?*) so they refract rays less and allow them to focus farther back on the retina.

d. The farsighted person can see (*near? far?*) well, but has difficulty seeing (*near? far?*) without the aid of corrective lenses. The farsighted (or

_____) person requires (*concave? convex?*) lenses. After about age 40, most persons lose the ability to see near objects clearly. This condition,

called _____, is due to loss of elasticity of the

_____ .

■ **D4.** Choose the correct answers to complete each statement.

a. Both accommodation of the lens and constriction of the pupil involve (*extrinsic? intrinsic?*) muscles, whereas convergence of the eyes involves (*extrinsic? intrinsic?*) muscles.

b. The pupil (*dilates? constricts?*) in the presence of bright light.

■ **D5.** Answer these questions about the image focused on the retina.

a. How would a letter *e* look as it is focused on the retina? _____

b. Explain why you see things right side up, even though refraction produces inverted images on your retina.

■ **D6.** Complete this exercise about what happens in the rods of your eyes after you walk into a dark movie theatre. Refer to Figure 17-7 (page 477) in your text.

a. In order to see in dim light (as in the dark theatre), you use (*rods? cones?*).

b. Rods produce a pigment called _____ (or visual purple). But in bright light (as in the lobby) this chemical is broken down very (*rapidly? slowly?*). So as you enter the theatre your rods lack sufficient rhodopsin to see in the dark.

c. During the short time that you are stumbling around in the darkness, rhodopsin can form in the following manner. An enzyme converts retinal from its all-trans

form to its _____ form, which fits well with a protein named

_____ to produce rhodopsin.

d. As soon as you have sufficient rhodopsin present in the rods, stimulation of the rods by the dim light in the theatre will begin breaking down rhodopsin. This is a rapid but stepwise process ultimately resulting in the cleavage of rhodopsin into

two parts: again the protein _____ and the (*cis? all-trans?*)

form of retinal. As a result of this process, a _____ impulse develops, and a nerve impulse then passes to your brain informing you of what you have seen in dim light, such as images of persons in seats near you.

e. Your vision becomes increasingly better after a few more minutes in the dim light since you continue to form rhodopsin. (See *c* above.) But should you step out into the lobby to buy popcorn, the bright light will (*produce? destroy?*) rhodopsin so rapidly and completely that you will have to start the dark-induced synthesis of rhodopsin all over again once you re-enter the dimly lit theatre.

f. Night blindness (also known as _____) is associated with

vitamin ____ deficiency since that vitamin is a precursor of (*scotopsin? retinal?*).

■ **D7.** Explain how cones help you to see in "living color" by completing this paragraph.

Cones contain pigments which (*do? do not?*) require bright light for breakdown (and therefore for a generator potential). Three different pigments are present, sensi-

tive to three colors: _____, _____, and

_____. A person who is red-green color-blind lacks some of the

cones receptive to two colors (_____ and

_____) and so cannot distinguish between these colors. This condition occurs more often in (*males? females?*).

■ **D8.** Describe the conduction pathway for vision by arranging these structures in sequence. Write the letters in correct order on the lines provided. Also, indicate the four points where synapsing occurs by placing an asterisk (*) between the letters.

___ ___ ___ ___ ___ ___ ___

   B. Bipolar cells
   C. Cerebral cortex (visual areas)
   G. Ganglion cells
OC. Optic chiasma

ON. Optic nerve
OT. Optic tract
  P. Photoreceptor cells (rods, cones)
  T. Thalamus

■ **D9.** Answer these questions about the visual pathway to the brain. It may be helpful to refer to Figure 17-8 (page 479) in the text.

a. Hold your left hand up high and to the left so you can still see it. Your hand is in the (*temporal? nasal?*) visual field of your left eye, and in the

_____ visual field of your right eye.

b. Due to refraction, the image of your hand will be projected on to the (*left? right?*) (*upper? lower?*) portion of the retinas of your eyes.

c. All nerve fibers from these areas of your retinas reach the (*left? right?*) side of your thalamus and cerebral cortex.

d. Damage to the right optic tract, right side of thalamus, or right visual cortex would result in loss of sight of the (*left? right?*) visual fields of each eye.

E. Auditory sensations and equilibrium (pages 480–490)

■ **E1.** Refer to Figure LG 17-2 and do the following exercise.

a. Color the three parts of the ear using color code ovals. Then write on lines next to ovals the letters of structures that are located in each part of the ear. One is done for you.

b. Label each lettered structure using leader lines on the figure.

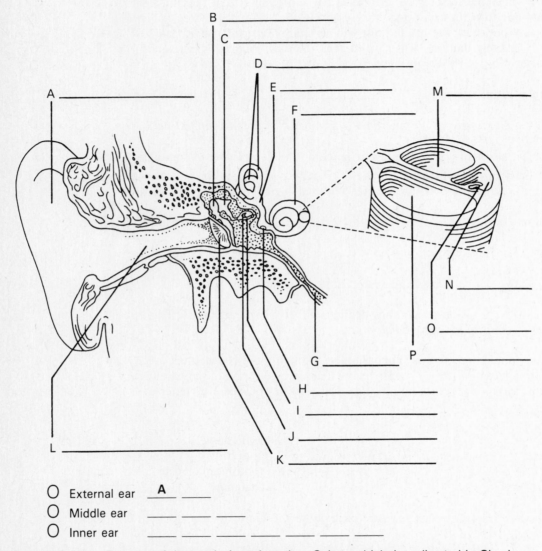

○ External ear __A__ ____
○ Middle ear ____ ____ ____
○ Inner ear ____ ____ ____ ____ ____ ____ ____

Figure LG 17-2 Diagram of the ear in frontal section. Color and label as directed in Checkpoint E1.

■ **E2.** *For extra review.* Select the ear structures in the box that fit the descriptions below. Not all answers will be used.

AT.	Auditory tube	P.	Pinna
I.	Incus	RW.	Round window
M.	Malleus	S.	Stapes
OW.	Oval window	TM.	Tympanic membrane

_____ a. Tube used to equalize pressure on
either side of tympanic membrane

_____ b. Eardrum

_____ c. Structure on which stapes exerts
pistonlike action

_____ d. Ossicle adjacent to eardrum

_____ e. Anvil-shaped ear bone

■ **E3.** Fill in the blanks and circle correct answers about sound waves.

a. Sound waves heard by humans range from frequencies of 20 to 20,000 cycles/sec

(Hz). Humans can best hear sounds in the range of _____ Hz.

b. A musical high note has a (*higher? lower?*) frequency than a low note. So frequency is (*directly? indirectly?*) related to pitch.

c. Sound intensity (loudness) is measured in units called _____.

Normal conversation is at a level of about _____ dB, while sounds at

_____ dB can cause pain.

■ **E4.** Summarize events in the process of hearing in this activity. It may help to refer to Figure 17-11 (page 486) in the text.

a. Sound waves travel through the _____ and strike the

_____ membrane. Sound waves are magnified by the action of

the three _____ in the middle ear.

b. The ear bone named (*malleus? incus? stapes?*) strikes the (*round? oval?*) window, setting up waves in (*endo-? peri-?*) lymph. This pushes on the floor of the upper *scala* (*vestibuli? tympani?*). As a result the cochlear duct is moved, and so is the perilymph in the lower canal, the *scala* (*vestibuli? tympani?*). The pressure of the perilymph is finally expended by bulging out the (*round? oval?*) window.

c. As the cochlear duct moves, tiny hair cells embedded in the floor of the duct are

stimulated. These hair cells are part of the _____ organ; its

name is based on its spiral arrangement on the _____ membrane all the way around the 2¾ coils of the cochlear duct.

d. As spiral organ hair cells are moved by waves in endolymph, hairs move against the

gelatinous _____ membrane. This movement generates a potential in the hair cells that excites nearby sensory neurons of the

_____ branch of cranial nerve _____.

The pathway continues to the brain stem, _____ (relay center),

and finally to the _____ lobe of the cerebral cortex.

■ **E5.** Check your understanding of the pathway of fluid conduction by placing the following structures in correct sequence. Write the letters on the lines provided.

___ ___ ___ ___ ___ ___ ___

    B. Basilar membrane
    C. Cochlear duct (endolymph)
    O. Oval window
RW. Round window

    ST. Scala tympani (perilymph)
    SV. Scala vestibuli (perilymph)
VM. Vestibular membrane

E6. Describe the roles of K^+ and Ca^{2+} in stimulating hair cells and generating nerve impulses related to hearing.

■ **E7.** Contrast receptors for hearing and for equilibrium in this summary of the inner ear.

    a. Receptors for hearing and equilibrium are all located in the (*middle? inner?*) ear.

       All consist of supporting cells and _____ cells that are covered by a _____ membrane.

    b. In the spiral organ, which senses _____, the gelatinous membrane is called the _____ membrane. Hair cells move against this membrane as a result of (*sound waves? change in body position?*).

    c. In the macula, located in the (*semicircular canals? vestibule?*), the gelatinous membrane is embedded with calcium carbonate crystals called

       _____. These respond to gravity in such a way that the macula is the receptor for (*static? dynamic?*) equilibrium. An example of such equilibrium occurs as you are aware of your (*position while standing on your head? change in position on a careening roller coaster?*).

    d. In the semicircular canals the gelatinous membrane is called the

       _____. Its shape is (*flat? like an inverted cup?*). The cupula is part of the (*crista? saccule?*) located in the ampulla. Change in direction (as in a

    roller coaster) causes _____ to bend hairs in the cupula. Cristae in semicircular canals are therefore receptors for (*static? dynamic?*) equilibrium.

F. Disorders, medical terminology (pages 490–492)

■ **F1.** Match the name of the disorder with the description.

Cat.	Cataract	NB.	Night blindness
Con.	Conjunctivitis	OM.	Otitis media
G.	Glaucoma	P.	Presbyopia
Men.	Ménière's syndrome	Str.	Strabismus
My.	Myopia	T.	Tinnitus

_____ a. Condition requiring corrective lenses to focus distant objects

_____ b. Excessive intraocular pressure resulting in blindness; second most common cause of blindness

_____ c. Ringing in the ears

_____ d. Crossed eyes

_____ e. Middle ear infection

_____ f. Associated with vitamin A deficiency

_____ g. Pinkeye

_____ h. Disturbance of the inner ear; cause unknown

_____ i. Loss of transparency of the lens

_____ j. Farsightedness due to loss of elasticity of lens, especially after age 40

F2. Discuss causes and treatments of:

a. Glaucoma

b. Trachoma

c. Motion sickness

F3. Contrast *sensineural deafness* with *conduction deafness*.

A1. (a) Superior, bi-, supporting, hairs. (b) Produce mucus that acts as a solvent for odoriferous substances. (c) Chemical, generator potential and nerve impulse. (d) Rapidly.

A2. (a) I, bulb, frontal. (b) Tracts, cerebral cortex, does not. (c) Special, V.

B1. (a) Gustatory, capsule, hairs, 10 days. (b) Papillae, circumvallate, fungiform, filiform, anterior, do not.

B2. (a) Sweet, sour, salt, and bitter; sweet and salty, bitter. (b) Olfactory, smell. (c) Low, only a small, bitter, sweet and salt. (d) VII, IX and X, thalamus.

C2. (a) Palpebra. (b) Tarsal plate. (c) Eyebrow. (d) Eyelashes. (e) Medial commissure. (f) Lacrimal glands. (g) Tarsal glands. (h) Conjunctiva.

C4. Fibrous tunic: G, K; vascular tunic: D, F, M; nervous tunic: E.

C5. (a) C. (b) R. (c) C. (d) R.

C6. (a) Can; health of blood vessels can be readily assessed, for example, in persons with diabetes. (b) Optic disc.

C7. (a) Photoreceptor, bipolar, ganglion. (b) Nerve, optic, blind spot.

C9. (a) Protein, clear, cataract. (b) Cavities, vitreous, jellylike, retina, is not. (c) Iris, anterior chamber, aqueous, ciliary. (d) Both are formed from blood vessels (choroid plexuses) and the fluid finally returns to venous blood. (e) Provides nutrients and oxygen and removes wastes since cornea and lens are avascular; also provides pressure to separate cornea from lens. (f) 16, increases.

C10. (a) S. (b) L. (c) OD. (d) CF. (e) Cor. (f) R. (g) Cho. (h) P. (i) I. (j) CM. (k) SVS.

D1. (a) Formation of image on retina. (b) Stimulation of rods and cones. (c) Conduction of nerve impulses to brain.

D3. (a) More, contraction, ciliary. (b) Emmetropic, 6 (20 ft). (c) Myopic, in front of, long, concave. (d) Far, near, hypermetropic, convex, presbyopia, lens.

D4. (a) Intrinsic, extrinsic. (b) Constricts.

D5. (a) "ɘ" ("e" is inverted 180° and is much smaller). (b) Brain learns to "turn" images so that you see things right side up.

D6. (a) Rods. (b) Rhodopsin, rapidly. (c) Cis, scotopsin. (d) Scotopsin, all-trans, generator. (e) Destroy. (f) Nyctalopia, A, retinal.

D7. Do; red, green, blue; red, green; males.

D8. P * B * G ON OC OT * T * C. (Note that some crossing but no synapsing occurs in the optic chiasma.)

D9. (a) Temporal, nasal. (b) Right, lower. (c) Right. (d) Left.

E1.

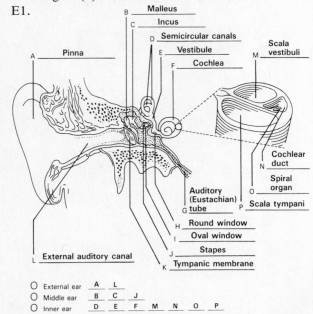

Figure LG 17-2A Diagram of the ear in frontal section.

E2. (a) AT. (b) TM. (c) OW. (d) M. (e) I.

E3. (a) 1,000–4,000. (b) Higher, directly. (c) Decibels (dB), 45, 115–120.

E4. (a) External auditory meatus, tympanic, ossicles. (b) Stapes, oval, peri-, vestibuli, tympani, round. (c) Spiral, basilar. (d) Tectorial, cochlear, VIII, thalamus, temporal.

E5. O SV VM C B ST RW.

E7. (a) Inner, hair, gelatinous. (b) Hearing, tectorial, sound waves. (c) Vestibule, otoliths, static, position while standing on your head. (d) Cupula, like an inverted cup, crista, endolymph, dynamic.

F1. (a) My. (b) G. (c) T. (d) Str. (e) OM. (f) NB. (g) Con. (h) Men. (i) Cat. (j) P.

Questions 1–5: Circle T (true) or F (false). If the statement is false, change the underlined word or phrase so that the statement is correct.

T F 1. Decomposition of photopigments in rods and cones leads to <u>increased</u> influx of Na^+ into the photoreceptor cells and therefore results in <u>depolarization.</u>

T F 2. <u>Convergence and accommodation are both results</u> of contraction of smooth muscle of the eye.

T F 3. <u>Crista, macula, otilith, and spiral organ</u> are all structures located in the inner ear.

T F 4. <u>Both aqueous and vitreous humors are</u> replaced constantly throughout your life.

T F 5. The receptor organs for special senses are <u>less</u> complex structurally than those for general sense.

Questions 6–12: Arrange the answers in correct sequence.

_____ _____ _____   6. Layers of the eye, from superficial to deep:

    A. Sclera
    B. Retina
    C. Choroid

_____ _____ _____   7. From anterior to posterior:

    A. Vitreous humor
    B. Optic nerve
    C. Cornea
    D. Lens

_____ _____ _____ _____   8. Pathway of aqueous humor, from site of formation to destination:

    A. Anterior chamber
    B. Scleral venous sinus
    C. Ciliary body
    D. Posterior chamber

_____ _____ _____ _____ _____   9. Pathway of sound waves and resulting mechanical action:

    A. External auditory canal
    B. Stapes
    C. Malleus and incus
    D. Oval window
    E. Tympanic membrane

_____ _____ _____ _____ _____   10. Pathway of tears, from site of formation to entrance to nose:

    A. Lacrimal gland and ducts
    B. Lacrimal sac
    C. Nasolacrimal duct
    D. Surface of conjunctiva
    E. Lacrimal punctae and lacrimal canals

_____ _____ _____ _____ _____ 11. Order of impulses along conduction pathway for smell;

A. Olfactory bulb
B. Olfactory hairs
C. Olfactory nerves
D. Olfactory tract
E. Primary olfactory area of cortex

_____ _____ _____ _____ _____ 12. From anterior to posterior:

A. Anterior chamber
B. Iris
C. Lens
D. Posterior cavity
E. Posterior chamber

Questions 13–20: Circle the letter preceding the one best answer to each question.

13. Which of the following answers is a screening test for glaucoma and involves checking pressure in the eye?

A. Audiometry   D. Radial keratotomy
B. Tonometry   E. Epikeratoplasty
C. Snellen test

14. Infections in the throat (pharynx) are most likely to lead to ear infections in the following manner. Bacteria spread through the:

A. External auditory meatus to the external ear
B. Auditory (Eustachian) tube to the middle ear
C. Oval window to the inner ear
D. Round window to the inner ear

15. Choose the FALSE statement about rods.

A. There are more rods than cones in the eye.
B. Rods are concentrated in the fovea and are less dense around the periphery.
C. Rods enable you to see in dim (not bright) light.
D. Rods contain rhodopsin.
E. No rods are present at the optic disc.

16. Choose the FALSE statement about the lens of the eye.

A. It is biconvex.
B. It is avascular.
C. It becomes more rounded (convex) as you look at distant objects.
D. Its shape is changed by contraction of the ciliary muscle.
E. Change in the curvature of the lens is called accommodation.

17. Choose the FALSE statement about the middle ear.

A. It contains three ear bones called ossicles.
B. Infection in the middle ear is called otitis media.
C. It functions in conduction of sound from the external ear to the inner ear.
D. The cochlea is located here.

18. Choose the FALSE statement about the semicircular canals.

A. They are located in the inner ear.
B. They sense acceleration or changes in position.
C. Nerve impulses begun here are conveyed to the brain by the vestibular branch of cranial nerve VIII.
D. There are four semicircular canals in each ear.
E. Each canal has an enlarged portion called an ampulla.

19. Destruction of the left optic tract would result in:

A. Blindness in the left eye
B. Loss of left visual field of each eye
C. Loss of right visual field of each eye
D. Loss of lateral field of view of each eye ("tunnel vision")
E. No effects on the eye

20. Mr. Frederick has a detached retina of the lower right portion of one eye. As a result he is unable to see objects in which area in the visual field of that eye?

A. High in the left   C. Low in the left
B. High in the right   D. Low in the right

Questions 21–25: fill-ins. Complete each sentence with the word or phrase that best fits.

_____ 21. ___ is the most common disorder leading to blindness.

_____ 22. Name the four processes necessary for formation of an image on the retina: ___ of light rays, ___ of the lens, ___ of the pupil, and ___ of the eyes.

_____ 23. ___ is the study of the structure, functions, and diseases of the eye.

_____ 24. The names of three ossicles are ___.

_____ 25. Two functions of the ciliary body are ___.

ANSWERS TO MASTERY TEST: ■ Chapter 17

True-False

1. F. Decreased, hyperpolarization
2. F. Accommodation, but not convergence, is a result.
3. T
4. F. Aqueous humor, but not vitreous humor, is
5. F. More

True-False

6. A C B
7. C D A B
8. C D A B
9. A E C B D
10. A D E B C
11. B C A D E
12. A B E C D

Multiple Choice

13. B   17. D
14. B   18. D
15. B   19. C
16. C   20. A

Fill-ins

21. Cataract
22. Refraction, accommodation, constriction, convergence
23. Ophthalmology
24. Malleus, incus, and stapes
25. Production of aqueous humor and alteration of lens shape for accommodation

FRAMEWORK 18

The Endocrine System

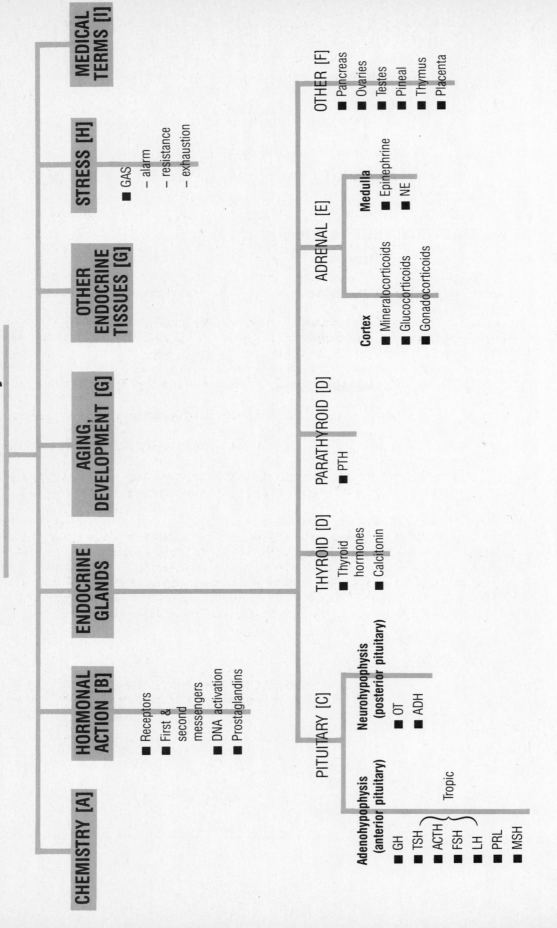

CHEMISTRY [A]

HORMONAL ACTION [B]
- Receptors
- First & second messengers
- DNA activation
- Prostaglandins

ENDOCRINE GLANDS

AGING, DEVELOPMENT [G]

OTHER ENDOCRINE TISSUES [G]

STRESS [H]
- GAS
  - alarm
  - resistance
  - exhaustion

MEDICAL TERMS [I]

PITUITARY [C]

Adenohypophysis (anterior pituitary)
- GH
- TSH
- ACTH
- FSH
- LH } Tropic
- PRL
- MSH

Neurohypophysis (posterior pituitary)
- OT
- ADH

THYROID [D]
- Thyroid hormones
- Calcitonin

PARATHYROID [D]
- PTH

ADRENAL [E]

Cortex
- Mineralocorticoids
- Glucocorticoids
- Gonadocorticoids

Medulla
- Epinephrine
- NE

OTHER [F]
- Pancreas
- Ovaries
- Testes
- Pineal
- Thymus
- Placenta

The Endocrine System

CHAPTER

18

Hormones produced by about a dozen endocrine glands exert widespread effects on just about every body tissue. Hormones are released into the bloodstream, which serves as the vehicle for distribution throughout the body. In fact, analysis of blood levels of hormones can provide information about the function of specific endocrine glands. How do hormones "know" which cells to affect? And how does the body "know" when to release more or less of a hormone? Mechanisms of action and regulation of hormone levels are discussed in this chapter. The roles of all major hormones and effects of excesses or deficiencies in those hormones are also included. The human body is continually exposed to stressors, such as rapid environmental temperature change, a piercing sound, or serious viral infection. Adaptations to stressors are vital to survival; however, certain adaptations lead to negative consequences. A discussion of stress and adaptation concludes Chapter 18.

One way to alleviate the stress associated with learning about stress is to consult the Chapter 18 Framework and to study the key terms for each section.

TOPIC OUTLINE AND OBJECTIVES

A. Endocrine glands, chemistry of hormones

1. Discuss the functions of the endocrine system in maintaining homeostasis.

B. Mechanisms of homonal action, control of hormonal secretion

2. Explain the mechanisms of hormonal action by interaction with plasma membrane receptors and intracellular receptors.
3. Describe the control of hormonal secretions via feedback cycles and give several examples.

C. Pituitary (hypophysis)

4. Describe the release of hormones stored in the neurohypophysis.

D. Thyroid and parathyroids

E. Adrenals

F. Other endocrine glands: pancreas, gonads, pineal (epiphysis cerebri), thymus

5. Describe the location, histology, hormones, and functions of the following endocrine glands: pituitary, thyroid, parathyroids, adrenals (suprarenals), pancreas, ovaries, testes, pineal, and thymus.

G. Aging and developmental anatomy of the endocrine system; other endocrine tissues

6. Describe the effects of aging on the endocrine system.
7. Describe the development of the endocrine system.

H. Stress and homeostasis

8. Define the general adaptation syndrome (GAS) and compare homeostatic responses and stress responses.

I. Summary of hormones, medical terminology

9. Discuss the symptoms of pituitary dwarfism, giantism, acromegaly, diabetes insipidus, cretinism, myxedema, exophthalmic goiter, simple goiter, tetany, osteitis fibrosa cystica, aldosteronism, Addison's disease, Cushing's syndrome, adrenogenital syndrome, pheochromocytomas, diabetes mellitus, and hyperinsulinism.

10. Define medical terminology associated with the endocrine system.

WORDBYTES

Now study the following parts of words that may help you better understand the terminology in this chapter.

Wordbyte	Meaning	Example
adeno-	gland	*adeno*hypophysis
crin-	to secrete	endo*crine*
endo-	within	*endo*crine
horm-	excite, urge on	*horm*one
insipid-	without taste	diabetes *insipid*us

Wordbyte	Meaning	Example
mellit-	sweet	diabetes *mellit*us
oxy(s)-	swift	*oxy*tocin
para-	around	*para*thyroid
-tocin	childbirth	pi*tocin*
trop-	turn	thyro*trop*in

Checkpoints

A. Endocrine glands, chemistry of hormones (pages 496–499)

■ **A1.** Compare the ways in which the nervous and endocrine systems exert control over the body. Write N (nervous) or E (endocrine) next to descriptions that fit each system.

_____ a. Sends messages to muscles, glands, and neurons only.

_____ b. Sends messages to virtually any part of the body.

_____ c. Effects are generally faster and shorter-lived.

■ **A2.** Next to each letter on Figure LG 18-1, label each endocrine gland. Then color each gland and identify the approximate location of each gland on your own body.

■ **A3.** Complete the table contrasting exocrine and endocrine glands.

	Products Secreted Into	Examples
a. Exocrine		
b. Endocrine		

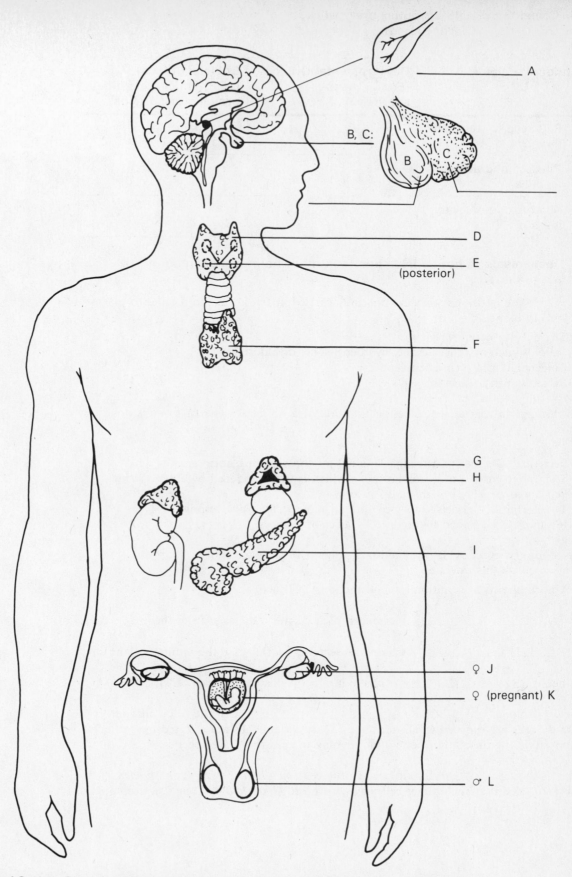

Figure LG 18-1 Diagram of endocrine glands. Label and color as directed in Checkpoints A2 and C1.

A

B, C:

B

C

D

E

(posterior)

F

G

H

I

♀ J

♀ (pregnant) K

♂ L

■ **A4.** Complete the table contrasting principal classes of hormones based on chemistry and origin of the endocrine glands.

Endocrine gland	Developmental Origin (*Endo-, meso-, ectoderm*)	Chemical Class (*APP*) amines, peptides, proteins or *L* (lipids)
a. Pancreas, thyroid, and parathyroid		
b. Pituitary and adrenal medulla		
c. Adrenal cortex, ovary, testis		

B. **Mechanisms of hormonal action, control of hormonal secretion**
(pages 499–503)

■ **B1.** Circle the hormones that are normally carried in free form (not bound to plasma protein) in blood:

A. Thyroid hormone (thyroxin)
B. Steroids such as progesterone, hydrocortisone, or aldosterone
C. Catecholamines such as epinephrine
D. Proteins such as insulin

■ **B2.** Arrange in correct sequence events in life of a hormone. The first one is done

for you: __B__ ____ ____ ____
A. Secreting cells receive a feedback signal to stop secreting hormome.
B. The body senses a need for hormone which is then produced by an endocrine gland and travels through blood stream.
C. The hormone interacts with receptors on target cells which respond.
D. Hormone is degraded by target cells, liver or kidneys.

■ **B3.** Study the cyclic AMP mechanism in Figure 18-4a (page 500) in your text. Then complete this exercise.

a. A hormone such as antidiuretic hormone (ADH) acts as the

_____ messenger, as it carries a message from the

_____ where it is secreted (in this case the pituitary gland) to

the _____ where it acts (in this case kidney cells).

b. All hormones travel in the _____, with the result that all body cells are exposed to all hormones. Only certain cells are affected by particular hormones because a hormone attaches only to cells that have specific

_____ located in the cell (*membrane? cytoplasm? nucleus?*). It is estimated that each target cell may have up to (*100? 1,500? 100,000?*) receptors.

c. The attachment of the hormone increases activity of the enzyme

_____, located in the plasma membrane. This enzyme cata-

lyzes conversion of ATP to _____, known as the

_____ messenger.

d. Cyclic AMP then activates one or more enzymes known as

_____ to help transfer (*calcium? phosphate?*) from ATP to a
protein. The resulting chemical can then set off the target cell's response. In the
case of ADH, this is an increase in permeability of kidney cells, with a resulting

decrease in _____ production.

e. Effects of cyclic AMP are (*short? long?*)-lived. This chemical is rapidly degraded by

an enzyme named _____.

f. A number of hormones are known to act by means of the cyclic AMP mechanism.

Included are most of the _____-soluble hormones.
Name several.

g. Name one chemical besides cyclic AMP that may act as a second messenger.

_____ What is its relationship to *calmodulin*?

■ **B4.** Thyroid and steroid hormones act by (*cyclic AMP second messengers? activating
DNA to form enzymes?*)

■ **B5.** The hormonal interaction by which effects of progesterone on the uterus in prepa-
ration for pregnancy are enhanced by earlier exposure to estrogen is known as a(n)
(*antagonistic? permissive? synergistic?*) effect.

■ **B6.** Describe prostaglandins in this exercise.
a. Prostaglandins (PGs) are considered (*local or tissue? circulating?*) hormones. They
are synthesized by (*one specific endocrine gland? most tissues of the body?*).
b. Chemically, prostaglandins are (*proteins? lipids?*). Name three prostaglandins.

c. Prostaglindins seem to alter the amount of _____ production.
Since cyclic AMP activates so many responses, PGs have (*few? a wide range of?*)
effects on the body. PGs are considered (*extremely? only slightly?*) potent chemicals.
d. List four functions of prostaglandins.

e. Prostaglandin synthesis has been linked to increase in temperature, as in fever.

Aspirin and acetominophen (Tylenol) _____-crease PG synthesis and can therefore

_____-crease fever.

■ **B7.** Precise regulation of hormone levels is critical to homeostasis. Describe three types of control mechanisms in this exercise.

    a. Some endocrine glands respond directly to chemicals in blood passing through the gland. A decreased blood level of calcium serves as the stimulus for release of

       parathyroid hormone (PTH), which causes blood calcium level to _____-crease. An increase in blood glucose passing through the pancreas stimulates release of

       insulin, which _____-creases blood glucose level. These are both examples of (*positive? negative?*) feedback mechanisms, meaning that the response of the body is (*the same as? opposite?*) the stimulus.

    b. Name two hormones that are released in response to sympathetic nerve stimulation.

       _____ and _____ Name one other hor-

       mone that is regulated by nervous stimulation. _____

    c. A third type of control mechanism involves regulating factors synthesized by the

       _____. Some of these, called releasing factors, cause release

       of anterior pituitary hormones; others, called _____ factors, have the opposite effect. (More about regulating factors in Checkpoint C3.)

C. Pituitary (hypophysis) (pages 503–512)

■ **C1.** Complete this exercise about the pituitary gland.

    a. The pituitary is also known as the _____.

    b. Why is it sometimes called the "master gland"?

    c. Where is it located?

    d. Seventy-five percent of the gland consists of the (*anterior? posterior?*) lobe, called

       the _____-hypophysis. The posterior lobe (or

       _____) lobe is somewhat smaller.

    e. The anterior pituitary is known to secrete (*2? 5? 7?*) different hormones. Write abbreviations for each of these hormones next to the diagram of the anterior pituitary on Figure LG 18-1.

    f. Which anterior pituitary hormones are *tropic hormones*? _____
    Define tropic hormone.

■ **C2.** Refer to Figure LG 18-2 as you do this Checkpoint about growth hormone.

    a. The main function of growth hormone (_____) is to stimulate growth and

       maintain size of _____ and _____. The

       alternate name for GH, _____ hormone (STH), indicates its function: to "turn" (*trop-*) the body (*soma-*) to growth. Note that somatotropic hormone (*is? is not?*) considered a true tropic hormone since it (*does? does not?*) stimulate another endocrine gland to produce a hormone.

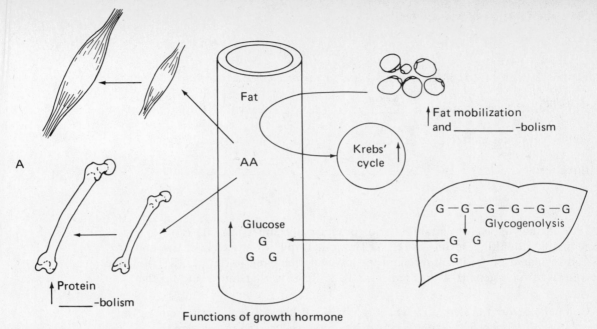

Figure LG 18-2 Functions of growth hormone in regulating metabolism. AA, amino acid; G-G-G-G-G-G, glycogen; G, glucose. Complete as directed in Checkpoint C2.

b. GH stimulates growth by (*accelerating? inhibiting?*) entrance of amino acids into

cells where they can be used for _____ synthesis. Therefore, growth hormone stimulates protein (*anabolism? catabolism?*). Indicate this on Figure LG 18-2.

c. GH promotes breakdown of glycogen in liver to _____.
It also promotes release of stored body fats, as well as stimulates

_____-bolism of fats. Show this on the figure.

d. Since cells now turn to fats for energy (by means of the Krebs cycle), and liver

glycogen is broken down, the level of glucose in cells and blood _____-creases, a condition known as (*hyper? hypo?*)-glycemia. In fact, excess GH can mimic diabetes

mellitus, and for this reason GH is said to be diabeto-_____.

e. GH secretion is controlled by two regulating factors: _____

and _____. One stimulus that promotes release of GH is low blood sugar (for example, during growth when cells require much energy). This condition, (*hyper-? hypo-?*) glycemia, stimulates secretion of (*GHRH? GHIH?*), which then causes release of (*high? low?*) levels of GH.

f. State the function of somatomedins.

g. Hypersecretion of GH in early years leads to (*dwarfism? giantism? acromegaly?*),

whereas deficiency causes _____. Excess GH after closure of

epiphyseal plates causes _____, characterized by bones in

three areas: _____, _____ and

_____ .

■ **C3.** Recall that release of certain anterior pituitary hormones like GH can be decreased by inhibiting hormones (IHs). The four tropic hormones lack IHs and are instead regulated by feedback mechanisms involving releasing hormones (RHs). These are shown in four steps (1–4) in Figure LG 18-3. Refer to that figure and do this exercise.

a. Color the arrow for step 1, in which a low blood level of target hormone, in this example, adrenocorticoid such as cortisone produced by the target gland

_____, triggers the hypothalamus to secrete a (*high? low?*) level of releasing hormone (CRH).

b. Color the two arrows for step 2. First the high level of releasing hormone (CRH)

passes through blood vessels to the _____ and stimulates *release* of ACTH. Then a high level of this (*tropic? target?*) hormone in blood passing through the adrenal cortex stimulates secretion of a (*high? low?* level of target hormone.

c. Color steps 3 and 4, and note thickness of arrows. Both steps 1 and 3 are (*positive? negative?*) feedback mechanisms, whereas steps 2 and 4 are (*positive? negative?*) feedback mechanisms.

■ **C4.** Complete the table listing three other releasing-tropic-target hormone relationships.

Regulating	TRH		
Tropic			LH
Target		Estrogen	♀ ♂

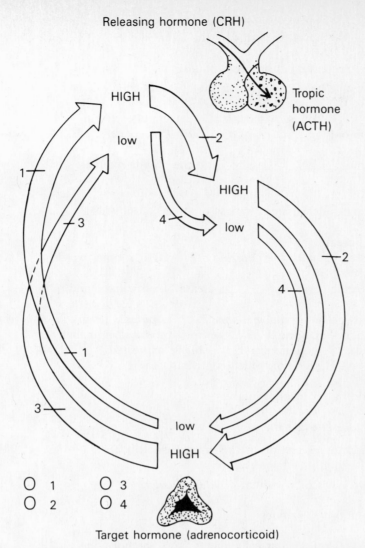

Releasing hormone (CRH)

Tropic hormone (ACTH)

HIGH

low

HIGH

low

low

HIGH

○ 1 ○ 3
○ 2 ○ 4

Target hormone (adrenocorticoid)

Figure LG 18-3 Control of hormone secretion: example of regulating, tropic, and target hormones. Thickness of arrows indicates amounts of hormones. Color arrows as directed in Checkpoint C3.

■ **C5.** Do this exercise about hormonal control of the mammary glands.

a. Milk is produced in mammary glands following stimulation by the hormone

_____ (_____) which is secreted by the (*anterior? posterior?*) pituitary. A different hormone causes ejection of milk from the glands at the time of the baby's suckling. this hormone, named

_____, is released by the (*anterior? posterior?*) pituitary.

b. During pregnancy, prolactin levels (*increase? decrease?*) due to increased levels of (*PIF? PRF?*). During most of a woman's lifetime prolactin levels are low as a result of high level of (*PIF? PRF?*). An exception is the time just prior to each menstrual flow when PIF levels (*increase? decrease?*). Due to this lack of inhibition prolactin secretion increases, activating breast tissues (and causing tenderness).

■ **C6.** Describe posterior pituitary hormone function by writing the correct terms on the lines provided on Figure LG 18-4. Use the list of terms on the figure.

■ **C7.** Answer these questions about posterior pituitary hormones.

a. The time period from the start of a baby's suckling at the breast to the delivery of

milk to the baby is about _____ seconds. Although afferent impulses (from breast to hypothalamus) require only (*a fraction of a second? 45 seconds?*), passage

of the oxytocin through the _____ to the breasts requires

about _____ seconds.

b. On a day when your body becomes dehydrated (by loss of sweat), your ADH

production is likely to _____-crease.

c. Effects of diuretic medications are (*similar to? opposite of?*) ADH. In other words,

diuretics _____-crease urine production and _____-crease blood volume and blood pressure.

d. Inadequate ADH production, as occurs in diabetes (*mellitus? insipidus?*) results in (*enormous? minute?*) daily volumes of urine. Unlike the urine produced in diabetes mellitus, this urine (*does? does not?*) contain sugar. It is bland or *insipid*.

e. Alcohol (*stimulates? inhibits?*) ADH secretion which contributes to

_____-creased urine production after alcohol intake.

D. Thyroid and parathyroids (pages 512–520)

■ **D1.** Describe the histology and hormone secretion of the thyroid gland in this exercise.

a. The thyroid is composed of two types of glandular cells. (*Follicular?*

Parafollicular?) cells produce the two hormones _____

and _____. Parafollicular cells manufacture the hormone

_____.

b. The ion (*Cl^-? I^-? Br?*) is highly concentrated in the thyroid since it is an essential component of T_3 and T_4. Each molecule of thyroxine (T_4) contains four of the I^-

ions, whereas T_3 contains _____ ions of I^-.

c. These hormones are combined with a (*lipid? glycoprotein?*) to form thyroglobulin (TGB) and (*released immediately? may be stored for months?*).

d. Thyroxine travels in the bloodstream bound to plasma proteins, and in this form

the hormone is often called PBI, or_____.

■ **D2.** Fill in blanks and circle answers about functions of thyroid hormone contrasted to those of growth hormone.

a. Like growth hormone, thyroid hormone increases protein _____-bolism.

b. Unlike GH, thyroid hormone increases most aspects of carbohydrate

_____-bolism. Therefore, thyroid hormone (*increases? decreases?*) basal metabolic rate (BMR), releasing heat and (*increasing? decreasing?*) body temperature.

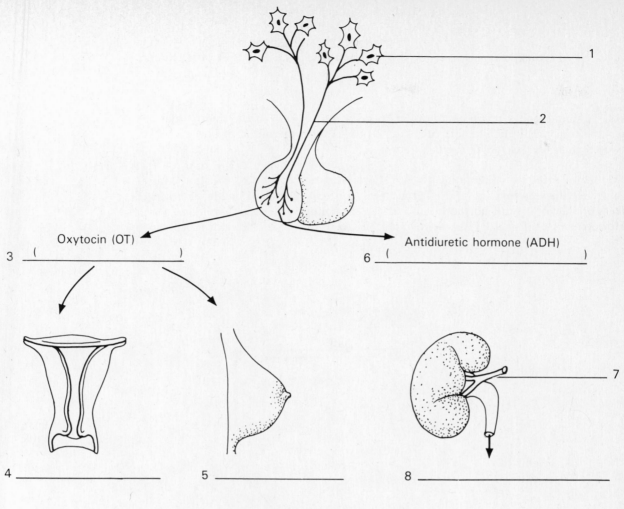

Oxytocin (OT)

Antidiuretic hormone (ADH)

1

2

3 (_____)

6 (_____)

4 _____

5 _____

8 _____

7

LIST OF TERMS

Axons to posterior pituitary	Pitocin
Blood volume ↑	Urine ↓
Breast milk letdown	Uterine contraction
Neurons in hypothalamus	Vasopressin

Figure LG 18-4 Posterior pituitary hormones. Label as directed in Checkpoint C6.

c. Thyroid hormone (*increases? decreases?*) heart rate and blood pressure and (*increases? decreases?*) nervousness.

d. This hormone is important for development of normal bones and muscles, as well as brain tissue. Therefore, deficiency in thyroid hormone during childhood (a condition called _____) (*does? does not?*) lead to retardation as well as small stature. Note that in childhood GH deficiency, retardation of mental development (*occurs also? does not occur?*).

D3. Release of thyroid hormone is under influence of releasing factor

_____ and tropic hormone _____ . Review
the table in Checkpoint C4. List factors that stimulate TRF and TSH production.

■ **D4.** A simple goiter is (*an enlargement? a decrease in size?*) of the thyroid gland. It results from the futile attempts of the gland to produce thyroid hormone when iodine is present in (*excessive? deficient?*) quantities.

■ **D5.** Color arrows on Figure LG 18-5 to show hormonal control of calcium. This figure emphasizes that CT and PTH are (*antagonists? synergists?*) with regard to effects on calcium. Both hormones also affect distribution of (*potassium? phosphate?*). However both CT and PTH act synergistically to (*raise? lower?*) blood phosphate levels.

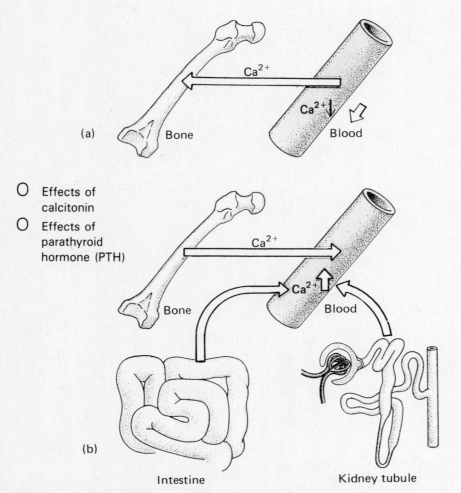

Figure LG 18-5 Hormonal control of calcium ion (Ca^{2+}). (a) Increased calcium storage in bones lowers blood calcium (hypocalcemic effect). (b) Calcium from three sources increases blood calcium level (hypercalcemic effect). Arrows show direction of calcium flow. Color arrows as directed in Checkpoint D5.

320

■ **D6.** *A clinical challenge.* Consider calcium imbalances in this exercise.

    a. Hypocalcemia can result from (*hypo? hyper?*)-parathyroid conditions. Decreased

    blood calcium results in abnormal _____-crease in nerve impulses to muscles,

    and the condition known as _____ .

    b. Parathyroid tumors or certain types of cancer cells may lead to hyper-parathyroidism. Explain why persons with those conditions are more subject to fractures.

E. Adrenals (pages 520–526)

■ **E1.** Fill in blank lines on Figure LG 18-6 to show mechanisms that control fluid balance and blood pressure (BP).

■ **E2.** *A clinical challenge.* Determine whether each of the following conditions is likely to *increase* or *decrease* blood pressure (BP).

    a. Excessive renin production: _____
    b. Use of an angiotensin converting enzyme (A.C.E.)-inhibitor medication:

    _____

    c. Aldosteronism (such as tumor of zona glomerulosa): _____

    E3. Name three glucocorticoids. Circle the one that is most abundant in the body.

■ **E4.** In general, glucocorticoids are involved with metabolism and resistance to stress. Describe their effects in this exercise.

    a. These hormones promote _____-bolism of proteins. In this respect their effects are (*similar to? opposite?*) those of both growth hormone and thyroid hormone.
    b. However, glucocorticoids may stimulate conversion of amino acids and fats to glu-

    cose, a process known as _____ . In this way glucocorticoids (like GH) (*increase? decrease?*) blood glucose.
    c. Refer to Figure 18-25 (page 535) in your text. Note that glucocorticoid (and also GH) production (*increases? decreases?*) during times of stress. By raising blood glucose levels, these hormones make energy available to cells most vital for a response to stress. During stress glucocorticoids also act to (*increase? decrease?*) blood pressure and to (*enhance? limit?*) the inflammatory process. (They are *anti-inflammatory.*)

    E5. How are glucocorticoid hormones regulated? (For help, review the table in Checkpoint C4, page LG 316.)

■ **E6.** A third category of adrenal cortex hormones are the _____ hormones. Name two.

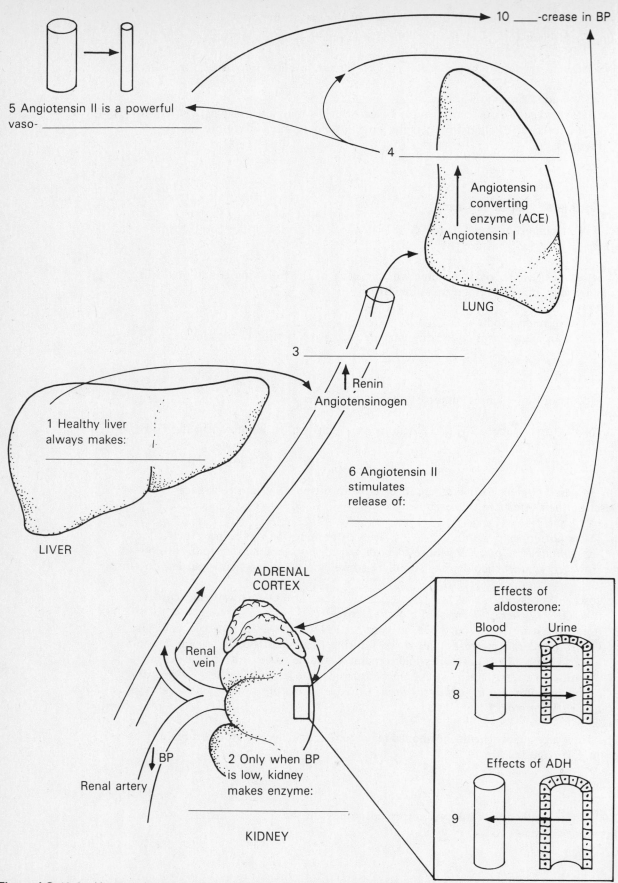

10 _____-crease in BP

5 Angiotensin II is a powerful
vaso-_____

4 _____

Angiotensin
converting
enzyme (ACE)

Angiotensin I

LUNG

3 _____

↑ Renin
Angiotensinogen

1 Healthy liver
always makes:

_____

LIVER

6 Angiotensin II
stimulates
release of:

_____

ADRENAL
CORTEX

Renal
vein

↓ BP

Renal artery

2 Only when BP
is low, kidney
makes enzyme:

_____

KIDNEY

Effects of
aldosterone:

Blood   Urine

7

8

Effects of ADH

9

Figure LG 18-6 Hormonal control of fluid balance and blood pressure (BP): renin-angiotensin, aldosterone and ADH. Fill in blanks 1–10 as directed in Checkpoint E1.

E7. Name the two principal hormones secreted by the adrenal medulla. Circle the name of the hormone that accounts for 80 percent of adrenal medulla secretion.

E8. List three effects of these hormones.

For extra review. See Chapter 16, Checkpoints C1 and C2, page LG 282.

F. Other endocrine glands: pancreas, gonads, pineal (epiphysis cerebri), thymus (pages 527–531)

■ **F1.** The islets of Langerhans are located in the _____. Do this exercise describing hormones produced there.

a. The name *islets of Langerhans* suggests that these clusters of

_____-crine cells lie amidst a "sea" of exocrine cells within the pancreas. Name the three types of islet cells.

b. Refer to Figure LG 18-7a. Glucagon, produced by (*alpha? beta? delta?*) cells, (*increases? decreases?*) blood sugar in two ways. First, it stimulates the breakdown

of _____ to glucose, a process known as (*glycogenesis? glycogenolysis? gluconeogenesis?*). Second, it stimulates the conversion of amino acids and other compounds to glucose. This process is called

_____.

c. Is glucagon controlled by an anterior pituitary *tropic* hormone? (*Yes? No?*) In fact,

control is by effect of _____ directly upon the pancreas. When blood glucose is low, then a (*high? low?*) level of glucagon will be produced. This will raise blood sugar.

d. *For extra review.* Define the terms glycogen and glucagon, noting their differences in spelling and pronunciation.

e. Now look at Figure LG 18-7b. The action of insulin is (*the same as? opposite?*) that of glucagon. In other words, *insulin decreases blood sugar.* (Repeat that statement three times; it is important!) Insulin acts by several mechanisms. Two are shown in the figure. First, insulin (*helps? hinders?*) transport of glucose from blood into cells. Second, it accelerates conversion of glucose to

_____, the process called _____.

f. Hormones that raise blood glucose levels are called (*hypo? hyper?*)-glycemic hormones. The one hormone that lowers blood glucose level is said to be

_____-glycemic. To show this, color arrows in Figure LG 18-7.

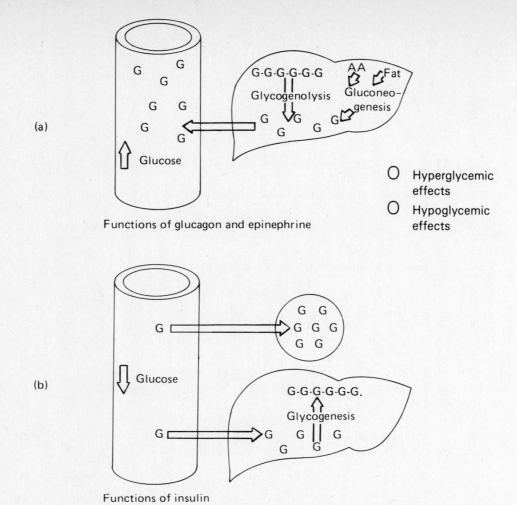

(a)

Functions of glucagon and epinephrine

○ Hyperglycemic effects

○ Hypoglycemic effects

(b)

Functions of insulin

Figure LG 18-7 Hormones that regulate blood glucose level. AA, amino acid; G, glucose; G-G-G-G-G-G, glycogen. Color according to directions in Checkpoint F1.

■ **F2.** Contrast Type I and Type II diabetes mellitus DM by writing *I* or *II* next to related descriptions.

_____ a. Also known as maturity-onset DM

_____ b. The more common type of DM

_____ c. Related to insensitivity of body cells to insulin, rather than to insulin deficiency

_____ d. Also known as insulin-dependent diabetes mellitus (IDDM)

_____ e. More likely to lead to serious complications such as ketosis

■ **F3.** *A clinical challenge.* Mrs. Jefferson has diabetes mellitus. In the absence of sufficient insulin, her cells are deprived of glucose. Explain why she experiences the following symptoms.

   a. Hyperglycemia and glycosuria

   b. Increased urine production (poly-_____) and increased water

   intake (poly-_____)

   c. Ketosis, a form of _____-osis

   d. Atherosclerosis

■ **F4.** Sex hormones are produced in several locations in the body. List four. (More about these in Chapter 28.)

   F5. Where is the pineal gland located?

   Does this gland appear to atrophy with age? (*Yes? No?*)
   Describe the probable action of melatonin.

■ **F6.** What is the general function of the thymus?

   G. Aging and developmental anatomy of the endocrine system; other endocrine tissues (pages 531–533)

■ **G1.** Name two endocrine organs that are perhaps most likely to cause problems

   among the elderly. _____ _____

   G2. To categorize major endocrine glands according to their primary germ layer organs, review Checkpoint A4, page LG 312.

■ **G3.** Now match the endocrine gland with the correct description of its embryological origin.

AC. Adrenal cortex	P. Pancreas
AM. Adrenal medulla	PT. Parathyroid
AP. Anterior pituitary	T. Thymus

_____ a. Develops from the foregut area that later becomes part of small intestine

_____ b. Derived from the roof of the stomodeum (mouth) called the hypophyseal (Rathke's) pouch

_____ c. Originates from the neural crest which also produces sympathetic ganglia

_____ d. Derived from tissue from the same region that forms gonads

_____ e. Arise from pharyngeal pouches (2 answers)

■ **G4.** As you continue your study of anatomy and physiology, you will consider hormones produced in three other parts of the body: those from the

_____ tract (Chapter 24), hormones made in the

_____ during pregnancy (Chapter 28), and also the hormone

erythropoietin, which affects _____ production (Chapter 19).

■ **G5.** Match growth factors with descriptions.

IGF. Insulinlike growth factor	TAF. Tumor angiogenesis factors
L. Lymphokines	

_____ a. Produced by normal and tumor cells; stimulate growth of new blood vessels

_____ b. Made by certain white blood cells, these help to provide immunity

_____ c. Made in the liver (and also in some cancer cells), this factor stimulates growth of cartilage and fibroblasts

H. Stress and homeostasis (pages 533–537)

H1. Define and give two examples of *stressors*.

■ **H2.** Summarize the body's response to stressors in this exercise.

a. The part of the brain that senses the stress and initiates response is the

_____. It responds by two main mechanisms.

b. First is the (*alarm? resistance?*) reaction involving the adrenal (*medulla? cortex?*)

and the _____ division of the autonomic nervous system.

c. Second is the _____ reaction. Playing key roles in this

response are three anterior pituitary hormones: _____,

_____, and _____ and their target

hormones. Therefore the adrenal (*medulla? cortex?*) hormones are activated.

■ **H3.** Now describe events that characterize the first (or alarm) stage of response to stress. Complete this exercise.

a. The alarm reaction is sometimes called the _____ response.

b. During this stage blood glucose (*increases? decreases?*) by a number of hormonal mechanisms, including those of adrenal medulla hormones

_____ and _____. Glucose must be available for cells to have energy for the stress response.

c. Oxygen must also be available to tissues; the respiratory system (*increases? decreases?*) its activity. Heart rate and blood pressure (*increase? decrease?*). Blood

is shunted to vital tissues such as the _____ and

_____ and is directed away from reservoir organs such as the

_____ and _____.

d. Nonessential activities such as digestion (*increase? decrease?*). Sweating (*increases? decreases?*) in order to control body temperature and eliminate wastes.

H4. Describe the effects of each of these hormones in the resistance stage.

a. Mineralocorticoids

b. Glucocorticoids

c. Thyroxin

d. GH

H5. When the resistance stage fails to combat the stressor, the body moves into the

_____ stage. Explain how this is related to:

a. Loss of potassium

b. Depletion of glucocorticoids

c. Weakening of organs

I. Summary of hormones, medical terminology (page 537)

■ **I1.** *For extra review.* Test your understanding of these hormones by writing the name (or abbreviation where possible) of the correct hormone after each of the following descriptions.

a. Stimulates release of growth hormone: _____

b. Stimulates testes to produce testosterone: _____

c. Promotes protein anabolism, especially of bones and muscles:

   _____

d. Tropic hormone for thyroxin: _____

e. Stimulates development of ova in females and sperm in males:

   _____

f. Present in bloodstream of nonpregnant women to prevent lactation:

   _____

g. Stimulates ovary to release egg and change follicle cells into corpus luteum:

   _____

h. Stimulates uterine contractions and also milk ejection: _____

i. Inhibited by MIF: _____

j. Stimulates kidney tubules to produce small volume of concentrated urine:

   _____

k. Target hormone of ACTH: _____

l. Increases blood calcium: _____

m. Decreases blood calcium and phosphate: _____

n. Contains iodine as an important component: _____

o. Raises blood glucose (3): _____, _____,

   _____

p. Lowers blood glucose: _____

q. Serves antiinflammatory functions: _____

r. Mimics many effects of sympathetic nerves: _____

I2. *For extra review.* Match the disorder with the hormonal imbalance.

Ac. Acromegaly	M. Myxedema
Ad. Addison's disease	PC. Pheochromocytoma
C. Cretinism	PD. Pituitary dwarfism
DI. Diabetes insipidus	S. Simple goiter
DM. Diabetes mellitus	T. Tetany
E. Exophthalmic goiter	

_____ a. Deficiency of GH in child; slow bone growth

_____ b. Excess of GH in adult; enlargement of hands, feet, and jawbones

_____ c. Deficiency of ADH; production of enormous quantities of "insipid" (nonsugary) urine

_____ d. Deficiency of insulin; hyperglycemia and glycosuria (sugary urine)

_____ e. Deficiency of thyroxin in child; short stature and mental retardation

_____ f. Deficiency of thyroxin in adult; edematous facial tissues, lethargy

_____ g. Excess of thyroxin; protruding eyes, "nervousness," weight loss

_____ h. Deficiency of thyroxin due to lack of iodine; most common reason for enlarged thyroid gland

_____ i. Result of deficiency of PTH; decreased calcium in blood and fluids around muscles resulting in abnormal muscle contraction.

_____ j. Deficiency of adrenocorticoids; increased K^+ and decreased Na^+ resulting in low blood pressure and dehydration

_____ k. Tumor of adrenal medulla causing sympathetic-like responses, such as increased pulse and blood pressure, hyperglycemia, and sweating

ANSWERS TO SELECTED CHECKPOINTS

A1. (a) N. (b) E. (c) N.

A2. A, pineal; B, posterior pituitary; c, anterior pituitary; D, thyroid; E, parathyroid; F, thymus; G, adrenal cortex; H, adrenal medulla; I, pancreas; J, ovary; K, placenta; L, testis.

A3.

	Products Secreted Into	Examples
a. Exocrine	Ducts	Sweat, oil, mucus, digestive juices
b. Endocrine	Bloodstream	Hormones

A4. (a) Endoderm, APP. (b) Ectoderm, APP. (c) Mesoderm, L.

B1. C, D.

B2. B C A D

B3. (a) First, endocrine gland, target cell. (b) Bloodstream, receptors, membrane, cytoplasm or nucleus; 100,000. (c) Adenylate cyclase, cyclic AMP, second. (d) Protein kinases, phosphate, urine. (e) Short; phosphodiesterase. (f) Water; ADH, OT, FSH, LH, TSH, ACTH, CT, NE and epinephrine. (g) Calcium or GMP; calcium binds to a protein named calmodulin which, in turn, activates or inhibits enzymes like protein kinases.

B4. Activating DNA to form enzymes.

B5. Permissive.

B6. (a) Local or tissue, most tissues of the body. (b) Lipids, examples: PGA, PGB, PGE_1, PGE_2, PGI. (c) Cyclic AMP, a wide range of, extremely. (d) See list on Page 502 of text. (e) De, de.

B7. (a) In, de, negative, opposite. (b) Epinephrine and norepinephrine (NE), OT and ADH. (c) Hypothalamus, inhibiting.

C1. (a) Hypophysis. (b) It regulates so many body activities. (c) In the sella turcica just inferior to the hypothalamus. (d) Anterior, adeno-, neuro. (e) 7; GH, PRL, MSH, ACTH, TSH, FSH, LH. (f) ACTH, TSH, FSH, LH; hormones that stimulate other endocrine glands to produce hormones.

C2. (a) GH, bones, skeletal muscles, somatotropic, is not, does not. (b) Accelerating, protein, anabolism. (c) Glucose, cata. (d) In, hyper, -genic. (e) GHRH, GHIH, hypo-, GHRH, high. (f) They mediate effects of GH. (g) Giantism; dwarfism; acromegaly; hands, feet, face.

C3. (a) Adrenal cortex, high. (b) Anterior pituitary, tropic, high. (c) Negative, positive.

C4.

Regulating	TRH	GnRH	GnRH
Tropic	**TSH**	**FSH**	LH
Target	**Thyroid hormone**	Estrogen	♀ **Estrogen, progesterone** ♂ **Testosterone**

C5. (a) Prolactin (or lactogenic hormone) (PRL), anterior, oxytocin (OT), posterior. (b) Increase, PRF, PIF, decrease.

C6. 1, neurons in hypothalamus; 2, axons to posterior pituitary; 3, pitocin; 4, uterine contraction; 5, breast milk letdown; 6, vasopressin; 7, blood volume ↑; 8, urine ↓.

C7. (a) 30–60, a fraction of a second, bloodstream, 30–60. (b) In. (c) Opposite of, in, de. (d) Insipidus, enormous, does not. (e) Inhibits, in.

D1. (a) Follicular; thyroxine (T_4) and triiodothyronine (T_3); calcitonin. (b) I^-; 3. (c) Glycoprotein; may be stored for months. (d) Protein-bound iodine.

D2. (a) Ana. (b) Cata, increases, increasing. (c) Increases, increases. (d) Cretinism, does, does not occur.

D4. An enlargement, deficient.

D5. Figure LG 18-5A: (a) Calcitonin (CT). (b) Parathyroid hormone (PTH). Antagonists, phosphate, lower.

D6. (a) Hypo, in, tetany. (b) Demineralization of bones occurs since PTH stimulates osteoclasts to destroy bone.

E1. 1, The protein angiotensinogen; 2, renin; 3, angiotensin I; 4, angiotensin II; 5, constrictor; 6, aldosterone; 7, Na^+ (and H_2O by osmosis, and Cl^- and HCO_3^- to achieve electrochemical balance); 8, K^+ and H^+; 9, H_2O; 10, In.

E2. (a) Increase (known as *high renin hypertension*). (b) Decrease, such as the antihypertensive captopril (Capoten). (c) Increase.

E4. (a) Cata-, opposite. (b) Gluconeogenesis, increase. (c) Increases, increase, limit.

E6. Gonadocorticoids or sex hormones; estrogens and androgens.

F1. Pancreas. (a) Endo-; alpha, beta, delta. (b) Alpha, increases, glycogen, glycogenolysis, gluconeogenesis. (c) No, glucose, high. (d)

Glycogen is a storage form of glucose; *glucagon* is a hormone that breaks down glycogen to glucose. (e) Opposite, helps, glycogen, glycogenesis. (f) Hyper, hypo; glucagon and epinephrine are hyperglycemic; insulin is hypoglycemic.

F2. (a–c) II. (d–e) I.

F3. (a) Lack of insulin prevents glucose from entering cells, so more is left in blood and spills into urine. (b) -Uria, -dipsia; glucose in urine draws water (osmotically), and as result of water loss, patient is thirsty. (c) Acid-; since cells have little glucose to use, they turn to excessive fat catabolism which leads to keto-acid formation (discussed in Chapter 25). (d) As fats are transported in extra amounts (for catabolism), some fats are deposited in walls of vessels; fatty (= *athero*) thickening (= *sclerosis*) results.

F4. Ovaries, testes, adrenal cortex and placenta.

F6. Immunity.

G1. Pancreas and thyroid.

G3. (a) P. (b) AP. (c) AM. (d) AC. (e) PT, T.

G4. Gastrointestinal, placenta, red blood cell.

G5. (a) TAF. (b) L. (c) IGF.

H2. (a) Hypothalamus. (b) Alarm, medulla, sympathetic. (c) Resistance; ACTH, GH, TSH, cortex.

H3. (a) Fight-or-flight. (b) Increases, norepinephrine, epinephrine. (c) Increases; increase; skeletal muscles, heart, or brain; spleen, skin. (d) Decrease, increases.

I1. (a) GHRH. (b) LH. (c) GH (also testosterone). (d) TSH. (e) FSH. (f) PIF. (g) LH. (h) OT. (i) MSH. (j) ADH. (k) Adrenocorticoid (or glucocorticoid, such as cortisol). (l) PTH. (m) CT. (n) Thyroid hormone. (o) GH, glucagon, epinephrine, norepinephrine, adrenocorticoids (or ACTH indirectly). (p) Insulin. (q) Adrenocorticoid. (r) Epinephrine, norepinephrine.

I2. (a) PD. (b) Ac. (c) DI. (d) DM. (e) C. (f) M. (g) E. (h) S. (i) T. (j) Ad. (k) PC.

MASTERY TEST: Chapter 18

Questions 1-2: Arrange the answers in correct sequence.

_____ _____ _____ _____ 1. Steps in action of TSH upon a target cell:

   A. Adenylate cyclase breaks down ATP.
   B. TSH attaches to receptor site.
   C. Cyclic AMP is produced.
   D. Second messenger causes the specific effect mediated by TSH, namely, stimulation of thyroid hormone production.

_____ _____ _____ _____ 2. Steps in renin-angiotensin mechanism to increase blood pressure:

   A. Angiotensin I is converted to angiotensin II.
   B. Renin converts angiotensinogen, a plasma protein, to angiotensinogen I.
   C. Angiotensin II causes vasoconstriction and stimulates aldosterone production which causes water conservation and raises blood pressure.
   D. Low blood pressure causes secretion of enzyme renin from cells in kidneys.

Questions 3-9: Circle the letter preceding the one best answer to each question.

3. All of these compounds are synthesized in the hypothalamus EXCEPT:

   A. ADH   D. PIF
   B. GHRH   E. Oxytocin
   C. CT

4. Choose the FALSE statement about endocrine glands.

   A. They secrete chemicals called hormones.
   B. Their secretions enter extracellular spaces and then pass into blood.
   C. Sweat and sebaceous glands are endocrine glands.
   D. Endocrine glands are ductless.

5. All of the following correctly match hormonal imblance with related signs or symptoms EXCEPT:

   A. Deficiency of ADH—excessive urinary output
   B. Excessive aldosterone—high blood pressure
   C. Deficiency of aldosterone—low blood level of potassium and muscle weakness
   D. Deficiency of parathyroid production—demineralization of bones; low blood level of calcium and tetany

6. All of the following hormones are secreted by the anterior ptiuitary EXCEPT:

   A. ACTH   D. Oxytocin
   B. FSH   E. GH
   C. Prolactin

7. Which one of the following is a function of adrenocorticoid hormones?

   A. Lower blood pressure
   B. Raise blood level of calcium
   C. Convert glucose to amino acids
   D. Lower blood level of sodium
   E. Antiinflammatory

8. All of these hormones lead to increased blood glucose EXCEPT:

   A. ACTH   D. Growth hormone
   B. Insulin   E. Epinephrine
   C. Glucagon

9. Choose the FALSE statement about the anterior pituitary.

   A. It secretes at least seven hormones.
   B. It develops from mesoderm.
   C. It secretes tropic hormones.
   D. It is stimulated by releasing factors from the hypothalamus.
   E. It is also known as the adenohypophysis.

10. Choose the FALSE statement.

   A. Both ADH and aldosterone tend to lead to retention of water and increase in blood pressure.
   B. PTH activates vitamin D which enhances calcium absorption.
   C. PTH activates osteoclasts, leading to bone destruction.
   D. Tetany is a sign of hypocalcemia.
   E. ADH and OT pass from hypothalamus to posterior pituitary via pituitary portal veins.

Questions 11–20: Circle T (true) or F (false). If the statement is false, change the underlined word or phrase so that the statement is correct.

T F 11. Polyphagia, polyuria, and hypoglycemia are all symptoms associated with insulin deficiency.

T F 12. Regulating factors are all secreted by the anterior pituitary and affect the hypothalamus.

T F 13. The hormones of the adrenal medulla mimic the action of parasympathetic nerves.

T F 14. Prostaglandins are hormones made in large quantities in the prostate gland.

T F 15. Epinephrine and cortisol are both adrenocorticoids.

T F 16. Secretion of growth hormone, insulin, and glucagon is controlled (directly or indirectly) by blood glucose level.

T F 17. Most of the feedback mechanisms that regulate hormones are positive (rather than negative) feedback mechanisms.

T F 18. The alarm reaction occurs before the resistance reaction to stress.

T F 19. Most hormones studied so far appear to act by a mechanism involving cyclic AMP rather than by the gene activation mechanism.

T F 20. A high level of thyroxine circulating in the blood will tend to lead to a low level of TRF and TSH.

Questions 21–25: fill-ins. Complete each sentence with the word or phrase that best fits.

_____ 21. ____ is a hormone used to induce labor since it stimulates contractions of smooth muscle of the uterus.

_____ 22. ____ and ____ are two hormones that are classified chemically as amines.

_____ 23. ____ is a term that refers to the control mechanism by which the number of receptors for a hormone will be decreased when that hormone is present in excessive amounts.

_____ 24. Type ____ diabetes mellitus is a non-insulin-dependent condition in which insulin level is adequate but cells have decreased sensitivity to insulin.

_____ 25. Three factors that stimulate release of ADH are ____.

ANSWERS TO MASTERY TEST: ■ Chapter 18

Arrange

1. B A C D
2. D B A C

Multiple Choice

3. C   7. E
4. C   8. B
5. C   9. B
6. D   10. E

True-False

11. F. Polyphagia and polyuria are both
12. F. Hypothalamus and affect the anterior pituitary
13. F. Sympathetic
14. F. Small quantities by nearly every body cell.
15. F. Cortisol is
16. T
17. F. Negative (rather than positive)
18. T
19. T
20. T

Fill-ins

21. Oxytocin (OT or pitocin)
22. Thyroid hormones (T_3 and T_4), epinephrine or norepinephrine
23. Down-regulation
24. II (maturity onset)
25. Decreased extracellular water, pain, stress, trauma, anxiety, nicotine, morphine, tranquilizers.

Maintenance of the Human Body

FRAMEWORK 19
Blood

INTRODUCTION [A]

TYPES OF ECF
- Blood
- Interstitial fluid [I]
- Lymph [I]

CHARACTERISTICS

FUNCTIONS

COMPONENTS

TYPES, ORIGIN [B]
- Hemopoiesis

FORMED ELEMENTS
- CBC

PLASMA [F]
- Plasma proteins

Erythrocytes [C]
- Hemoglobin
- Transport of gases
- 4.8–5.4 million/mm^3
- Erythropoiesis

Leucocytes [D]
- Defense
- 5–10,000/mm^3
- Types

Granular

Agranular

Thrombocytes [E]
- Blood clotting
- 250,000–400,000/mm^3

HEMOSTASIS [G]
- Vascular spasm
- Platelet plug
- Coagulation
  - prothrombin activator
  - thrombin formation
  - fibrin formation
- Retraction
- Fibrinolysis

GROUPING [H]
- Agglutinogens (isoantigens)
- Agglutinins (isoantibodies)
- ABO
- Rh
  - erythroblastosis fetalis

DISORDERS, TERMS
- RBCs
  - anemia
  - polycythemia
- WBCs
  - mononucleosis
  - leukemia

The Cardiovascular System: The Blood

CHAPTER 19

The major role of the blood is transportation through an intricate system of channels—blood vessels—that reach virtually every part of the body. Blood constantly courses through the vessels, carrying gases, fluids, nutrients, electrolytes, hormones, and wastes to and from body cells. Most of these chemicals "ride" within the fluid portion of blood (plasma). Some, such as oxygen, piggyback on red blood cells. White blood cell "passengers" take short trips in the blood and depart from the vessels at sites calling for defense. Platelets and other clotting factors congregate to plug holes in the network of vessels.

To explore this marvelous transport system, start by examining the Chapter 19 Framework and the list of key words for each section.

TOPIC OUTLINE AND OBJECTIVES

A. Types of body fluids, characteristics and functions of blood

1. Contrast the general roles of blood, lymph, and interstitial fluid in maintaining homeostasis.
2. Define the principal physical characteristics of blood and its functions in the body.

B. Components of blood: types, origin

3. Compare the origins, histology, and functions of the formed elements in blood.

C. Formed elements: erythrocytes

D. Formed elements: leucocytes

4. Explain the significance of a differential (DIF) count.

E. Formed elements: thrombocytes

F. Plasma

5. List the components of plasma and explain their importance.

G. Hemostasis

6. Identify the stages involved in blood clotting and explain the various factors that promote and inhibit blood clotting.

H. Grouping (typing)

7. Explain ABO and Rh blood grouping.

I. Interstitial fluid, lymph

J. Disorders, medical terminology

8. Contrast the causes of nutritional, pernicious, hemorrhagic, hemolytic, aplastic, and sickle-cell anemia (SCA).
9. Identify the clinical symptoms of infectious mononucleosis (IM), chronic fatigue syndrome, and leukemia.
10. Define medical terminology associated with blood.

WORDBYTES

Now study the following parts of words that may help your better understand terminology in this chapter.

Wordbyte	Meaning	Example	Wordbyte	Meaning	Example
a-, an-	not, without	*an*emia	leucos-	white	*leuco*cytosis
crit	to separate	hemato*crit*	mega-	large	*mega*karyocyte
-cyte	cell	leuco*cyte*	-osis	increase	leucocy*tosis*
dia-	through	*dia*phragm	-pedesis	moving	dia*pedesis*
-emia	blood	leuk*emia*	-penia	want, lack	leucocyto*penia*
erythro-	red	*erythro*poiesis	-poiesis	formation	hemo*poiesis*
heme-, hemo-	blood	*hemo*stasis	thromb-	clot	pro*thromb*in

CHECKPOINTS

A. **Types of body fluids, characteristics and functions of blood**
(pages 546–547)

A1. Review relationships among these three body fluids: blood, interstitial fluid, and lymph. Refer to Figure LG 1-2, page 13.)

■ **A2.** Name the structures that comprise the:

a. Cardiovascular system

_____   _____   _____

b. Lymphatic system

_____   _____   _____

■ **A3.** Circle the answers that correctly describe characteristics of blood.

a. Viscosity:

    2.5× thicker than water   4.5× thicker than water

b. Salt (NaCl) concentration:

    0.85–0.9 percent   8.5–9.0 percent

c. Temperature:

    36°C (96.8°C)   38°C (100.4°C)
    37°C (98.6°C)

d. pH:

    6.8   7.35–7.45
    7.0–7.1

e. Volume of blood in average adult:

    1.5–3 liters   8–10 liters
    4–6 liters

A4. Describe the functions of blood in this Checkpoint.

a. Blood transports many substances. List six.

_____   _____   _____

_____   _____   _____

b. List three characteristics of the body that are regulated by blood:

_____   _____   _____

c. List two protective mechanisms of blood.

B. Components of blood: types, origin   (pages 547–548)

■ **B1.** Describe blood components in this exercise.

a. Blood consists of about ____ percent plasma and ____ percent cells.

b. The three types of formed elements (or cells) are: red blood cells (_____-cytes),

white blood cells (_____-cytes), and platelets (_____-cytes).

■ **B2.** Blood formation is a process known as _____. Most blood cells are formed in (*lymphoid tissue? red bone marrow?*). Name six bones in which hemopoiesis occurs in adults.

_____   _____   _____

_____   _____   _____

C. Formed elements: erythrocytes   (pages 548–552)

■ **C1.** Refer to Figure LG 19-1. Which diagram represents a mature erythrocyte? ____ Note that it (*does? does not?*) contain a nucleus. The chemical named

_____ accounts for the color of red blood cells (RBCs). Label and color the RBC.

■ **C2.** Describe hemoglobin in this exercise.

a. The hemoglobin molecule consists of a central portion, which is the protein (*heme? globin?*) with four side groups called (*hemes? globins?*).

b. Each heme contains one _____ atom on which a molecule of (*oxygen? carbon dioxide?*) can be transported. So each hemoglobin molecule can carry (*1? 4?*) oxygen molecule(s). Almost all oxygen is transported in this manner.

c. Carbon dioxide has one combining site on hemoglobin; it is the amino acid portion

of the (*heme? globin?*), thereby forming _____-hemoglobin. About (*23? 70? 97?*) percent of carbon dioxide is carried in this state. Most carbon dioxide is transported (*in blood cells? in plasma?*) within the

_____ (HCO_3^-) ion.

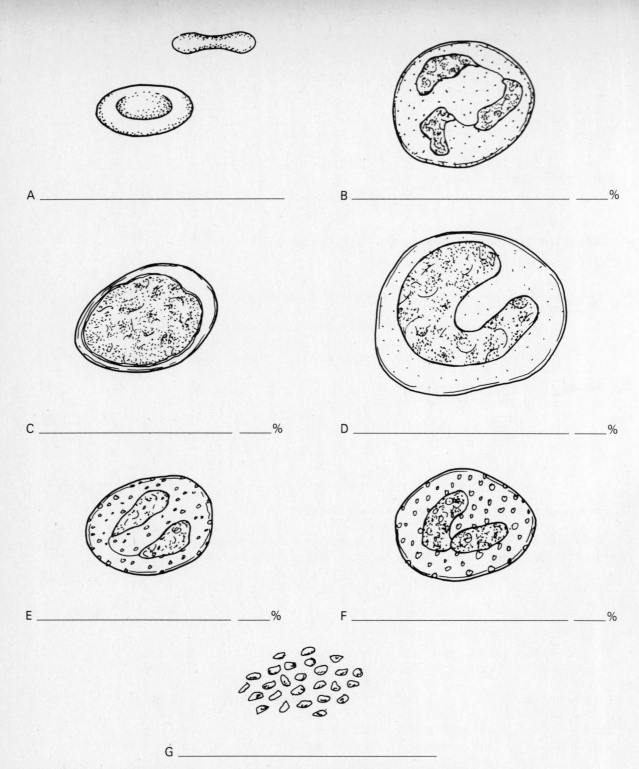

A _____

B _____ _____ %

C _____ _____ %

D _____ _____ %

E _____ _____ %

F _____ _____ %

G _____

Figure LG 19-1 Diagrams of blood cells. Label, color, and complete as directed in Checkpoints C1, D1, and E2.

■ **C3.** Circle the most normal blood values. (Note that values vary slightly according to age and sex.)

a. Average life of a red blood cell: 4 hours   4 days   4 months   4 years

b. RBC count in cube this size: □ (mm^3)

   500   5,000   250,000   5 million   250 million

c. Hemoglobin in adults (g/100 ml blood):   1   8   15   27   41

■ **C4.** Which product of hemoglobin is secreted by liver to form bile, which contributes color to fecal material?

A. Globin   C. Hemosiderin

B. Bilirubin

■ **C5.** Write both the correct term and value after each description.

> Terms: hematocrit, reticulocyte count, sed rate (ESR)
> Values: 1   8   15   27   41

a. Percentage of RBCs that are not quite mature in the circulating blood; this value

   increases during rapid erythropoiesis: _____ ____%

b. Percentage of blood that consists of RBCs; value is usually about 3× the person's

   hemoglobin value: _____ ____%

c. A measure of the rate at which RBCs fall towards the bottom of a test tube, this value often increases in infections, inflammation, or cancer:

   _____ ____ mm/hr

■ **C6.** Explain the roles of the terms in the box in red blood cell production by matching terms with descriptions.

> B_{12}. Vitamin B_{12} H. Hypoxia
>   E. Erythropoietin   IF. Intrinsic factor
>   Fe. Iron   REF. Renal erythropoietic factor

_____ a. Decrease of oxygen in cells; serves as a signal that erythropoiesis is needed (to help provide more oxygen to tissues)

_____ b. Enzyme produced by kidneys when they are hypoxic; causes production of erythropoietin

_____ c. Hormone that travels to marrow to stimulate red blood cell production

_____ d. Component of the heme of hemoglobin

_____ e. Vitamin necessary for normal erythropoiesis

_____ f. Substance produced by the stomach lining and necessary for normal vitamin B_{12} absorption

■ **C7.** *A clinical application.* Explain the rationale behind each of the following practices of some athletes in attempts to increase oxygen-carrying capacity of blood, and, ultimately, athletic performance.

  a. Training in high altitude city such as Mexico City or Denver for a competition to be held in one of those cities.

  b. *Induced erythrocytemia:* removal of RBCs a certain period before competition with reinjection of the cells shortly before the competition.

D. Formed elements: leucocytes (pages 552–555)

■ **D1.** Refer to Figure LG 19-1 and do this exercise about leucocytes.

  a. Label each leucocyte and indicate what percent of the total leucocyte count is accounted for by each type of leucocyte. Such a breakdown of WBCs is known

   as a _____ blood count.

  b. Leucocytes (*have? lack?*) hemoglobin, and so these cells are known as (*red? white?*) blood cells.

  c. Each white blood cell (WBC) (*has? lacks?*) a nucleus. Which type of WBC has a

   large S- or C-shaped nucleus? _____ Which WBC has a

   nucleus that occupies most of the cell? _____. Which three

   WBCs are polymorphonuclear? ____ ____ ____

  d. Which three WBCs are granular? ____ ____ ____
  e. *For extra review.* Color nucleus, cytoplasm, and granules of all WBCs.

■ **D2.** Check your understanding of types of leucocytes by matching names of leucocytes with descriptions. Answers may be used more than once.

B. Basophils	M. Monocytes
E. Eosinophils	N. Neutrophils
L. Lymphocytes	

_____ a. Constitute the largest percentage of leucocytes

_____ b. Contain defensins with antibiotic activity

_____ c. Involved in immunity; may form plasma cells for antibody production

_____ d. Involved in allergic response; become mast cells that release serotonin, heparin and histamine

_____ e. Involved in allergic response; release antihistamines

_____ f. Form wandering macrophages that clean up sites of infection

_____ g. Important in phagocytosis (two answers)

_____ h. Classified as agranular leucocytes (two answers)

■ **D3.** Complete this exercise about antigens and antibodies.

a. Antigens are defined as substances that _____. Most are composed of (*carbohydrate? lipid? protein?*). Most (*are? are not?*) synthesized by the body. Several examples of antigens are:

b. In response to antigens, (*B? T?*)-lymphocytes become plasma cells that secrete

_____. These chemicals are all globulin-type (*lipids? proteins?*). They function to (*activate? inactivate?*) antigens.

c. Other lymphocytes are known as (*A? L? T?*)-lymphocytes. One group of T-cells are activated by antigens to kill these antigens either directly or with help of other lymphocytes and macrophages. Such T-cells are appropriately called

_____ T-cells.

■ **D4.** A normal leucocyte count is _____ mm³. In other words, a typical ratio of RBCs to WBCs is about:

A. 700 to 1   C. 2:1
B. 30:1   D. 1:1

■ **D5.** *A clinical challenge.* Mrs. Doud arrives at a health clinic with a suspected acute infection. Complete this Checkpoint about her.

a. During infection, it is likely that Mrs. Doud's leucocyte count will (*in? de?*)-crease. A count of (*4,000? 8,000? 15,000?*) leucocytes/mm³ blood is most likely. This condition is known as (*leucocytosis? leucopenia?*).

b. A differential increase in the number of WBCs named _____ is most indicative of enhanced phagocytic activity during infection. Neutrophils are most likely to account for (*48? 62? 76?*) percent of the total white count in Mrs. Doud's blood. A sign of a chronic infection is increase in the percent of cells that

can become macrophages; these are _____.

D7. Production of leucocytes is stimulated by chemicals known as (*erythropoietin? colony-stimulating factors [CSFs]?*). List three of these chemicals.

■ **D8.** ____ is a term that relates to success of organ transplants based on presence of HLA antigens on WBCs and other cells.

A. Histocompatibility   C. Chemotaxis
B. Fluosol-DA   D. Diapedesis

E. Formed elements: thrombocytes (page 555)

■ **E1.** Circle correct answers related to thrombocytes.

a. Thromboctyes are also known as:

antibodies   red blood cells
platelets   white blood cells

b. Thrombocytes are formed in:

red bone marrow
tonsils and lymph nodes
spleen

c. Platelets are:

entire cells   chips off the old
                   megakaryocytes

d. A normal range for platelet count is ____/mm³

5,000–1000   4.5–5.5 million
250,000–400,000

e. The primary function of platelets is related to:

O₂ and CO₂ transport   blood clotting
defense   blood typing

■ **E2.** Label platelets on Figure LG 19-1.

■ **E3.** List the components of a complete blood count (CBC).

F. Plasma (pages 556–557)

■ **F1.** Match the names of components of plasma with their descriptions.

> A. Albumin   GAF. Glucose, amino acids, and fats
> E. Electrolytes   HE. Hormones and enzymes
> F. Fibrinogen   N. Nonprotein nitrogen (NPN) substances
> G. Globulin   W. Water

_____ a. Makes up about 92 percent of plasma

_____ b. Regulatory substances carried in blood

_____ c. Cations and anions carried in plasma

_____ d. Constitutes about 55 percent of plasma protein

_____ e. Made by liver; a protein used in clotting

_____ f. Antibody protein

_____ g. Wastes carried to kidneys or sweat glands

_____ h. Food substances carried in blood

■ **F2.** _____ refers to temporary withdrawal of plasma with selective removal of unwanted components, such as certain antibodies.

G. Hemostasis (pages 557–562)

■ **G1.** List the three basic mechanisms of hemostasis:

a. Vascular _____

b. _____ plug formation

c. _____ (clotting)

■ **G2.** Check your understanding of three phases of platelet plug formation in this Checkpoint.

a. When platelets snag on the inner lining of a damaged blood vessel, their shape is altered. They become (*small, disc-shaped? large, irregular-shaped?*). Additional platelets aggregate on the uneven projections. This phase is known as platelet (*adhesion? release reaction? plug?*).

b. Platelets then release chemicals. Some chemicals (*dilate? constrict?*) vessels. How does this help hemostasis?

    Name two of these chemicals. _____ _____

c. State two roles of ADP that is released by platelets.

d. The resulting platelet plug is useful in preventing blood loss in (*large? small?*) vessels.

■ **G3.** The third step in hemostasis, coagulation, requires clotting factors which are derived from three sources.

a. Some of these factors are made in the liver and constantly circulate in blood (*plasma? serum?*).

b. Others are released by (*red blood cells? platelets?*).

c. One important factor, called _____ factor (TF) or thromboplastin, is released from damaged cells. If these cells are within blood or in the lining of blood vessels, then the (*extrinsic? intrinsic?*) pathway to clotting is initiated. If other tissues are traumatized and release tissue factor, then the (*extrinsic? intrinsic?*) pathway is initiated.

■ **G4.** Summarize the three major steps in the cascade of events in coagulation by filling in blanks A–E in Figure LG 19-2. Choose from the list of terms on the figure.

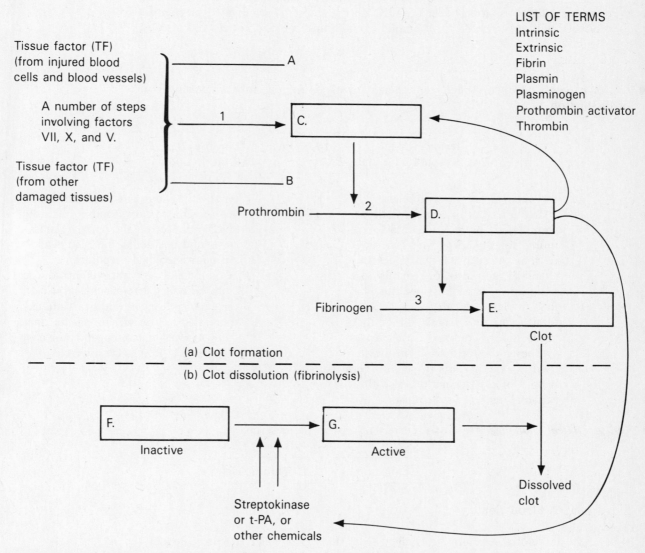

Figure LG 19-2 Summary of steps in clot formation and dissolution. Complete as directed in Checkpoints G4, G5, and G7.

■ **G5.** Refer to Figure LG 19-2 again. The figure indicates that thrombin exerts a (*positive? negative?*) feedback effect upon prothrombin activator, causing the clot to (*enlarge? shrink?*). The ion (Ca^{2+}? K^+? Na^+?) plays a critical role in many aspects of coagulation. Write this ion on Figure LG 19-2 next to each of the three major steps.

■ **G6.** Although there are many forms of hemophilia, all involve deficiency of

_____. Write three signs or symptoms of hemophilia.

■ **G7.** Following clot formation, clots undergo two changes. Summarize these.

a. First clot _____ or syneresis occurs.
b. The lower portion of Figure LG 19-2 shows the next step: clot dissolution or

_____. An inactive enzyme (F on the figure) is activated to (G) which dissolves clots. Label F and G on the figure. F is activated by chemicals produced within the body, for example, from damaged blood vessels and also by

(D) _____.

■ **G8.** *A clinical challenge.* Match each chemical with its role in clot formation or dissolution.

> H. Heparin   Vit K. Vitamin K
> S, t-PA. Streptokinase and t-PA   W. Warfarin (Coumadin)

_____ a. Required for synthesis of factors II, VII, IX, and X in clot formation. Commercial preparations may be required for newborns who lack the intestinal bacteria that synthesize this chemical, and also for persons with malabsorption of fat since this is a fat-soluble chemical.

_____ b. Slow-acting anticoagulant that is antagonistic to vitamin K so decreases synthesis of prothrombin in liver; used to prevent clots in persons with history of clot formation; also the active ingredient in rat poison.

_____ c. Fast-acting anticoagulant that blocks conversion of prothrombin to thrombin; produced in the body by mast cells and basophils, but also available in pharmacologic preparations

_____ d. Commercial preparations introduced into coronary arteries can limit clot size and prevent severe heart attacks; classified as fibrinolytic agents since they activate plasminogen (shown on Figure LG 19-2) to dissolve clot.

■ **G9.** *A clinical challenge.* Match the correct term with the description.

> E. Embolus   T. Thrombus

_____ a. A blood clot

_____ b. A "clot-on-the-run," dislodged from the site at which it formed (usually a deep vein of the leg); also fat from broken bone, bubble of air, or amniotic fluid traveling through blood, possibly to lung (pulmonary) vessels

■ **G10.** *A clinical challenge.* Mr. Eisen has normal coagulation function. Match each of his lab values with the correct test:

12 sec	40 sec	4 min

a. Bleeding time: _____

c. Prothrombin time (PT or Protime): _____

b. Partial thromboplastin time (PTT): _____

■ **G11.** Mr. B's chronic alcoholism has altered his PT and PTT times. These values are likely to be (*higher? lower?*) than normal. Explain why.

H. Grouping (typing) (pages 562–565)

■ **H1.** Contrast *agglutination* with *coagulation.*

a. Coagulation (or _____) involves coagulation factors found in platelets, plasma, or other tissue fluids. The process (*requires? can occur in absence of?*) red blood cells.

b. Agglutination (or "clumping") of erythrocytes is an

antigen-_____ process that (*does? does not?*) require red

blood cells since these are sites of _____ used in the agglutination reaction.

■ **H2.** Do this exercise about factors responsible for blood groups.

a. Agglutinogens are (*antigens? antibodies?*) located (*in plasma? on surface of red blood cells?*).

b. Type A blood has (*A? B? both? neither?*) antigen on RBCs.

c. Type A has (*a? b? both? neither?*) antibody or _____. These are also called *anti-B* antibodies since they attack agglutinogen B.

d. If a person with type A blood receives type B blood, the recipient's (*anti-A? anti-B?*) antibodies will attack the (*A? B?*) antigens on the donor's cells.

H3. Complete the table contrasting ABO blood types.

Blood Group	Percent of White Popu- lation	Percent of Black Popu- lation	Sketch of Blood Showing Correct Agglutinogens and Agglutinins	Can Donate Safely to	Can Receive Blood Safe- ly From
a. Type A					A, O
b.	10				
c. Type AB		7			
d.				A, B, AB, O	

■ **H4.** Type O is known as the universal (*donor? recipient?*) with regard to the ABO

group since type O blood lacks _____ of the ABO group. Type

_____ is known as the universal recipient. Explain why.

H5. Explain what problems can result from transfusion of an incompatible blood type.

■ **H6.** Complete this exercise about the Rh group.
   a. The Rh (*positive? negative?*) group is more common. Rh (*positive? negative?*) blood has Rh agglutinogens on the surfaces of RBCs.
   b. Under normal circumstances plasma of (*Rh positive blood? Rh negative blood? both Rh groups? neither Rh group?*) contains anti-Rh agglutinins.
   c. Rh (*positive? negative?*) persons can develop these agglutinins when they are ex- posed to Rh (*positive? negative?*) blood.
   d. An example of this occurs in fetal-maternal incompatibility when a mother who is Rh (*positive? negative?*) has a baby who is Rh (*positive? negative?*) and some of the baby's blood enters the mother's bloodstream. The mother develops anti-Rh ag- glutinins which may cross the placenta in future pregnancies and hemolyze the

   RBCs of Rh (*positive? negative?*) babies. Such a condition is known as _____.

■ **H7.** Explain how administration of anti-Rh agglutinins to a mother just after the birth of an Rh positive baby can prevent Rh problems in future pregnancies.

I. Interstitial fluid and lymph (pages 565–566)

■ **I1.** Contrast these two body fluids in the table.

	Interstitial Fluid and Lymph	Plasma
a. Amount of protein (more or less)		
b. Formed elements present		Red blood cells, white blood cells, platelets
c. Location	Between cells and in lymph vessels	

J. Disorders, medical terminology (pages 566–569)

■ **J1.** Match names of types of anemia with descriptions below.

A. Aplastic	N. Nutritional
Hl. Hemolytic	P. Pernicious
Hr. Hemorrhagic	S. Sickle cell

_____ a. Condition resulting from inadequate diet, such as deficiency of iron, vitamin B_{12}, or amino acids

_____ b. Condition in which intrinsic factor is not produced, so absorption of vitamin B_{12} is inadequate

_____ c. Inherited condition in which hemoglobin forms stiff rodlike structures causing erythrocytes to assume sickle shape and rupture, reducing oxygen supply to tissues

_____ d. Rupture of red blood cell membranes due to variety of causes, such as parasites, toxins, or antibodies

_____ e. Condition due to excessive bleeding, as from wounds, gastric ulcers, heavy menstrual flow

_____ f. Indequate erythropoiesis as a result of destruction or inhibition of red bone marrow

J2. Persons with one gene for sickle cell anemia are said to have (*sickle cell trait? sickle cell anemia?*). These individuals have (*greater? less?*) resistance to malaria. Explain why.

■ **J3.** *A clinical challenge.* Amy (21 years old) has a red blood count of 7 million/mm³. She has the condition known as (*anemia? polycythemia?*). Amy has a hemotocrit done. It is more likely to be (*under 32? over 45?*). Her blood is (*more? less?*) viscous than normal, which is likely to cause (*high? low?*) blood pressure.

■ **J4.** Infectious mononucleosis is caused by the Epstein-Barr (*bacteria? virus?*). It affects mostly (*teenagers? elderly?*). The name of this infection is based upon the fact that B lymphocytes become abnormal in appearance, resembling

_____. A differential count shows a great increase in

_____-cytes. Permanent damage usually (*does? does not?*) result.

■ **J5.** Leukemia is a form of cancer involving abnormally high production of (*erythrocytes? leucocytes?*). Briefly explain why the following symptoms are likely to occur.

a. Anemia

b. Hemorrhage

c. Infection

J6. Define and explain clinical significance of each of these terms.

a. Gamma globulin

b. Thrombocytopenia

c. Septicemia

ANSWERS TO SELECTED CHECKPOINTS

A2. (a) Blood, heart, and blood vessels. (b) Lymph, lymph vessels, and lymph nodes and organs (tonsils, spleen).

A3. (a) 4.5× thicker than water. (b) 0.85–0.9 percent. (c) 38°C (100.4°C). (d) 7.35–7.45. (e) 4–6 liters.

B1. (a) 55, 45. (b) Erythro, leuco, thrombo.

B2. Hemopoiesis, red bone marrow (myeloid tissue); proximal epiphyses of femur and humerus, flat bones of skull, sternum, ribs, vertebrae, pelvis.

C1. A, does not, hemoglobin, see Figure 19-2, page 549.

C2. (a) Globin, hemes. (b) Iron (Fe), oxygen, 4. (c) Globin, carbamino-, 23, in plasma, bicarbonate.

C3. (a) 4 months. (b) 5 million. (c) 15.

C4. B.

C5. (a) Reticulocyte, 1. (b) Hematocrit, 41. (c) Sed rate, 15 is in the normal range for males under 50 years; 27 is normal for females over 50 years.

C6. (a) H. (b) REF. (c) E. (d) Fe. (e) B_{12}. (f) IF.

C7. Mild hypoxia stimulates RBC formation. Hypoxia is induced in (a) by a move to high altitudes where the air is "thinner" (lower pressure, so lower in oxygen). Residents of high-altitude cities develop higher hematocrits. Removal of RBCs (in b) induces hypoxia and erythropoiesis, giving this athlete an (illegal) boost of hematocrit once original RBCs are reintroduced.

D1. (a) B, neutrophil (60–70); C, lymphocyte (20–25); D, monocyte (3–8); E, eosinophil, 2–4; F, basophil (0.5–1); differential (DIF). (b) Lack, white. (c) Has, monocyte, lymphocyte; B, E, F. (d) B, E, F. (e) Refer to Figure 19-2, page 549 of text.

D2. (a) N. (b) N. (c) L. (d) B. (e) E. (f) M. (g) M, N. (h) L, M.

D3. (a) Stimulate production of antibodies; protein; are not; pollen, toxin released by bacteria, structural parts of microorganisms. (b) B, antibodies, proteins, inactivate. (c) T, killer or cytotoxic.

D4. 5,000–10,000, A.

D5. (a) In, 15,000, leucocytosis. (b) Neutrophils (or polymorphs or "polys"), 76, monocytes.

D8. A.

E1. (a) Platelets. (b) Red bone marrow. (c) Chips off the old megakaryocytes. (d) 250,000–400,000. (e) Blood clotting.

E2. G.

E3. RBC, WBC, and platelet counts; DIF; hemoglobin and hematocrit (H & H).

F1. (a) W. (b) HE. (c) E. (d) A. (e) F. (f) G. (g) N. (h) GAF.

F2. Plasmapheresis or therapeutic plasma exchange (TPE).

G1. (a) Spasm. (b) Platelet. (c) Coagulation.

G2. (a) Large, irregular-shaped; adhesion. (b) Constrict, reduces blood flow into injured area; serotonin, thromboxane 2. (c) Activates platelets and makes them stickier, thereby inviting more platelets to the scene. (d) Small.

G3. (a) Plasma. (b) Platelets. (c) Tissue, intrinsic, extrinsic.

G4. A, Intrinsic. B, extrinsic. C, Prothrombin activator. D, Thrombin. E, Fibrin.

G5. Positive, enlarge; Ca^{2+}.

G6. Some clotting factor; hemorrhaging, blood in urine, joint damage and joint pain.

G7. (a) Retraction. (b) Fibrinolysis; F, plasminogen; G, plasmin; D, thrombin.

G8. (a) Vit K. (b) W. (c) H. (d) S, t-PA.

G9. (a) T. (b) E.

G10. (a) 4 min. (b) 40 sec. (c) 12 sec.

G11. Higher (longer times to clot blood). Most coagulation factors are normally synthesized in the liver, which may be damaged by cirrhosis induced by alcohol.

H1. (a) Clotting, can occur in absence of. (b) Antibody, does, antigens.

H2. (a) Antigens, on surface of red blood cells. (b) A. (c) b, agglutinin (or isoantibody). (d) Anti-B (or b), B.

H4. Donor; agglutinogens or isoantigens; AB since AB plasma lacks both anti-A and anti-B antibodies (agglutinin a and b).

H6. (a) Positive, positive. (b) Neither Rh group. (c) Negative, positive. (d) Negative, positive, positive, erythroblastosis fetalis.

H7. These antibodies tie up any fetal RBCs (with Rh antigens on them) so that the Rh-negative mother's immune system will not respond and make anti-Rh antibodies against future Rh-positive babies.

I1. (a) Less, more. (b) Contain some leucocytes, but no red blood cells or platelets, -. (c) -, within blood vessels.

J1. (a) N. (b) P. (c) S. (d) Hl. (e) Hr. (f) A.

J3. Polycythemia, over 45, more, high.

J4. Virus, teenagers, monocytes, lympho, does not.

J5. Focus on production of only immature leukocytes in abnormal bone marrow limits production of RBCs which leads to (a) anemia; platelet production also decreases (b: hemorrhage) and WBC production declines (c: infection).

MASTERY TEST: Chapter 19

Questions 1–2: Arrange the answers in correct sequence.

_____ _____ _____ 1. Events in the coagulation process:

A. Retraction or tightening of fibrin clot
B. Fibrinolysis or clot dissolution by plasma
C. Clot formation

_____ _____ _____ 2. Stages in the clotting process:

A. Formation of prothrombin activator
B. Conversion of prothrombin to thrombin
c. Conversion of fibrinogen to fibrin

3. The normal red blood cell count in healthy females is ____ RBCs/mm^3.

   A. 4.8 million   D. 250,000
   B. 2 million   E. 8,000
   C. 0.5 million

4. All of the following types of formed elements are produced in bone marrow EXCEPT:

   A. Neutrophils   D. Erythrocytes
   B. Basophils   E. Lymphocytes
   C. Platelets

5. Normal prothrombin time is about:

   A. 12 sec   C. 10 to 15 min
   B. 1 to 4 min   D. 3 hr

6. Megakaryocytes are involved in formation of:

   A. Red blood cells D. Platelets
   B. Basophils   E. Neutrophils
   C. Lymphocytes

7. Choose the FALSE statement about Norine, who has type O blood.

   A. Norine has neither A nor B agglutinogens on her red blood cells.
   B. She has both anti-A and anti-B agglutinins in her plasma.
   C. She is called the universal donor.
   D. She can receive blood safely from both type O and type AB persons.

8. Choose the FALSE statement about blood.

   A. Blood is thicker than water.
   B. It normally has a pH of 7.0.
   C. The human body normally contains about 5 to 6 liters of it.
   D. It normally consists of more plasma than cells.

9. Choose the FALSE statement about factors related to erythropoiesis.

   A. Erythropoietin stimulates RBC formation.
   B. Oxygen deficiency in tissues serves as a stimulus for erythropoiesis.
   C. Intrinsic factor, which is necessary for RBC formation, is produced in the kidneys.
   D. A reticulocyte count of over 1.5 percent of circulating RBCs indicates that erythropoiesis is occurring rapidly.

10. Choose the FALSE statement about plasma.

    A. It is red in color.
    B. It is composed mainly of water.
    C. Its concentration of protein is almost four times that of interstitial fluid.
    D. It contains plasma proteins, primarily albumin.

11. Choose the FALSE statement about neutrophils.

    A. They are actively phagocytic.
    B. They are the most abundant type of leucocyte.
    C. Neutrophil count decreases during most infections.
    D. An increase in their number would be a form of leucocytosis.

12. All of the following correctly match parts of blood with principal functions EXCEPT:

    A. RBC: carry oxygen and CO_2
    B. Plasma: carries nutrients, wastes, hormones, enzymes
    C. WBCs: defense
    D. Platelets: determine blood type

13. Which of the following chemicals is an enzyme that converts fibrinogen to fibrin?

    A. Heparin
    B. Thrombin
    C. Prothrombin
    D. Coagulation factor IV
    E. Tissue factor

14. A type of anemia that is due to a deficiency of intrinsic factor production, for example, caused by alterations in the lining of the stomach, is:

    A. Aplastic   C. Hemolytic
    B. Sickle cell   D. Pernicious

15. A person with a hematocrit of 66 and hemoglobin of 22 is most likely to have the condition named:

    A. Anemia
    B. Polycythemia
    C. Infectious mononucleosis
    D. Leukemia

T   F   16. Plasmin is an enzyme that facilitates clot <u>formation.</u>

T   F   17. Erythroblastosis fetalis is most likely to occur with an Rh <u>positive mother and her Rh negative babies.</u>

T   F   18. In general, values for hemoglobin, hematocrit, and red blood cell count are <u>higher</u> among males than females.

T   F   19. In both black and white populations in the U.S., type O is <u>most</u> common and type AB is <u>least</u> common of the ABO blood groups.

T   F   20. In 100 ml of blood there are usually about <u>15 ml</u> of formed elements, most of which are red blood cells.

Questions 21–25: fill-ins. Answer questions or complete sentences with the word or phrase that best fits.

_____ 21. Name five types of substances transported by blood.

_____ 22. Name three types of plasma proteins.

_____ 23. Name the type of leucocyte that is transformed into plasma cells which then produce antibodies.

_____ 24. Write a value for a normal leucocyte count: ____/mm^3.

_____ 25. All blood cells and platelets are derived from ancestor cells called ____.

ANSWERS TO MASTERY TEST: ■ Chapter 19

Arrange

1. C A B   2. A B C

Multiple Choice

3. A	10. A
4. E	11. C
5. A	12. D
6. D	13. B
7. D	14. D
8. B	15. B
9. C	

True-False

16. F. Dissolution or fibrinolysis
17. F. Negative mother and her Rh-positive babies
18. T
19. T
20. F. 47 ml (males) to 42 ml (females)

Fill-ins

21. Oxygen, carbon dioxide, nutrients, wastes, sweat, hormones, enzymes
22. Albumin, globulins, and fibrinogen
23. Lymphocyte
24. 5,000–9,000
25. Hemocytoblasts

FRAMEWORK 20

The Heart

ANATOMY

INTRODUCTION [A]
- Location, borders

STRUCTURE [B]

Coverings, Layers
- Pericardium
- Myocardium
- Endocardium

- Chambers
- Vessels
- Valves

BLOOD SUPPLY [C]
- Coronary arteries
- Coronary sinus
- Angina, MI, enzyme studies

PHYSIOLOGY

CONDUCTION SYSTEM [D]
- SA node
- AV node
- AV bundle
- Purkinje fibers
- ECG: P, QRS, T

CARDIAC CYCLE [E]
- Duration: 0.8 sec
  - atrial systole, 0.1 sec
  - atrial diastole, 0.7 sec
  - ventricular systole, 0.3 sec
  - ventricular diastole, 0.5 sec
- Sounds
- Pressure changes

CARDIAC OUTPUT [F]

SV × **Heart Rate**
(75 ml/beat × **75 beat/min)**

- EDV, ESV
- Starling's law
- Cardiac reserve

- Regulation by medulla
- Baroreceptor reflexes

ADVANCES IN CARDIOLOGY [G]
- Artificial heart
- Diagnostics
- Risk factors

DEVELOPMENT [H]

DISORDERS [I]

DISORDERS [I]
- CAD
- Atherosclerosis
- Congenital defects
- Arrhythmias
- CHF
- CP

The Cardiovascular System: The Heart

CHAPTER **20**

Make a fist, then open and squeeze tightly again. Repeat this about once a second as you read the rest of this Overview. Envision your heart as a muscular pump about the size of your fist. Unfailingly, your heart exerts pressure on your blood, moving it onward through the vessels to reach all body parts. The heart has one job; it is simply a pump. But this function is critical. Without the force of the heart, blood would come to a standstill, and tissues would be deprived of fluids, nutrients, and other vital chemicals. To serve as an effective pump, the heart requires a rich blood supply to maintain healthy muscular walls, a specialized nerve conduction system to synchronize actions of the heart, and intact valves to direct blood flow corrrectly. Heart sounds and ECG recordings as well as a variety of more complex diagnostic tools provide clues to the status of the heart.

Is your fist tired yet? The heart ordinarily pumps 24 hours a day without complaint and seldom reminds us of its presence. As you complete this chapter on the heart, keep in mind the value of this virtually irreplaceable organ. Start by studying the Chapter 20 Framework and the key terms for each section.

TOPIC OUTLINE AND OBJECTIVES

A. Introduction

B. Heart structure

1. Describe the structure and functions of the chambers, great vessels, and valves of the heart.

C. Blood supply

D. Conduction system, electrocardiogram

2. Explain the structural and functional features of the conduction system of the heart.

E. Cardiac cycle

3. Describe the principal events of a cardiac cycle.
4. Contrast the sounds of the heart and their clinical significance.

F. Cardiac output

5. Explain the various factors that affect heart rate and cardiac output (CO).

G. Advances in cardiology: prevention, diagnosis, and treatment

6. List the risk factors involved in heart disease.

H. Developmental anatomy

I. Disorders, medical terminology

7. Describe how atherosclerosis and coronary artery spasm contribute to coronary artery disease (CAD).
8. Describe coarctation of the aorta, patent ductus arteriosus, septal defects, valvular stenosis, and tetralogy of Fallot as congenital heart defects and atrioventricular (AV) block, atrial flutter, atrial fibrillation, and ventricular fibrillation as abnormalities of the conduction system of the heart (arrhythmias).
9. Define congestive heart failure (CHF) and cor pulmonale (CP).
10. Define medical terminology associated with the heart.

Now study the following parts of words that may help you better understand terminology in this chapter.

Wordbyte	Meaning	Example	Wordbyte	Meaning	Example
cardi-	heart	*cardi*ologist	myo-	muscle	*myo*cardial infarction
coron-	crown	*coron*ary arteries			
endo-	within	*endo*carditis	peri-	around	*peri*cardium
			vascul-	small vessel	cardio*vascular*

CHECKPOINTS

A. Introduction (page 573)

■ **A1.** What is the main function of the heart? _____

■ **A2.** Closely examine Figure 20-1, page 573 in your text. Consider the location, size, and shape of your heart as you do this exercise. Trace its outline on your body.

a. Your heart lies in the _____ portion of your thorax, between

your two _____. About (*one-third? one-half? two-thirds?*) of the mass of your heart lies to the left of the midline of your body.

b. Your heart is about the size and shape of your _____.

c. The pointed part of your heart, called the _____, lies in the

_____ intercostal space, about _____ cm (_____ inches) from the midline of your body.

d. The region of the heart that lies directly on top of the diaphragm is the right (*atrium? ventricle?*).

B. Heart structure (pages 573–580)

■ **B1.** Arrange in order from most superficial to deepest. ___ ___ ___ ___ ___ ___

E. Endocardium	PC. Pericardial cavity
FP. Fibrous pericardium	SP. Serous pericardium
M. Myocardium	VP. Visceral pericardium (epicardium)

For extra review of the pericardium, look again at Checkpoint D2 in Chapter 4, page LG 76.

■ **B2.** Inflammation of the pericardium is known as _____. This

condition may lead to cardiac tamponade which means _____.

■ **B3.** Refer to Figures LG 20-1 and LG 20-2 to do the following activity.

   a. Identify all structures with letters (A–I) by writing labels on lines A–I.

   b. Draw arrows on Figure LG 20-1 to indicate direction of blood flow.

   c. Color red the chambers of the heart and vessels that contain highly oxygenated blood; color blue the regions in which blood is low in oxygen and high in carbon dioxide.

   d. On Figure LG 20-1, label the four valves that control blood flow through the heart.

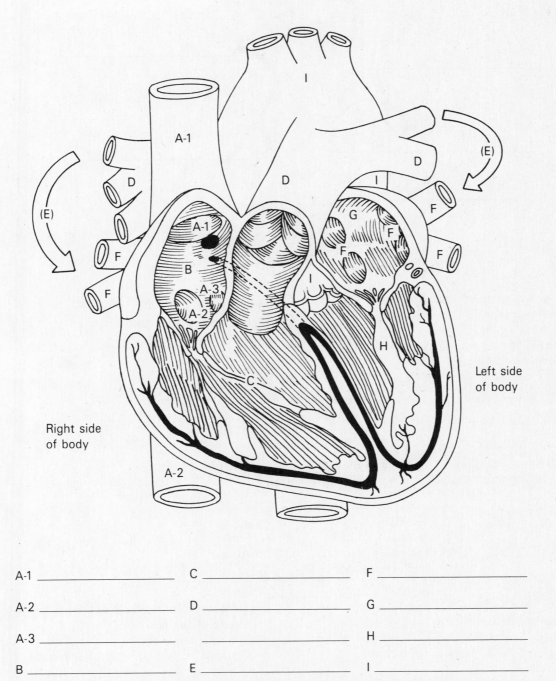

A-1 _____ C _____ F _____

A-2 _____ D _____ G _____

A-3 _____ _____ H _____

B _____ E _____ I _____

Figure LG 20-1 Diagram of a frontal section of the heart. Letters follow the path of blood through the heart. Label, color, and draw arrows as directed in Checkpoints B3 and D2.

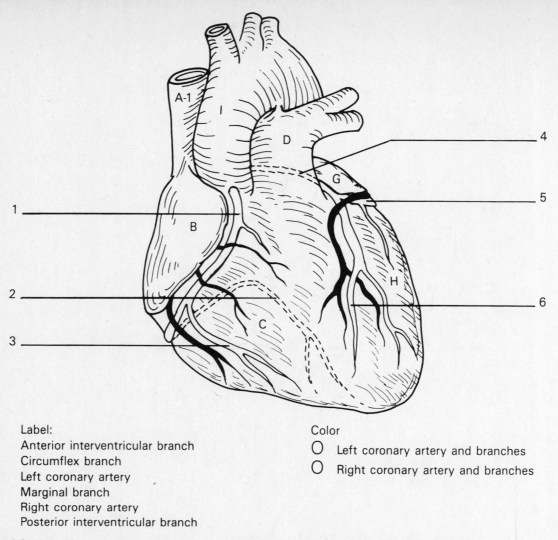

Label:
Anterior interventricular branch
Circumflex branch
Left coronary artery
Marginal branch
Right coronary artery
Posterior interventricular branch

Color
O   Left coronary artery and branches
O   Right coronary artery and branches

Figure LG 20-2   Anterior view of the heart. Label and color as directed in Checkpoints B3 and C2.

■ **B4.** Check your understanding of heart structure and the pathway of blood through the heart by selecting terms that fit descriptions below. Not all answers will be used.

A. Atria	PV. Pulmonary vein
CT. Chordae tendiniae	S. Septum
PA. Pulmonary artery	V. Ventricles
PM. Papillary muscles	

_____ a. Thin-walled chambers that receive blood from veins

_____ b. Blood vessel that carries blood that is rich in oxygen from lungs to left atrium

_____ c. Wall separating right side from left side of the heart

_____ d. Strong tendons that anchor atrioventricular valves to ventricular muscle

_____ e. Sites of most of the myocardium

■ **B5.** Check your understanding of the valves by doing this matching exercise. More than one answer may be required for each description.

| A. Aortic semilunar | P. Pulmonary semilunar |
| B. Bicuspid | T. Tricuspid |

_____ a. Also called the mitral valve

_____ b. Prevents backflow of blood from right ventricle to right atrium

_____ c. Prevents backflow from pulmonary trunk to right ventricle

_____ d. Prevents backflow of blood into left atrium

_____ e. Have half-moon-shaped leaflets or cusps (2 answers)

_____ f. Also called atrioventricular (AV) valves (2 answers)

■ **B6.** Using Figure 20-5 (page 580) of your text, locate the sites on the surface of your own body where heart sounds related to each valve are best heard. Remember that these sounds are created by turbulence of blood when valves close, so these sites (*do? do not?*) necessarily lie directly over the valve itself.

C. Blood supply (pages 580–582)

C1. Defend or dispute this statement: "The myocardium receives all of the oxygen and nutrients it needs from blood that is passing through its four chambers."

■ **C2.** Refer to Figure LG 20-2 and complete this Checkpoint.

a. Using color code ovals on the figure, color the two coronary arteries and their main branches.

b. Label each of the vessels using list of labels on the figure.

■ **C3.** The coronary sinus functions as (*an artery? a vein?*). It collects blood that has

passed through coronary arteries and capillaries into _____ veins. The coronary sinus finally empties this blood into the (*right? left?*) (*atrium? ventricle?*).

C4. Define *ischemia* and explain how it is related to *angina pectoris*.

C5. A "heart attack" is the common name for a myocardial

_____ (*MI*). Answer the following questions about this problem.

a. What does the term *infarction* mean? _____

b. Explain how either a *thrombus* or *free radicals* may lead to an MI.

c. Measurement of CPK level of blood is one test used to diagnose heart attacks.

Write abbreviations for two other enzyme tests: _____ and

_____. Explain why elevation of these enzyme levels may indicate that myocardial infarction has occurred.

D. Conduction system, electrocardiogram (pages 582–585)

D1. Define *intercalated discs* and *gap junctions,* and explain how these are related to the contraction of atria as a unit and contraction of ventricles as a unit.

■ **D2.** Summarize the conduction system of the heart by doing this learning activity.

a. The (*autonomic? somatic?*) nervous system innervates the heart. These nerves function to (*initiate? alter rate of?*) contraction of the heart. In other words, the heart (*can? cannot?*) contract in the absence of ANS stimulation. (More about ANS control in Chapter 21.)

b. The heart's intrinsic regulating system is known as its _____ system. Cells forming this system are developed embryologically from cardiac muscle cells; however, their specialty is (*contraction? impulse transmission?*).

c. Label the parts of the conduction system on Figure LG 20-1.

d. What structural feature of the heart makes the AV node and the AV bundle necessary for conduction to the entire heart?

D3. Label the following parts of a typical ECG recording on Figure LG 20-3c: *P wave, P–R interval, QRS wave (complex), S–T segment, T wave.*

E. Cardiac cycle (pages 586–588)

■ **E1.** Complete the following overview of movement of blood through the heart.

a. Blood moves through the heart as a result of _____ and

_____ of cardiac muscle, as well as

_____ and _____ of valves. Valves open

or close due to _____ within the heart.

b. Contraction of heart muscle is known as _____, whereas

relaxation of myocardium is called _____.

■ **E2.** Now refer to Figure LG 20-3 and consider details of the cardiac cycle in this Checkpoint.

a. The duration of one average cycle is ____ sec; (*54? 60? 75?*) complete cardiac cycles (or heart beats) occur per minute (if pulse rate is 75).

b. As the P wave of the ECG occurs, nerve impulses spread across the

_____. During this initial (*0.1? 0.3?*) sec of the cycle, the atria begin their contraction. As a result, pressure within the atria (*increases? decreases?*), as shown at point A in Figure LG 20-3b.

c. The _____ wave then signals contraction (or

_____) of the ventricles. Note that although the QRS wave itself happens over a very brief period, the ventricular contraction that follows occurs over a period of (*0.3? 0.5? 0.7?*) sec.

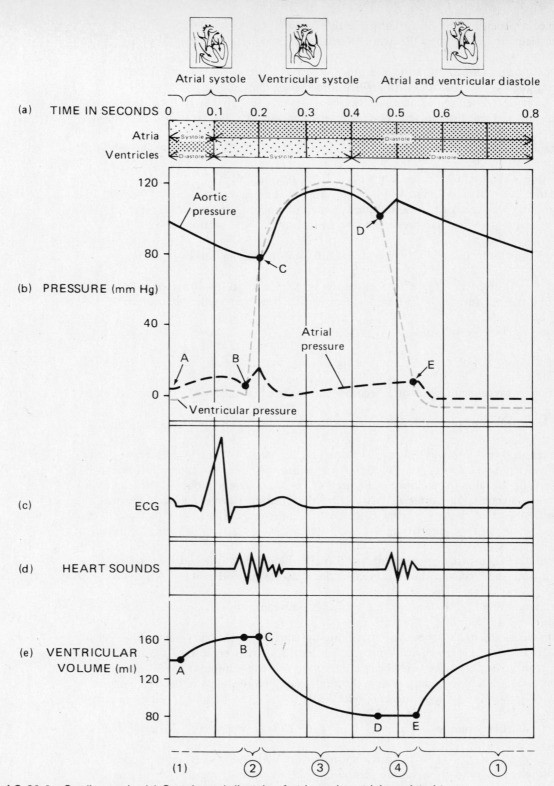

Figure LG 20-3 Cardiac cycle. (a) Systole and diastole of atria and ventricles related to time. (b) Pressures in aorta, atria and ventricles. Points labeled *A* to *E* refer to Checkpoint E2 and E3. (c) ECG related to cardiac cycle. Label as directed in Checkpoint D3. (d) Heart sounds related to cardiac cycle. Label according to Checkpoint E2m. (e) Volume of blood in ventricles. Refer to Checkpoint E2j–l.

d. Trace a pencil lightly along the curve that shows changes in pressure within the ventricles following the QRS wave. Notice that pressure there (*increases? decreases?*)

(*slightly dramatically?*). In fact it quickly surpasses atrial pressure (at point ____ in the figure). As a result, AV valves are forced (*open? closed?*).

e. A brief time later ventricular pressure becomes so great that it even surpasses pressure in the great arteries (pulmonary artery and aorta). This pressure forces blood

against the undersurface of the _____ valves, (*opening?*

closing?) them. This occurs at point _____ on the figure.

f. Continued ventricular systole ejects blood from the heart into the great vessels. In the aorta this creates a systolic blood pressure of about (*15? 80? 120?*) mm Hg.

g. What causes ventricular pressure to begin to drop? The cessation of

_____-polarization (and contraction) followed by repolarization of these

chambers (after the ____ wave) causes the ventricles to go into (*systole? diastole?*).

h. When the pressure within ventricles drops just below that in the great arteries,

blood in these vessels fills the _____ valves and closes them.

This occurs at point _____ in the figure. Ventricular pressure now drops at a rapid rate.

i. When intraventricular pressure becomes lower than that in atria, the force of blood

within the atria causes the AV valves to (*open? close?*). This happens at point ____ in the figure.

j. Note that the AV valves now remain open all the way to Point *B* in the next cycle. This permits adequate filling of ventricles before their next contraction. Trace a pencil lightly along the curve on Figure LG 20-2e starting at point *E*. Note that the curve is steepest (*just after E? just before A of the next cycle?*). In other words, the first part of the curve from *E* to *A* of the next cycle (ventricular diastole) is the time when ventricles are filling (*rapidly? slowly?*), whereas (*rapid filling? diastasis?*)

occurs later. In fact, ____ percent of all of the blood that will enter ventricles does so during the first two thirds of ventricular diastole (period *E–A*).

k. The remaining ____ percent enters ventricles between points *A* and *B,* that is, as a result of atrial (*systole? diastole?*).

l. Summarize changes in ventricular volume by labeling regions 1–4 below Figure LG 20-2e. In part 1, the ventricles are relaxed and they are in a state of (*filling? ejection?*). In part 2, ventricles are contracting but the semilunar valves have not yet opened. So blood cannot leave the ventricles yet. Thus this is the brief period named isovolumetric (*contraction? relaxation?*). Once the semilunar valves open,

the ventricular volume ____-creases. This period (3) is ventricular

_____. In the short period 4, the ventricles have stopped contracting so pressure is rapidly dropping, but AV valves have not yet opened. So

ventricles cannot fill yet. This is the period of _____. Note that *iso-* means *same.*

m. The first heart sound (*lubb? dupp?*) is created by turbulence of blood at the

(*opening? closing?*) of the _____ valves. What causes the second sound?

Write *first* and *second* next to the parts of Figure LG 20-2d showing each of these heart sounds.

■ **E3.** *For extra review.* Label points *B, C, D,* and *E* on Figure LG 20-3 to indicate which valves open or close at those points.

E4. Explain what causes heart murmurs.

F. Cardiac output, cardiac reserve, regulation of heart rate (pages 588–592)

■ **F1.** Determine the average cardiac output in a resting adult.
Cardiac output = stroke volume × heart rate

$$= \text{_____} \text{ ml/stroke} \times \text{_____} \text{ strokes/min}$$

$$= \text{_____} \text{ ml/min (_____ liter/min)}$$

■ **F2.** Do this exercise about volumes of blood in the heart at different times during the cardiac cycle.

a. During diastole ventricles (*eject? fill with?*) blood. At the end of their diastole the

ventricular volume is about _____ ml of blood. This is known as the EDV, or

_____ volume.

b. As ventricles contract, they eject about _____ ml, known as their

_____ volume (SV).

c. So at the end of systole ventricles contain _____ ml, called their

_____ volume (or _____).

d. In summary, which statement is true? (Circle one.)

   A. ESV − SV = EDV   C. EDV − SV = ESV
   B. ESV − EDV = SV

e. *A clinical challenge.* Mrs. G. has an EDV of 120. But a myocardial infarction (MI) has weakened her ventricles so that they can eject a stroke volume (SV) of only 40 ml. Calculate Mrs. G's ESV.

■ **F3.** Describe factors affecting stroke volume by doing this exercise.

  a. During exercise skeletal muscles surrounding blood vessels squeeze (*more? less?*) blood back toward the heart. So venous return (*increases? decreases?*), causing heart muscle fibers to stretch (*more? less?*).

  b. A stretched muscle contracts with (*greater? less?*) force than a muscle that is only

  slightly stretched. This is a statement of _____ law of the heart. To get an idea of this, blow up a balloon slightly and let it go. Then blow it up quite full of air (much like the heart when stretched out by large venous return) and now let the balloon go. In which stage do the walls of the balloon/heart compress the air/blood with greater force?

  c. A heart that is overstretched too often is much like a balloon that has been greatly expanded 5000 times. Now the heart fibers are beyond the optimal length according to Starling's law. So now the force of contraction of the balloon/heart

  is ____-creased. This heart is now capable of a(n) ____-creased stroke volume. State two reasons why blood backs up in heart chambers, causing such overdilation of the heart.

  d. As a result of Starling's law, during exercise the ventricles of the normal heart contract (*more? less?*) forcefully, and stroke volume (*increases? decreases?*).

  e. State several reasons why venous return might be decreased, leading to reduced cardiac output.

■ **F4.** At rest Dave has a cardiac output of 5 liters per minute. During a strenuous cross country run, Dave's maximal cardiac output is 25 liters per minute. In other words, his

heart has the reserve to pump out ____ liters per minute more during maximal exercise than at rest. Write your calculations here:

Cardiac reserve = maximal CO − resting CO

Dave's cardiac reserve is ____ percent of his resting cardiac output.
Note that cardiac reserve is usually (*higher? lower?*) in trained athletes than in sedentary persons.

■ **F5.** Complete this exercise describing control of heart rate and blood pressure (BP).

a. Recall the two factors that determine cardiac output: _____

   and _____. If one of these is altered from normal, the other
   one may compensate. For example, if SV is decreased (as by an MI), then the heart

   rate will normally _____-crease in an attempt to maintain adequate cardiac
   output.

b. Cardiac output is directly related to blood pressure (BP). This means that when the
   heart increases its output of blood, more blood is present in vessels, exerting
   (*increased? decreased?*) BP. Cardiac output must be kept at an appropriate level,
   however, or else BP will become too low or too high.

c. Cardiac and blood pressure control centers are located in the

   _____ of the brain. Sensory nerves must constantly inform
   the medulla of the need for adjustments. Cranial nerves IX and X carry these messages from receptors sensitive to BP. These baroreceptors are located in the walls of

   the _____ and the _____.

d. When baroreceptors indicate that BP is increased slightly, BP must then be
   (*decreased slightly? increased even more?*) in order to maintain homeostasis. This

   can be accomplished by ____-creasing heart rate.

e. Consider the cardiac center as comparable to the control center of an automobile.
   The center has an accelerator portion (like a gas pedal) and an inhibitor portion

   (like the _____ of a car).

f. When BP is reported to be too high (as if the car is running too fast), the
   cardioacceleratory center must be (*stimulated? inhibited?*); in the car analogy, you
   would (*step on the gas? take your foot off the gas?*). At the same time the
   cardioinhibitory center (brakes) would be (*stimulated? inhibited?*).

g. Dual control is carried out by autonomic nerve fibers. When BP is too high,
   sympathetic (accelerator) nerve messages would (*increase? decrease?*), while

   parasympathetic nerve impulses along the _____ nerve to the
   heart would (*increase? decrease?*).

h. As a result of these nerve messages heart rate would (*increase? decrease?*) and
   thereby (*raise? lower?*) BP in response to each occasion of slightly increased BP.

   This is a statement of _____ law. The mechanism is an example of (*positive? negative?*) feedback.

■ **F6.** The Bainbridge reflex works by a (*positive? negative?*) feedback mechanism.
Describe this reflex.

■ **F7.** Draw arrows to indicate whether each of the following factors increases (↑) or
decreases (↓) heart rate.

a. Epinephrine ____ heart rate.

b. Increased temperature, such as in fever, ____ heart rate.

c. Fear and anger ____ heart rate.

G1. Implantation of the first permanent artificial heart in a human took place in De-

cember, 19____. In this surgery patient Dr. Barney Clark's own atria were (*removed?
left in place?*). Briefly describe the structure of the implanted heart and its power
source.

G2. Discuss *echocardiography.* Tell how the procedure is performed and for what purposes it may be used.

■ **G3.** In each case indicate whether a catheter should be placed into an artery (A),
such as the femoral or brachial artery, or into a vein (V), such as the femoral or
brachial vein.

_____ a. To record pressure in the right ventricle

_____ b. To record pressure in the left ventricle

_____ c. To determine oxygen content in a pulmonary artery

_____ d. To insert dye into coronary arteries in order to diagnose narrowing of
these vessels

G4. Explain how each of the following techniques aids open-heart surgery.
a. Hypothermia

b. Heart–lung machine

G5. List five risk factors related to heart attacks. Explain why each factor increases
the likelihood of an attack.

H. Developmental anatomy (pages 594–595)

■ **H1.** The heart is derived from _____-derm. The heart begins to develop during the (*third? fifth? seventh?*) week. Its initial formation consists of two

endothelial tubes which unite to form the _____ tube.

■ **H2.** Match the regions of the primitive heart at right with the related parts of a mature heart listed below.

A. Atrium	SV. Sinus venosus
BCTA. Bulbus cordis and	V. Ventricle
truncus arteriosus	

_____ a. Superior and inferior vena cava

_____ b. Right and left ventricles

_____ c. Parts of the fetal heart connected by foramen ovale

_____ d. Aorta and pulmonary trunk

I. Disorders, medical terminology (pages 595–601)

■ **I1.** Outline the sequence of changes in atherosclerosis in this learning activity.

a. Atherosclerosis involves damage to the walls of arteries. State several factors that can lead to damage of this lining.

b. Next _____-cytes stick to the damaged lining. These cells soon

become _____ which are capable of ingesting cholesterol.

Name one other type of cell that takes in cholesterol: _____.

c. The resulting fatty lesion in the arterial lining is known as atherosclerotic

_____. The rough surface of plaque may snag platelets. These

may then cause _____ formation at the site, possibly leading to unwanted emboli. Platelets may also release a chemical called

_____.

d. Name two other types of cells that release PDGF: _____ and

_____. What is the effect of this chemical?

■ **I2.** Distinguish classes of chemicals related to atherosclerosis by doing this exercise.

a. HDL and LDL both consist of complexes of _____ combined

with _____. HDL refers to (*high? low?*) density lipoprotein,

whereas LDL means _____.

b. (*HDLs? LDLs?*) are considered culprits in causing atherosclerosis since they may carry cholesterol and deposit this lipid in walls of arteries. So a (*high? low?*) blood level of LDL is desirable.

c. High levels of LDL may occur in blood of persons who have too (*many? few?*) LDL receptors on their cells. Consequently, LDL remains in blood and may travel to blood vessel walls and take up residence there.

d. A high level of HDL tends to (*lead to? prevent?*) atherosclerosis. List two ways that HDLs may be increased in your own blood.

I3. List several factors that may lead to a coronary artery spasm.

■ **I4.** Match each congenital heart disease in the box with the related description below.

> C. Coarctation of aorta PDA. Patent ductus arteriosus
> IASD. Interatrial septal defect VS. Valvular stenosis
> IVSD. Interventricular septal defect

_____ a. Connection between aorta and pulmonary artery is retained after birth, allowing backflow of blood to right ventricle.

_____ b. Foramen ovale fails to close.

_____ c. Septum between ventricles does not develop properly.

_____ d. Valve is narrowed, limiting forward flow and causing increase in workload of heart.

_____ e. Aorta is abnormally narrow.

■ **I5.** Which of the conditions in Checkpoint I4 result in left-to-right shunts? Explain why this is so.

■ **I6.** Do this exercise on arrhythmias.

a. (*All? Not all?*) arrhythmias are serious.

b. Conduction failure across the AV node results in an arrhythmia known as

_____ .

c. Which is more serious? (*Atrial? Ventricular?*) fibrillation. Explain why.

d. Excitation of a part of the heart other than the normal pacemaker (SA node) is

known as a(n) _____ focus. Such excitations, as by caffeine or lack of sleep, or by more serious problems such as ischemia, may lead to an extra

early ventricular systole known as _____ .

I7. Briefly describe each of these cardiac conditions.

a. Congestive heart failure (CHF)

b. Cor pulmonale

c. Cardiomegaly

ANSWERS TO SELECTED CHECKPOINTS

A1. It is a pump, pumping about 1,000 gal/day.

A2. (a) Mediastinal, lungs, two-thirds. (b) Fist. (c) Apex, fifth, 8, 3. (d) Ventricle.

B1. FP SP PC VP M E

B2. Pericarditis; compression of heart due to accumulation of fluid in the pericardial space.

B3.

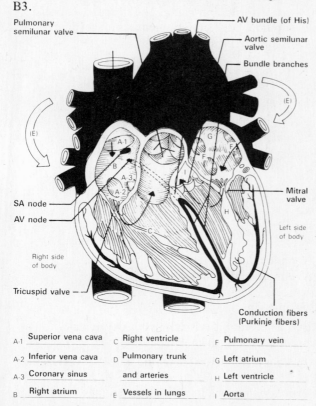

A-1 Superior vena cava   **C** Right ventricle   **F** Pulmonary vein

A-2 Inferior vena cava   **D** Pulmonary trunk   **G** Left atrium

A-3 Coronary sinus   and arteries   **H** Left ventricle

B Right atrium   **E** Vessels in lungs   **I** Aorta

Figure LG 20-1A Diagram of a frontal section of the heart.

(a, b, d) See Figure LG 20-1A. (c) Red: A–D and first part of E; blue: last part of E; F–I.

B4. (a) A. (b) PV. (c) S. (d) CT. (e) V.

B5. (a) B. (b) T. (c) P. (d) B. (e) A, P. (f) B, T.

B6. Do not.

C2. (a) Left coronary artery: 4–6; right coronary artery: 1–3. (b) 1, right coronary artery; 2, posterior interventricular branch; 3, marginal branch; 4, left coronary artery; 5, circumflex branch; 6, anterior interventricular branch.

C3. Vein, cardiac, right atrium.

D2. (a) Autonomic, alter rate of, can. (b) Conduction, impulse transmission. (c) See Figure LG 20-1A above. (d) The fibrous cardiac skeleton (including AV valves) completely separates atrial myocardium from ventricular myocardium.

E1. (a) Contraction, relaxation; opening, closing; pressure changes. (b) Systole, diastole.

E2. (a) 0.8, 75. (b) Atria, 0.1, increases. (c) QRS, systole, 0.3. (d) Increases, dramatically, *B*, closed. (e) Semilunar, opening, *C*. (f) 120. (g) De-, T, diastole. (h) Semilunar, *D*. (i) Open, *E*. (j) Just after *E*, rapidly, diastasis, 70. (k) 30, systole. (l) Filling, contraction, de, ejection, isovolumetric relaxation. (m) Lubb, closing, AV; closing of the semilunar valves; see Figure 20-8d, page 587 of text.

E3. B, AV valves close; C, SL valves open; D, SL valves close; E, AV valves open.

F1. CO = 70 ml/stroke × 72 strokes (beats)/min = 5,000 ml/min = 5 l/min.

F2. (a) Fill with, 120–130, end-diastolic. (b) 70, stroke. (c) 50, end-systolic, ESV. (d) C. (e) 120 − 40 = 80.

F3. (a) More, increases, more. (b) Greater, Starling's, the more expanded balloon/heart exerts greater force on air/blood. (c) De, de, examples: hypertension of the arteries of the body causes backflow into left ventricle, dilating it; mitral valve malfunction backlogs blood into the left atrium, dilating it; pulmonary artery thickening (stenosis) leads to dilation of the right ventricle. (d) More, increases. (e) Lack of exercise, MI, loss of blood (hemorrhage).

F4. 20, 20 = 25 − 5; 400 (= 20/5 × 100%), higher.

F5. (a) Stroke volume and heart rate, in. (b) Increased. (c) Medulla, carotid sinus, aorta. (d) Decreased slightly, de. (e) Brakes. (f) Inhibited, take your foot off the gas, stimulated. (g) Decrease, vagus (X), increase. (h) Decrease, lower, Marey's, negative.

F6. Positive: increase in venous return to right atrium (as in exercise) increases heart rate, allowing heart to more quickly pump the blood, for example, to skeletal muscles.

F7. (a) ↑. (b) ↑. (c) ↑.

G3. (a) V (since all veins except pulmonary lead back to the right side of the heart). (b) A (catheter moves retrograde, that is, backward, through the aorta to the left side of the heart). (c) V (since pulmonary artery can be reached via right side of heart). (d) A (since coronary arteries branch off of the aorta).

H1. Meso, third, primitive heart.

H2. (a) SV. (b) V. (c) A. (d) BCTA.

I1. (a) Carbon monoxide from smoking, diabetes mellitus, diet high in cholesterol and certain triglycerides. (b) Mono; macrophages; smooth muscle cells. (c) Plaque; clot; platelet-derived growth factor (PDGF). (d) Macrophages and endothelial cells; promote growth of smooth muscle cells, thus narrowing arterial lumen.

I2. (a) Lipids (like cholesterol and triglycerides), proteins, high, low density lipoproteins. (b) LDLs, low. (c) Few. (d) Prevent; exercise, diets low in animal fats and not smoking.

I4. (a) PDA. (b) IASD. (c) IVSD. (d) VS. (e) C.

I5. IASD, IVSD, and PDA. In all three cases, blood is forced from high pressure area in left side of heart or aorta back to lower pressured right side of heart.

I6. (a) Not all. (b) Heart block. (c) Ventricular; atrial fibrillation reduces effectiveness of the heart by only about 30 percent, but ventricular fibrillation causes the heart to fail as a pump. (d) Ectopic, premature ventricular contractions (PVCs).

MASTERY TEST: Chapter 20

Questions 1–3: Arrange the answers in correct sequence.

_____ _____ _____ _____ 1. Pathway of the conduction system of the heart:

A. AV node
B. AV bundle and bundle branches
C. SA node
D. Purkinje fibers

_____ _____ _____ _____ _____ 2. Route of a red blood cell supplying oxygen to myocardium of left atrium:

A. Arteriole, capillary, and venule within myocardium
B. Branch of great cardiac vein
C. Coronary sinus leading to right atrium
D. Left coronary artery
E. Circumflex artery

_____ _____ _____ _____ _____ 3. Route of a red blood cell now in the right atrium:

A. Left atrium
B. Left ventricle
C. Right ventricle
D. Pulmonary artery
E. Pulmonary vein

4. All of the following are correctly matched EXCEPT:

    A. Myocardium—heart muscle
    B. Visceral pericardium—epicardium
    C. Endocardium—forms heart valves which are especially affected by rheumatic fever
    D. Pericardial cavity—space between fibrous pericardium and parietal layer of serous pericardium

5. Which of the following factors will tend to decrease heart rate?

    A. Stimulation by cardiac nerves
    B. Release of the transmitter substance norepinephrine in the heart
    C. Activation of neurons in the cardioaccelerator center
    D. Increase of vagal nerve impulses
    E. Increase of sympathetic nerve impulses

6. The average cardiac output for a resting adult is about ____/min.

    A. 1 qt   D. 0.5 liter
    B. 5 pt   E. 2000 ml
    C. 1.25 gal (5 liters)

7. When ventricular pressure exceeds atrial pressure, what event occurs?

    A. AV valves open
    B. AV valves close
    C. Semilunar valves open
    D. Semilunar valves close

8. The second heart sound is due to turbulence of blood flow as a result of what event?

    A. AV valves opening
    B. AV valves closing
    C. Semilunar valves opening
    D. Semilunar valves closing

9. Choose the FALSE statement about the circumflex artery.

    A. It is a branch of the left coronary artery.
    B. It provides the major blood supply to the right ventricle.
    C. It lies in a groove between the left atrium and left ventricle.
    D. Damage to this vessel would leave the left atrium with virtually no blood supply.

10. Choose the FALSE statement about heart structure.

    A. The heart chamber with the thickest wall is the left ventricle.
    B. The apex of the heart is more superior in location than the base.
    C. The heart has four chambers
    D. The left ventricle forms the apex and most of the left border of the heart.

11. All of the following are defects involved in Tetralogy of Fallot EXCEPT:

    A. Ventricular septal defect
    B. Enlarged right ventricle
    C. Stenosed mitral valve
    D. Aorta emerging from both ventricles

12. Choose the TRUE statement.

    A. The T wave is associated with atrial depolarization.
    B. The normal P-R interval is about 0.4 sec.
    C. Myocardial infarction means strengthening of heart muscle.
    D. Myocardial infarction is commonly known as a "heart attack" or a "coronary."

13. Choose the FALSE statement.

    A. Pressure within the atria is known as intraarterial pressure.
    B. During most of ventricular diastole the semilunar valves are closed.
    C. During most of ventricular systole the semilunar valves are open.
    D. Diastole is another name for relaxation of heart muscle.

14. Which of the following structures are located in ventricles?

    A. Papillary muscles
    B. Fossa ovalis
    C. Ligamentum arteriosum
    D. Pectinate muscles

15. The heart is composed mostly of:

    A. Epithelium
    B. Muscle
    C. Dense connective tissue

Questions 16–20: Circle T (true) or F (false). If the statement is false, change the underlined word or phrase so that the statement is correct.

T F 16. The blood in the left chambers of the heart contains <u>higher</u> oxygen content than blood in the right chambers.

T F 17. At the point when intraarterial pressure surpasses ventricular pressure, semilunar valves <u>open</u>.

T F 18. The pulmonary <u>artery carries</u> blood from the lungs to the left atrium.

T F 19. The normal cardiac cycle <u>does not</u> require direct stimulation by the autonomic nervous system.

T F 20. During <u>about half</u> of the cardiac cycle, atria and ventricles are contracting simultaneously.

Questions 21–25: fill-ins. Complete each sentence with the word or phrase that best fits.

_____ 21. Most ventricular filling occurs during atrial ____.

_____ 22. ____ is a cardiovascular disorder that is the leading cause of death in the U.S.

_____ 23. An ECG is a recording of ____ of the heart.

_____ 24. ____ is a term that means abnormality or irregularity of heart rhythm.

_____ 25. According to Marey's law of the heart, if blood pressure against baroreceptors decreases, heart rate ____-creases.

ANSWERS TO MASTERY TEST: ■ Chapter 20

Arrange

1. C A B D
2. D E A B C
3. C D E A B

Multiple Choice

4. D	10. B
5. D	11. C
6. C	12. D
7. B	13. A
8. D	14. A
9. B	15. B

True-False

16. T
17. F. Close
18. F. Veins carry
19. T
20. F. No part

Fill-ins

21. Diastole
22. Coronary artery disease
23. Electrical changes or currents that precede myocardial contractions
24. Arrhythmia or dysrhythmia
25. In

Circulation: Vessels & Routes

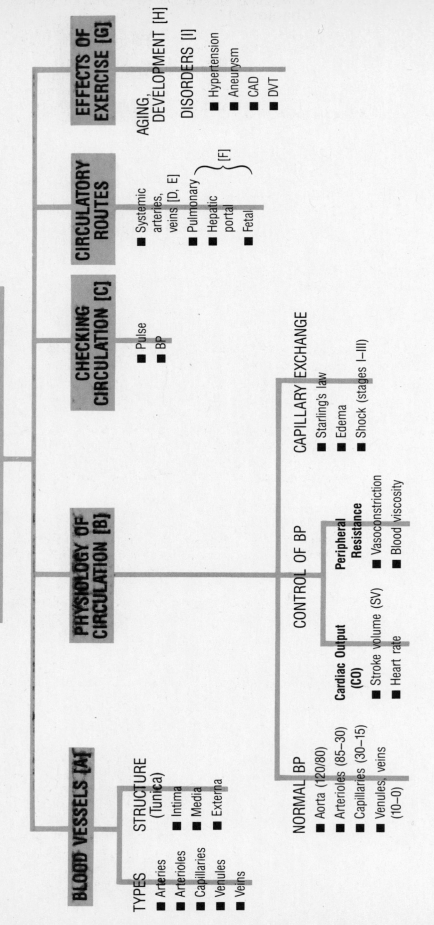

BLOOD VESSELS [A]

TYPES
- Arteries
- Arterioles
- Capillaries
- Venules
- Veins

STRUCTURE (Tunica)
- Intima
- Media
- Externa

PHYSIOLOGY OF CIRCULATION [B]

NORMAL BP
- Aorta (120/80)
- Arterioles (85–30)
- Capillaries (30–15)
- Venules, veins (10–0)

CONTROL OF BP

Cardiac Output (CO)
- Stroke volume (SV)
- Heart rate

Peripheral Resistance
- Vasoconstriction
- Blood viscosity

CAPILLARY EXCHANGE
- Starling's law
- Edema
- Shock (stages I–III)

CHECKING CIRCULATION [C]
- Pulse
- BP

CIRCULATORY ROUTES
- Systemic arteries, veins [D, E]
- Pulmonary
- Hepatic portal } [F]
- Fetal

EFFECTS OF EXERCISE [G]

AGING, DEVELOPMENT [H]

DISORDERS [I]
- Hypertension
- Aneurysm
- CAD
- DVT

The Cardiovascular System: Vessels and Routes

Blood is pumped by heart action to all regions of the body. The network of vessels is organized much like highways, major roads, side streets, and tiny alleys to provide interconnecting routes through all sections of the body. In city traffic, tension mounts when traffic is heavy or side streets are blocked. Similarly, pressure builds (arterial blood pressure) when volume of blood increases or when constriction of small vessels (arterioles) occurs. In a city, traffic may be backed up for hours, with vehicles unable to deliver passengers to destinations. A weakened heart may lead to similar and disastrous effects, backing up fluid into tissues (edema) and preventing transport of oxygen to tissues (ischemia, shock, and death of tissues).

Careful map study leads to trouble-free travel along the major arteries of a city. A similar examination of the major arteries and veins of the human body facilitates an understanding of the normal path of blood as well as clinical applications, such as the exact placement of a blood pressure cuff over the brachial artery or the expected route of a deep venous thrombus (pulmonary embolism) to the lungs.

Begin your route through circulatory pathways at the Chapter 21 Framework and familiarize yourself with the names of important routes.

TOPIC OUTLINE AND OBJECTIVES

A. Blood vessels

1. Contrast the structure and function of the various types of blood vessels.

B. Physiology of circulation; shock

2. Explain how blood pressure is developed and controlled.
3. Discuss the various pressures involved in the movement of fluids between capillaries and interstitial spaces.
4. Explain how the return of venous blood to the heart is accomplished.

C. Checking circulation

5. Define pulse and blood pressure (BP) and contrast the clinical significance of systolic, diastolic, and pulse pressure.

D. Circulatory routes: systemic arteries

E. Circulatory routes: systemic veins

F. Circulatory routes: hepatic portal, pulmonary and fetal circulation

6. Identify the principal arteries and veins of systemic, pulmonary, hepatic, portal, and fetal circulation.

G. Exercise and the cardiovascular system

H. Aging and developmental anatomy

7. Describe the effects of exercise and aging on the cardiovascular system.
8. Describe the development of blood vessels and blood.

I. Disorders and medical terminology

9. List the causes and symptoms of hypertension, aneurysms, coronary artery disease (CAD), and deep-venous thrombosis (DTV).
10. Define medical terminology associated with the cardiovascular system.

WORDBYTES

Now study the following parts of words that may help you better understand terminology in this chapter

Wordbyte	Meaning	Example	Wordbyte	Meaning	Example
hepat-	liver	*hepat*ic artery	sphygmo-	pulse	*sphygmo*manometer
med-	middle	tunica *med*ia	tunica	sheathe	*tunica* intima
phlebo-	vein	*phleb*itis	vaso-	vessel	*vaso*constriction
port-	carry	*port*al vein	ven-	vein	*ven*ipuncture
retro-	backward, behind	*retro*peritoneal			

Checkpoints

A. Blood vessels   (pages 606–610)

■ **A1.** Refer to Figure LG 21-1 and complete this Checkpoint on blood vessels.

a. Label vessels 1–7. Note that numbers are arranged in sequence of the pathway of blood flow. Use these labels:

aorta	venule
arteriole	small artery
capillary	small vein
inferior vena cava	

b. Then color vessels according to color code ovals.

■ **A2.** Use the names of vessels in A1 to answer the following descriptions.

_____ a. Site of gas, nutrient, and waste exchange with tissues.

_____ b. Plays primary role in regulating momentary distribution of blood and in regulating blood pressure.

_____ c. Sinusoids of liver are wider, more leaky versions of this type of vessel.

_____ d. Reservoirs for about 60 percent of the volume of blood in the body; vasoconstriction (due to sympathetic impulses) permits redistribution of blood stored here (3 answers).

■ **A3.** Match the terms to related descriptions:

A. Anastomosis	TI. Tunica intima
L. Lumen	TM. Tunica media
TE. Tunica externa	VV. Vasa vasorum

_____ a. The opening in blood vessels through which blood flows; narrows in atherosclerosis

_____ b. Formed of endothelium, it lines lumen; the only layer in capillaries.

_____ c. Junction of two or more vessels supplying one region of the body such as the cerebral arterial circle (of Willis) to the brain

_____ d. Composed of muscle and elastic tissue; contraction permits vasoconstriction

_____ e. Strong tunica adventitia

_____ f. Tiny blood vessels that carry nutrients to walls of blood vessels

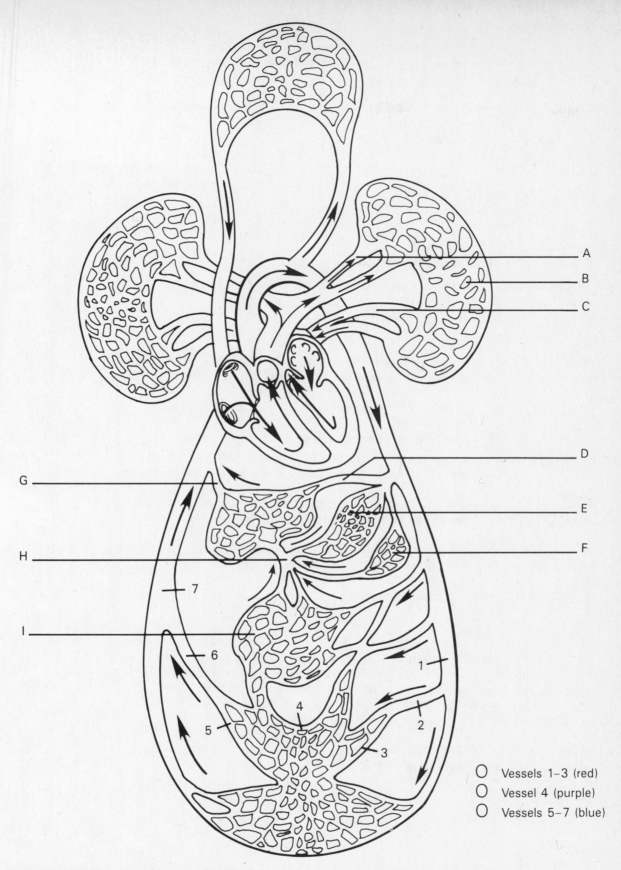

○ Vessels 1–3 (red)
○ Vessel 4 (purple)
○ Vessels 5–7 (blue)

Figure LG 21-1 Circulatory routes Numbers 1–7 follow the pathway of blood through the systemic blood vessels. Color and label numbered vessels according to directions in Checkpoint A1. Letters *A–I* refer to Checkpoints D1 and F1.

■ **A4.** Complete this exercise about structure of veins.

a. Veins have (*thicker? thinner?*) walls than arteries. This structural feature relates to the fact that the pressure in veins is (*more? less?*) than in arteries. The pressure difference is demonstrated when a vein is cut; blood leaves a cut vein in (*rapid spurts? an even flow?*)

b. Gravity exerts back pressure on blood in veins located inferior to the heart. To

counteract this, veins contain _____.

c. When valves weaken, veins become enlarged and twisted. This condition is called

_____. This occurs more often in (*superficial? deep?*) veins. Why?

B. Physiology of circulation; shock (pages 610–620)

■ **B1.** State the one main factor that explains why blood flows from one vessel to the next.

Two aditional factors that aid blood flow are the (*large? small?*) diameter of veins, which offers (*more? less?*) resistance to capillary blood, and also the contraction of

_____, which squeeze venous blood back toward the heart.

■ **B2.** Refer to Figure 21-4 (page 611) of the text, and complete this exercise.

a. In the aorta and brachial artery blood pressure (BP) is about 120 mm Hg immediately following ventricular contraction. This is called (*systolic? diastolic?*) BP. As

ventricles relax (or go into _____), blood is no longer ejected into these arteries. However, the normally (*elastic? rigid?*) walls of these vessels

recoil against blood, pressing it onward with a diastolic BP of _____ mm Hg.

b. Write the normal ranges of BP values for the following types of vessels:

arterioles, _____ mm Hg; capillaries, _____ mm Hg; venules and veins,

_____ mm Hg; venae cavae, _____ mm Hg.

c. The pressure in the right atrium is normally about _____ mm Hg. If the right side of the heart is weakened and cannot adequately pump blood into the lungs, or if pulmonary vessels are sclerosed, offering resistance to blood flow, blood will tend to backlog in the right atrium causing a (an) (*increase? decrease?*) in right atrial pressure.

■ **B3.** Complete Figure LG 21-2 which summarizes factors determining arterial blood pressure (BP). Indicate directions of arrows and add missing words.

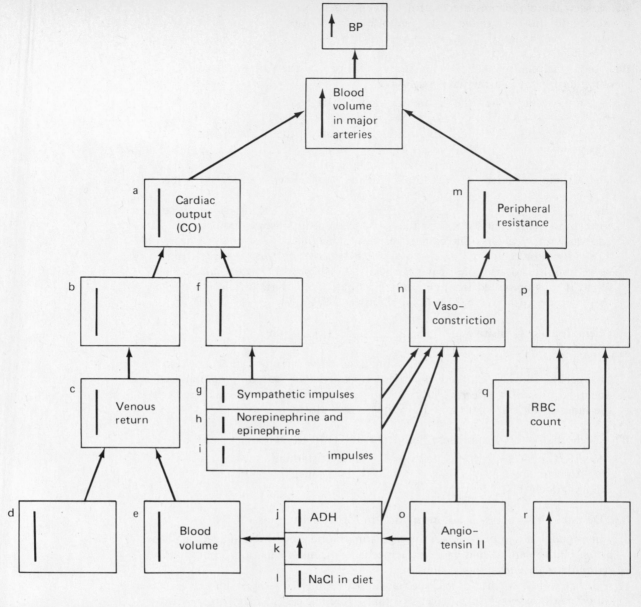

Figure LG 21-2 Summary of factors controlling blood pressure (BP). Arrows indicate increase (↑) or decrease (↓). Complete as directed in Checkpoint B3.

■ **B4.** If directions of arrows in Figure 21-2 are reversed, blood pressure can be lowered. Check your understanding of factors that affect arterial blood pressure by circling those that will tend to *decrease* it.

A. Increase in cardiac output, as by increased heart rate or stroke volume
B. Increase in (vagal) impulses from cardioinhibitory center
C. Decrease in blood volume, as following hemorrhage
D. Increase in blood volume, by excess salt intake and water retention
E. Increased peripheral resistance due to vasoconstriction of arterioles (blood prevented from leaving large arteries and entering small vessels)
F. Decrease in sympathetic impulses to the smooth muscle of arterioles
G. Decreased viscosity of blood via loss of blood protein or red blood cells, as by hemorrhage
H. Use of medications called dilators since they dilate arterioles, espcially in areas such as abdomen and skin

B5. Review the blood pressure control mechanisms involving afferent pathways from pressoreceptors to the *cardiac* center and efferent responses by the heart (Chapter 20, Checkpoint F5, page LG 363).

■ **B6.** Additional control of blood pressure is provided by the *vasomotor* center. Describe its function in this exercise.

a. Like the cardiac center, the vasomotor center is located in the

_____. Its function is to control the diameter of

_____ and thereby to alter BP.

b. The vasomotor center "is told" to alter BP by pressure sensitive receptors located in

the _____ and _____. When these pressoreceptors sense that BP is low, they send afferent impulses to both the cardiac and vasomotor centers. The cardiac response is to (*increase? decrease?*) heart rate.

c. The vasomotor center responds to low BP by (*increasing? decreasing?*) sympathetic nerve impulses to arterioles, causing them to (*vasoconstrict? vasodilate?*). Since blood is thus prevented from entering these narrowed arterioles, blood stays in major arteries, (*increasing? decreasing?*) arterial BP.

d. Chemoreceptors sensitive to _____ and

_____ can also stimulate the vasomotor center. Low oxygen or high carbon dioxide will tend to cause (*vasocontriction? vasodilation?*) and consequently an (*increase? decrease?*) in BP. As a result, blood can move more rapidly to the lungs to pick up oxygen and give up carbon dioxide.

■ **B7.** Circle the factor(s) that cause blood pressure (BP) to increase.

A. ADH   D. Histamine
B. Aldosterone   E. Epinephrine
C. Angiotensin II

■ **B8.** Describe the process of autoregulation in this activity.

a. Autoregulation is a mechanism for regulating blood flow through specific tissue areas. Unlike cardiac and vasomotor mechanisms, autoregulation occurs by (*local? autonomic nervous system?*) control.

b. In other words, when a tissue (such as an active muscle) is hypoxic, that is, has a (*high? low?*) oxygen level, the cells of that tissue release (*vasocontrictor? vaso-*

dilator?) substance. These may include _____ acid, built up by active muscle. Other products of metabolism that serve as dilator substances include

the gas _____ and the _____ ion.

c. The effect of these vasodilators is to cause precapillary sphincters to (*contract? relax?*), resulting in (*increased? decreased?*) blood flow into the active tissue.

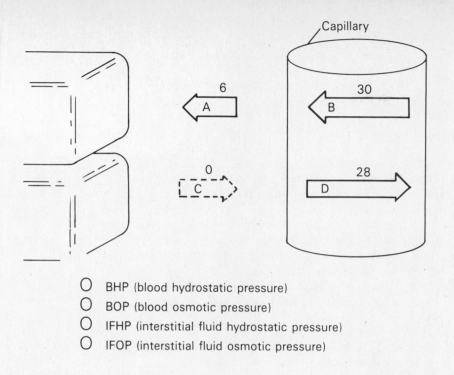

○ BHP (blood hydrostatic pressure)
○ BOP (blood osmotic pressure)
○ IFHP (interstitial fluid hydrostatic pressure)
○ IFOP (interstitial fluid osmotic pressure)

Figure LG 21-3 Practice diagram showing forces controlling capillary exchange and for calculation of *P*eff at arterial end of capillary. Numbers are pressure values in mm Hg. Color arrows as directed in Checkpoint B9.

■ **B9.** Recall that pressure is the principal factor causing blood to move through the vessels of the body. Pressure is also the major factor determining movement of substances across capillary membranes. Refer to Figure LG 21-3 and do this activity.

a. Color arrows A–D on Figure LG 21-3 to identify the four pressures involved in capillary exchange.

b. These pressures interact in a kind of tug-of-war that results in an effective filtration

pressure (*P*eff) of _____ mm Hg. In other words, at the arterial end of the capillary, substances tend to move (*into? out of?*) the vessel.

c. Which value is different at the venous end of capillaries? _____ Calculate the

*P*eff there: _____ mm Hg.

d. The near-equilibrium at the two ends of capillaries is based on _____'s law of the capillaries. Which system "mops up" the slight amount of fluid that is

not drawn back into veins? _____
e. Excessive amounts of fluid accumulating in interstitial spaces is the condition

known as _____. This may result from high blood pressure in which (*BOP? BHP?*) is increased, loss of plasma protein in which (*BOP? BHP?*) is decreased, or (*increase? decrease?*) in capillary permeability.

■ **B10.** Do this exercise about factors that influence blood flow back to the heart and therefore affect blood pressure.

a. Any factor that increases venous return to the heart will increase cardiac output

(due to _____ law of the heart). This will (*increase? decrease?*) blood pressure. (*For extra review,* return to Chapter 20, Checkpoint F3, page LG 362.)

b. Figure 21-8, page 616 of your text shows that velocity of blood flow is greatest in

(*arteries? capillaries? veins?*). Flow is slowest in _____. As a result, ample time is available for exchange between capillary blood and tissues.

c. As blood moves from capillaries into venules and veins, its velocity (*increases? decreases?*), enhancing venous return.

d. Although each individual capillary has a cross-sectional area which is microscopic, the body contains so many capillaries that their total cross-sectional area is enormous. (For comparision, look at the eraser end on one pencil; you are seeing the cross-sectional area of that one pencil. If you bind together 100 pencils with a strong rubber band and observe their eraser ends as a group, you see that the individual pencils contribute to a relatively large total cross-sectional area.) All of

the capillaries in the body have a total cross-sectional area of about _____ cm^2.

e. (*A direct? An indirect?*) relationship exists between cross-sectional area and velocity of blood flow. This is evident in the aorta also. This vessel has a cross-section of

about _____ cm^2 and has a very (*high? low?*) velocity.

f. The increased respiratory rate during active exercise also enhances venous return to the heart. During inspiration the size of the thorax enlarges causing (*increased? decreased?*) intrathoracic pressure, while intraabdominal pressure (*increases? decreases?*). A gradient is therefore established which pushes blood upward to the heart.

B11. How do skeletal muscle contractions and venous valves increase venous return?

■ **B12.** *Shock* is said to occur when tissues receive inadequate _____ supply to meet their demands for oxygen, nutrients, and waste removal. Shock related

to inadequate heart function is known as _____ shock, whereas

_____ shock results from decrease in circulating blood volume.

Widespread vasodilation leads to _____ shock, and

_____ shock is related to infection that causes vessels to dilate and increase permeability.

■ **B13.** *A clinical challenge.* Write arrows to indicate signs and symptoms of shock. (↑ = increase; ↓ = decrease)

a. Sweat ____   d. Blood pH ____

b. Heart rate and pulse ____   e. Mental status ____

c. BP ____   f. Urine volume ____

■ **B14.** Complete the following exercise about mechanisms during compensatory circulatory shock.

a. Blood vessels in nonessential areas are constricted so that blood is shunted into major vessels and BP is maintained. This occurs by means of powerful vasoconstrictors such as _____, released from the adrenal medulla, and

angiotensin II, formed when _____ is secreted from the

kidneys. Angiotensin II also stimulates release of the hormone _____.

b. Both aldosterone and _____ are hormones that cause water (*retention? elimination?*). Increase in fluid level of the body will (*increase? decrease?*) blood volume, and so will increase BP.

C. Checking circulation (pages 620–621)

C1. Define *pulse.*

C2. In general, where may pulse best be felt? List six places where pulse can readily be palpated. Try to locate pulses at these points on yourself or on a friend.

■ **C3.** Normal pulse rate is about _____ beats/min. *Tachycardia* means (*rapid? slow?*)

pulse rate; _____ means slow pulse rate.

■ **C4.** Describe the method commonly used for taking blood pressure by answering these questions about the procedure.

a. Name the instrument used to check BP. _____

b. The cuff is usually placed over the _____ artery. How can you tell that this artery is compressed after the bulb of the sphygmomanometer is squeezed?

c. As the cuff is deflated, the first sound heard indicates _____ pressure, since reduction of pressure from the cuff below systolic pressure permits blood to begin spurting through the artery.

d. Diastolic pressure is indicated by the point at which the sounds

_____. At this time blood can once again flow freely through the artery.

■ **C5.** *A clinical challenge.* Mrs. Kinney has a systolic pressure of 128 mm Hg and diastolic of 83 mm Hg. Write her blood pressure in the form in which BP is usually

expressed: _____. Determine her pulse pressure.

Would this be considered a normal pulse pressure? _____

D. Circulatory routes: systemic arteries (pages 621-632)

D1. On Figure LG 21-1, label the *pulmonary artery, pulmonary capillaries,* and a *pulmonary vein.* All blood vessels in the body other than pulmonary vessels are

known as _____ vessels since they supply a variety of body systems.

■ **D2.** Use Figure LG 21-4 to do this learning activity about systemic arteries.

a. The major artery from which all systemic arteries branch is the

_____. It exits from the chamber of the heart known as the (*right? left?*) ventricle. The aorta can be divided into three portions. They are the

_____ (*N*1 on the figure), the _____

(*N*2), and the _____ (*N*3).

b. The first arteries to branch off of the aorta are the _____ arteries. Label these (at *O*) on the figure. These vesssels supply blood to the

_____.

c. Three major arteries branch from the aortic arch. Write the names of these (at *K, L,* and *M*) on the figure, beginning with the vessel closest to the heart.

d. Locate the two branches of the brachiocephalic artery. The right subclavian artery

(letter _____ supplies blood to _____

_____.

Vessel *E* is named the (*right? left?*) _____ artery. It supplies the right side of the head and neck.

e. As the subclavian artery continues into the axilla and arm, it bears different names (much as a road does when it passes into new towns.) Label *G* and *H*. Note that *H* is the vessel you studied as a common site for measurement of

_____.

f. Label vessels *I* and *J*, and then continue the drawing of these vessels into the forearm and hand. Notice that they anastomose in the hand. Vessel (*I? J?*) is often used for checking pulse in the wrist.

g. Besides supplying the arms, subclavian arteries each send a branch that ascends the

neck through foramina in cervical vertebrae. This is the _____ artery. The right and left vertebral arteries join at the base of the brain to form the

_____ artery. (Refer to *D* and *C* on Figure LG 21-5.)

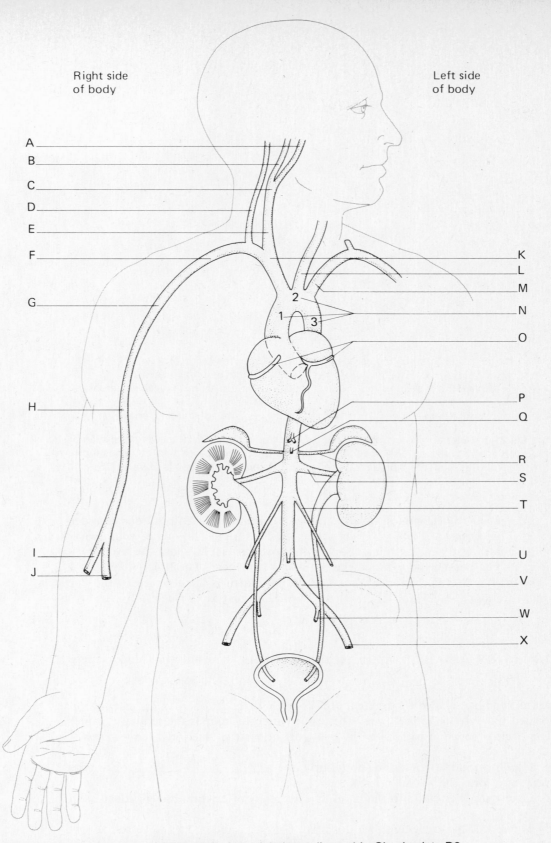

Figure LG 21-4 Major systemic arteries. Identify letter labels as directed in Checkpoints D2, D3, and D6–D8.

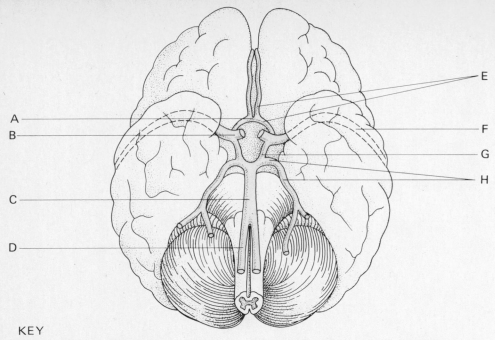

KEY

A. Anterior communicating cerebral artery (1)
B. Internal carotid arteries (cut) (2)
C. Basilar artery (1)
D. Vertebral arteries (2)

E. Anterior cerebral arteries (2)
F. Middle cerebral arteries (2)
G. Posterior communicating cerebral arteries (2)
H. Posterior cerebral arteries (2)

Figure LG 21-5 Arterial blood supply to the brain including vessels of the cerebral arterial circle (of Willis). View is from the undersurface of the brain. Numbers in parentheses indicate whether the arteries are single (1) or paired (2). Refer to Checkpoints D2 and D4.

■ **D3.** The right and left common carotid arteries ascend the neck in positions considerably (*anterior? posterior?*) to the vertebral arteries. Find a pulse in one of your own carotids. These vessels each bifurcate (divide into two vessels) at about the level of your mandible. Identify the two branches (*A* and *B*) on Figure LG 21-4. Which supplies the brain and eye? (*Internal? External?*) What structures are supplied by the external carotids? (See Exhibit 21-3, pages 625–626, in the text, for help.)

■ **D4.** Refer to Figure LG 21-5 and answer these questions about the vital blood supply to the brain.

a. An anastomosis, called the cerebral arterial circle of _____ is located just inferior to the brain. This circle closely surrounds the pituitary gland.

b. The circle is supplied with blood by two pairs of vessels: anteriorly, the (*internal?*

external?*) carotid arteries, and, posteriorly, the _____ arteries (joined to form the basilar artery).

c. The circle supplies blood to the brain by three pairs of cerebral vessels, identified

on Figure LG 21-5 by letter labels _____, _____, and _____. Vessels *A* and *G* complete the circle; they are known as anterior and posterior

_____ arteries. Now cover the key and correctly label each of the lettered vessels.

■ **D5.** *A clinical challenge.* Relate symptoms to the interruption of normal blood supply to the brain as described in each case. Explain your reasons.

a. A patient's basilar artery is sclerosed. Which part of the brain is most likely to be deprived of normal blood supply? (*Frontal lobes of cerebrum? Occipital lobes of cerebrum, as well as cerebellum and pons?*)

b. Paralysis and loss of some sensation are noted on a patient's *right* side. Angiography points to sclerosis in a common carotid artery. Which one is more likely to be sclerosed? (*Right? Left?*) Explain.

c. The middle cerebral artery which passes between temporal and frontal lobe (see broken lines on Figure LG 21-5) is blocked. Which sense is more likely to be affected (*Hearing? Vision?*) Explain.

d. Anterior cerebral arteries pass anteriorly from the circle of Willis and then arch backward along the medial aspect of each cerebral hemisphere. Disruption of blood flow in the right anterior cerebral artery will be most likely to cause effects in the (*right? left?*) (*hand? foot?*). (*Hint:* The affected brain area would include region *Z* in Figure LG 15-1 and/or region *R* in Figure LG 15-2, pages LG 265 and 267. Note that these regions are on the medial aspect of the right hemisphere.)

■ **D6.** Complete the table describing visceral branches of the abdominal aorta. Identify these on Figure LG 21-4 (vessels *P* to *U*). Refer also to Figure LG 24-4, page 449.

Artery	Letter	Structures Supplied
a. Celiac (three major branches) 1. 2. 3.		1. Liver 2. 3.
b.	Q	Small intestine and part of large intestine and pancreas
c.		Adrenal (suprarenal) glands
d. Renal	S	
e.		Testes (or ovaries)
f. Inferior mesenteric		

■ **D7.** The aorta ends at about level _____ vertebra by dividing into right and left

_____ arteries. Label these at *V* on Figure LG 21-4.

D8. Label the two branches (*W* and *X*) of each common iliac artery. Name structures supplied by each.

a. Internal iliac artery

b. External iliac artery

D9. Trace one possible route of a red blood cell from the common iliac artery to the little toe of the right foot. (Refer to Figure 21-16, page 632 of the text for help.) The pathway is started for you: Right common iliac artery → right external iliac artery →

■ **D10.** *For extra review.* Suppose that the right femoral artery were completely occluded. Describe an alternate (or collateral) route that would enable blood to pass from the right common iliac artery to the little toe of the right foot.

Right common iliac artery →

E. Circulatory routes: systemic veins (pages 633–642)

■ **E1.** Name the three main vessels that empty venous blood into the right atrium of the

heart. _____ _____ _____

■ **E2.** Do this exercise about venous return from the head.

a. Venous blood from the brain drains into vessels known as

_____. These are (*structurally? functionally?*) like veins.

b. Name the sinus that lies directly beneath the sagittal suture:

_____.

c. Blood from all of the cranial vascular sinuses eventually drains into the

_____ veins which descend in the neck. These veins are

positioned close to the _____ arteries.

d. Where are the jugular arteries and carotid veins located?

■ **E3.** Imagine yourself as a red blood cell currently located in the *left brachial vein.* Answer these questions.

a. In order for you to flow along in the bloodstream to the *left brachial artery,* which of these must you pass through? (*A brachial capillary? The heart and a lung?*)

b. In order for you to flow along to the *right arm,* must you pass through the heart? (*Yes? No?*) Both sides of the heart, that is, right and left? (*Yes? No?*) One lung? (*Yes? No?*)

■ **E4.** Once you are familiar with the pathways of systemic arteries, you know much about the veins that accompany these arteries. Demonstrate this by tracing routes of a drop of blood along the following deep vein pathways.

a. From the thumb side of the left forearm to the right atrium

b. From the medial aspect of the left knee to the right atrium

c. From the right kidney to the right atrium

■ **E5.** Contrast veins with arteries in this exercise.

a. Veins have (*higher? lower?*) blood pressure than arteries and a (*faster? slower?*) rate of blood flow.

b. Veins are therefore (*more? less?*) numerous, in order to compensate for slower blood flow in each vein. The "extra" veins consist of vessels located just under the skin (in subcutaneous fascia); these are known as _____ veins.

■ **E6.** On text Figures 21-19 to 21-21 (pages 638–642), locate the following veins. Identify the location at which each empties into a deep vein.

a. Basilic _____

c. Great saphenous _____

b. Median cubital _____

d. Small saphenous _____

E7. Veins connecting the superior and inferior venae cavae are named

_____ veins. Identify the azygos system of veins on Figure 21-20 (page 640) in the text. What parts of the thorax do they drain? Explain how these veins may help to drain blood from the lower part of the body if the inferior vena cava is obstructed.

■ **E8.** Name the visceral veins that empty into the abdominal portion of the inferior vena cava. Then color them blue on Figure LG 26-1 (page LG 489). Contrast accompanying systemic arteries by coloring them red on that figure. Note on Figure LG 26-1 that the inferior vena cava normally lies on the (*right? left?*) side of the aorta. One way to remember this is that the inferior vena cava is returning blood to the (*right? left?*) side of the heart.

F. Circulatory routes: pulmonary, hepatic portal, and fetal circulation
(pages 643-647)

■ **F1.** Refer to Figure LG 21-1 and do the following activity.

a. Label the following vessels: *gastric, splenic,* and *mesenteric* vessels. Note that all of these vessels drain directly into the vessel named (*7, inferior vena cava? H, portal vein to the liver?*). Label that vessel on Figure LG 21-1.

b. Label the vessel that carries oxygenated blood into the liver.

c. In the liver, blood from both the portal vein and hepatic artery mixes as it passes

through tortuous capillary-like vessels known as _____. On Figure LG 21-1 label a vessel that collects this blood and transports it to the inferior vena cava.

d. What special functions are served by the special hepatic portal circulation?

F2. What aspects of fetal life require the fetus to possess special cardiovascular structures not needed after birth?

■ **F3.** Complete the following table about fetal structures. (Note that the structures are arranged in order of blood flow from fetal aorta back to fetal aorta.)

Fetal Structure	Structures Connected	Function	Fate of Structure After Birth
a.		Carries fetal blood low in oxygen and nutrients and high in wastes.	
b. Placenta	(Omit)		
c.	Placenta to liver and ductus venosus		
d.		Branch of umbilical vein, bypasses liver	
e. Foramen ovale			
f.			Becomes ligamentum arteriosum

■ **F4.** For each of the following pairs of fetal vessels, circle the vessel with higher oxygen content.

    a. Umbilical artery/umbilical vein

    b. Femoral artery/femoral vein

    c. Pulmonary artery/pulmonary vein

    d. Aorta/thoracic portion of inferior vena cava

G. Exercise and the cardiovascular system   (page 647)

G1. List several examples of aerobic exercises.

■ **G2.** Write arrows to indicate effects of regular exercise (↑, increase; ↓, decrease).

    a. Cardiac output during exercise ____

    b. Endorphins ____

    c. HDLs ____

    d. Systolic BP in hypertensive persons ____

    e. Oxygen requirement of cardiac muscle ____

    f. Heart rate at rest ____

H. Aging and developmental anatomy   (pages 647–648)

■ **H1.** Draw arrows next to each factor listed below to indicate whether it increases (↑) or decreases (↓) with normal aging.

_____ a. Size and strength of cardiac muscle cells

_____ b. Cardiac output (CO)

_____ c. Deposits of cholesterol in blood vessels supplying brain

_____ d. Systolic blood pressure

H2. Write a paragraph describing embryonic development of blood vessels. Include the following key terms: *15–16 days, mesoderm, mesenchyme, blood islands, spaces, endothelial, tunica.*

H3. List four sites of blood formation in the embryo and fetus.

I. Disorders, medical terminology (pages 648–650)

I1. Contrast *primary hypertension* and *secondary hypertension.*

■ **I2.** Match each organ listed in the box with the related disorder that may cause secondary hypertension.

> AM. Adrenal medulla   K. Kidney
> AC. Adrenal cortex

_____ a. Excessive production of renin which catalyzes formation of angiotensin II, a powerful vasoconstrictor

_____ b. Pheochromocytoma, a tumor that releases large amounts of norepinephrine and epinephrine

_____ c. Release of excessive amounts of aldosterone which promotes salt and water retention

I3. Describe effects of hypertension upon the following body structures:

a. Heart

b. Cerebral blood vessels

c. Kidneys

I4. Explain how each of the following therapeutic measures can help control hypertension.

a. Sodium restriction

b. Cessation of smoking

c. Exercise

d. Diuretics

e. Vasodilators

I5. Define *aneurysm* and list four possible cause of aneurysms.

■ **I6.** *A clinical challenge.* Explain how a deep vein thrombosis (DVT) of the left femoral vein can lead to a pulmonary embolism.

A1. (a) 1, aorta; 2, small artery; 3, arteriole; 4, capillary; 5, venule; 6, small vein; 7, inferior vena cava. (b) See the key on Figure 21-1 and also Figure 21-11, (page 622) of the text.

A2. (a) Capillary. (b) Arteriole. (c) Capillary. (d) Venule, small vein, inferior vena cava (actually all veins and venules).

A3. (a) L. (b) TI. (c) A. (d) TM. (e) TE. (f) VV.

A4. (a) Thinner, less, an even flow. (b) Valves. (c) Varicose veins, superficial, skeletal muscles around deep veins prevent over-stretching.

B1. Continuous drop in blood pressure from one vessel to the next; large, less, skeletal muscles.

B2. (a) Systolic, diastole, elastic, 70 to 80. (b) 85 to 30, 30 to 10, 10 to 0, 0. (c) 0, increase.

B3. (a) ↑ CO. (b) ↑ Stroke volume. (c) ↑ Venous return. (d) ↑ exercise. (e) ↑ Blood volume. (f) ↑ Heart rate. (g) ↑ Sympathetic impulses. (h) ↑ Norepinephrine and epinephrine. (i) ↓ Vagal impulses. (j) ↑ ADH. (k) ↑ Aldosterone. (l) ↑ NaCl in diet. (m) ↑ Peripheral resistance. (n) ↑ Vasoconstriction. (o) ↑ Angiotensin II. (p) ↑ Viscosity of blood. (q) ↑ RBC count. (r) ↑ Plasma proteins.

B4. B, C, F, G, H.

B6. (a) Medulla, arterioles. (b) Aorta (arch), carotid sinus, increase. (c) Increasing, vasoconstrict, increasing. (d) Oxygen, carbon dioxide, vasoconstriction, increase.

B7. A B C E

B8. (a) Local. (b) Low, vasodilator, lactic, carbon dioxide, hydrogen. (c) Relax, increased.

B9. (a) A, IFOP; B, BHP; C, IFHP; D, BOP. (b) +8, out of. (c) BHP is about 15 mm Hg, −7. (d) Starling, lymphatic. (e) Edema, BHP, BOP, increase.

B10. (a) Starling's, increase. (b) Arteries, capillaries. (c) Increases. (d) 2,500. (e) An indirect, 2.5, high. (f) Decreased, increases.

B12. Blood; cardiogenic, hypovolemic, neurogenic, septic.

B13. (a–b) ↑. (c–f) ↓.

B14. (a) Epinephrine, renin, aldosterone. (b) ADH, retention, increase.

C3. 60–100, rapid, bradycardia.

C4. (a) Sphygmomanometer. (b) Left brachial; no pulse is felt or no sound is heard via a stethoscope. (c) Systolic. (d) Change or cease.

C5. 128/83, 45 (128-83); yes, it is in the normal range.

D1. See Figure 21-11, page 622 in the text; systemic.

D2. (a) Aorta; left; ascending aorta, arch of the aorta, descending aorta. (b) Coronary, heart. (c) K, brachiocephalic; L, left common carotid; M, left subclavian. (d) F; in general, the right extremity and right side of thorax, neck, and head; right common carotid. (e) G, Axillary; H, brachial; blood pressure. (f) I, radial; J, ulnar; I. (g) Vertebral, basilar.

D3. Anterior; internal; external portions of head (face, tongue, ear, scalp) and neck (throat, thyroid).

D4. (a) Willis. (b) Internal, vertebral, (c) E, F, H, communicating.

D5. (a) Occipital lobes of cerebrum, as well as cerebellum and pons. (b) Left, since this artery supplies the left side of the brain. (Remember that neurons in tracts cross to the opposite side of the body, but arteries do not.) (c) Hearing, since it is perceived in the temporal lobe. (Visual areas are in occipital lobes). (d) Left foot.

D6.

Artery	Letter	Structures Supplied

a. Celiac (three major branches)
  1. **Hepatic**
 	
  2. **Gastric**
  3. **Splenic**
	P	1. Liver, **gallbladder and parts of stomach, duodenum, pancreas**
		2. **Stomach and esophagus**
		3. **Spleen, pancreas, and stomach**

b. **Superior mesenteric**	Q	Small intestine and part of large intestine **and pancreas**
c. **Adrenal (suprarenal)**	R	Adrenal (suprarenal) glands
d. Renal	S	**Kidney**
e. **Gonadal (testicular or ovarian)**	T	Testes (or ovaries)
f. Inferior mesenteric	**U**	**Parts of colon and rectum**

D7. L-4, common iliac.

D10. Right external iliac artery → right lateral circumflex artery (right descending branch) → right anterior tibial artery → right dorsalis pedis artery → to arch and digital arteries in foot.

E1. Superior vena cava, inferior vena cava, coronary sinus.

E2. (a) Intracranial vascular (venous) sinuses, functionally. (b) Superior sagittal. (c) Internal jugular, common carotid. (d) There are no such vessels.

E3. (a) The heart and a lung. (b) Yes, Yes, Yes.

E4. (a) Left radial vein → left brachial vein → left axillary vein → left subclavian vein → left brachiocephalic vein → superior vena cava. (Note that there are left and right brachiocephalic veins, but only one artery of that name.) (b) Small veins in medial aspect of left knee → left popliteal vein → left femoral vein → left external iliac vein → left common iliac vein → inferior vena cava. (c) Right renal vein → inferior vena cava.

E5. (a) Lower, slower. (b) More, superficial.

E6. (a) Axillary. (b) Axillary. (c) Femoral. (d) Popliteal.

E8. See key, page LG 489. Right, right.

F1. (a) E, gastric; F, splenic, I, mesenteric; H, portal vein to the liver. (b) D, hepatic artery. (c) Sinusoids; G, hepatic vein (there are two but only one is shown on Figure LG 21-1). (d) While in liver sinusoids, blood is cleaned, modified, detoxified. Ingested nutrients and other chemicals are metabolized and stored.

F3.

Fetal Structure	Structures Connected	Function	Fate of Structure After Birth
a. Umbilical arteries (2)	Fetal internal iliac arteries to placenta	Carries fetal blood low in oxygen and nutrients and high in wastes.	Medial umbilical ligaments
b. Placenta	(Omit)	Site where maternal and fetal blood exchange gases, nutrients and wastes	Delivered as "afterbirth"
c. Umbilical vein (1)	Placenta to liver and ductus venosus	Carries blood high in oxygen and nutrients, low in wastes	Round ligament of the liver
d. Ductus venosus	Umbilical vein to inferior vena cava	Branch of umbilical vein, bypasses liver	Ligamentum venosum
e. Foramen ovale	Right and left atria	Bypasses lungs	Fossa ovalis
f. Ductus arteriosus	Pulmonary artery to aorta	Bypasses lungs	Ligamentum arteriosum

F4. (a) Umbilical vein. (b) Femoral artery. (c) Pulmonary artery (no oxygenation occurs in fetal lungs). (d) Thoracic portion of inferior vena cava (since umbilical vein blood enters it via ductus venosus).

G2. (a–c) ↑. (d–f) ↓.

H1. (a) ↓. (b) ↓. (c) ↑. (d) ↑.

I2. (a) K. (b) AM. (c) AC.

I6. Embolus (a "clot-on-the-run") travels through femoral vein to external and common iliac veins to inferior vena cava to right side of heart. It lodges in pulmonary arterial branches, blocking further blood flow into the pulmonary vessels.

MASTERY TEST: Chapter 21

Questions 1–2: Circle the letters preceding ALL correct answers to each question.

1. In the most direct route from the left leg to the left arm of an adult, blood must pass through all of these structures:

    A. Inferior vena cava
    B. Brachiocephalic artery
    C. Capillaries in lung
    D. Hepatic portal vein
    E. Left subclavian artery
    F. Right ventricle of heart
    G. Left external iliac vein

2. In the most direct route from the fetal right ventricle to the fetal left leg, blood must pass through all of these structures:

    A. Aorta
    B. Umbilical artery
    C. Lung
    D. Ductus arteriosus
    E. Ductus venosus
    F. Left ventricle
    G. Left common iliac artery

3. Choose the FALSE statement.

    A. In order for blood to pass from a vein to an artery, it must pass through chambers of the heart.
    B. In its passage from an artery to a vein a red blood cell must ordinarily travel through a capillary.
    C. The wall of the femoral artery is thicker than the wall of the femoral vein.
    D. Most of the smooth muscle in arteries is in the tunica interna.

4. Choose the FALSE statement.

    A. Arteries contain valves, but veins do not.
    B. Decrease in the size of the lumen of a blood vessel by contraction of smooth muscle is called vasoconstriction.
    C. Exercise tends to increase HDL levels.
    D. Sinusoids are wider and more tortuous (winding) than capillaries.

5. Choose the FALSE statement.

    A. Cool, clammy skin is a sign of shock that results from sympathetic stimulation of blood vessels and sweat glands.
    B. Nicotine is a vasodilator that helps control hypertension.
    C. Orthostatic hypotension refers to a sudden, dramatic drop in blood pressure upon standing or sitting up straight.
    D. Phlebitis is an inflammation of a vein.

6. Choose the FALSE statement.

    A. The vessels that act as the major regulators of blood pressure are arterioles.
    B. The only blood vessels that carry out exchange of nutrients, oxygen, and wastes are capillaries.
    C. Blood in the umbilical artery is normally more highly oxygenated than blood in the umbilical vein.
    D. At any given moment more than 50 percent of the blood in the body is in the veins.

7. Choose the TRUE statement.

    A. The basilic vein is located lateral to the cephalic vein.
    B. The great saphenous vein runs along the lateral aspect of the leg and thigh.
    C. The vein in the body most likely to become varicosed is the femoral.
    D. The hemiazygos and accessory hemiazygos veins lie to the left of the azygos.

8. Choose the TRUE statement.

    A. A normal blood pressure for the average adult is 160/100.
    B. During exercise, blood pressure will tend to increase.
    C. Sympathetic impulses to the heart and to arterioles tend to decrease blood pressure.
    D. Decreased cardiac output causes increased blood pressure.

9. Choose the TRUE statement.

    A. Cranial venous sinuses are composed of the typical three layers found in walls of all veins.
    B. Cranial venous sinuses all eventually empty into the external jugular veins.
    C. Internal and external jugular veins do unite to form common jugular veins.
    D. Most parts of the body supplied by the internal carotid artery are ultimately drained by the internal jugular vein.

10. Hepatic portal circulation carries blood from:

    A. Kidneys to liver
    B. Liver to heart
    C. Stomach, intestine, and spleen to liver
    D. Stomach, intestine, and spleen to kidneys
    E. Heart to lungs
    F. Pulmonary artery to aorta

11. Which of the following is a parietal (rather than a visceral) branch of the aorta?

    A. Superior mesenteric artery
    B. Bronchial artery
    C. Celiac artery
    D. Lumbar artery
    E. Esophageal artery

12. All of these vessels are in the leg or foot EXCEPT:
    A. Saphenous vein
    B. Azygos vein
    C. Peroneal artery
    D. Dorsalis pedis artery
    E. Popliteal artery

13. All of these are superficial veins EXCEPT:
    A. Median cubital   D. Great saphenous
    B. Basilic   E. Brachial
    C. Cephalic

14. Vasomotion refers to:
    A. Local control of blood flow to active tissues
    B. The Doppler method of checking blood flow
    C. Intermittent contraction of small vessels resulting in discontinuous blood flow through capillary networks
    D. A new aerobic dance sensation

Questions 15–20: Arrange the answers in correct sequence.

_____ _____ _____ _____ 15. Route of a drop of blood from the right side of the heart to the left side of the heart:
    A. Pulmonary artery
    B. Arterioles
    C. Capillaries
    D. Venules and veins

Questions 16–18, use the following answers:

| A. Aorta and other arteries | C. Capillaries |
| B. Arterioles | D. Venules and veins |

_____ _____ _____ _____ 16. Blood pressure in vessels, from highest to lowest:

_____ _____ _____ _____ 17. Total cross-sectional areas of vessels, from highest to lowest:

_____ _____ _____ _____ 18. Velocity of blood in vessels, from highest to lowest:

_____ _____ _____ _____ _____ 19. Route of a drop of blood from small intestine to heart:
    A. Superior mesenteric vein
    B. Hepatic portal vein
    C. Small vessels within the liver
    D. Hepatic vein
    E. Inferior vena cava

_____ _____ _____ 20. Layers of blood vessels, from most superficial to deepest:
    A. Tunica interna
    B. Tunica externa
    C. Tunica media

Questions 21–25: fill-ins. Complete each sentence or answer the question with the word(s) or phrase that best fits.

_____ 21. A weakened section of a blood vessel forming a balloonlike sac is known as a(n) ____.

_____ 22. Name the major factor that creates blood osmotic pressure (BOP) that prevents excessive flow of fluid from blood to interstitial areas.

_____ 23. Name the two blood vessels that are locations of baroreceptors and chemoreceptors for regulation of blood pressure.

_____ 24. In taking a blood pressure, the cuff is first inflated over an artery. As the cuff is then slowly deflated, the first sounds heard indicate the level of ____ blood pressure.

_____ 25. Most arteries are paired (one on the right side of the body, one on the left). Name five vessels that are unpaired.

ANSWERS TO MASTERY TEST: ■ Chapter 21

Multiple Answers

1. A C E F G
2. A D G

Multiple Choice

3. D	9. D
4. A	10. C
5. B	11. D
6. C	12. B
7. D	13. E
8. B	14. C

Arrange

15. A B C D
16. A B C D
17. C D B A
18. A B D C
19. A B C D E
20. B C A

Fill-ins

21. Aneurysm
22. Plasma protein such as albumin
23. Aorta and carotids
24. Systolic
25. Aorta, anterior communicating cerebral, basilar, brachio-cephalic, celiac, gastric, hepatic, splenic, superior and inferior mesenteric, middle sacral

Lymphatic System & Immunity

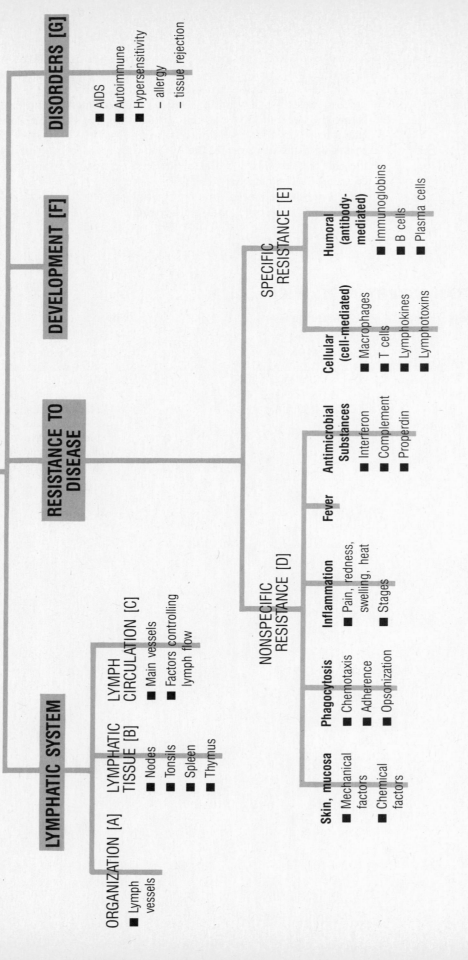

LYMPHATIC SYSTEM

ORGANIZATION [A]
- Lymph vessels

LYMPHATIC TISSUE [B]
- Nodes
- Tonsils
- Spleen
- Thymus

LYMPH CIRCULATION [C]
- Main vessels
- Factors controlling lymph flow

RESISTANCE TO DISEASE

NONSPECIFIC RESISTANCE [D]

Skin, mucosa
- Mechanical factors
- Chemical factors

Phagocytosis
- Chemotaxis
- Adherence
- Opsonization

Inflammation
- Pain, redness, swelling, heat
- Stages

Fever

Antimicrobial Substances
- Interferon
- Complement
- Properdin

SPECIFIC RESISTANCE [E]

Cellular (cell-mediated)
- Macrophages
- T cells
- Lymphokines
- Lymphotoxins

Humoral (antibody-mediated)
- Immunoglobins
- B cells
- Plasma cells

DEVELOPMENT [F]

DISORDERS [G]
- AIDS
- Autoimmune
- Hypersensitivity
  – allergy
  – tissue rejection

The Lymphatic System and Immunity

The human body is continually exposed to foreign (nonself) entities: bacteria or viruses, pollens, chemicals in a mosquito bite, or foods or medications that provoke allergic responses. Resistance to specific invaders is normally provided by a healthy immune system made up of antibodies and battalions of T-lymphocytes.

Nonspecific resistance is offered by a variety of structures and mechanisms, such as skin, mucus, white blood cells assisting in inflammation, and even by a rise in temperature (fever) that limits the survival of invading microbes. The lymphatic system provides sentinels at points of entry (as in tonsils) and at specific sites (lymph nodes) near entry to body cavities (inguinal and axillary regions). The lymphatic system also wards off the detrimental effects of edema by returning proteins and fluids to blood vessels.

Start your study of the physical defense mechanisms of the body by previewing Chapter 22 Framework and the key terms it contains.

TOPIC OUTLINE AND OBJECTIVES

A. Lymphatic organization

B. Lymphatic tissue

C. Lymphatic circulation

1. Describe the components of the lymphatic system and list their functions.
2. Discuss how edema develops.

D. Nonspecific resistance to disease

3. Discuss the roles of the skin and mucous membranes, phagocytosis, inflammation, fever, and antimicrobial substances as components of nonspecific resistance.

E. Immunity (specific resistance to disease)

4. Define immunity and explain the relationship between an antigen (Ag) and an antibody (Ab).
5. Contrast the role of T cells in cellular immunity and the role of B cells in humoral immunity.
6. Discuss the relationship of immunology to cancer.

F. Aging and developmental anatomy of the lymphatic system

7. Describe the effects of aging on the immune system.
8. Describe the development of the lymphatic system.

G. Disorders and medical terminology

9. Describe the clinical symptoms of the following disorders: acquired immune deficiency syndrome (AIDS), autoimmune diseases, severe combined immune deficiency (SCID), hypersensitivity (allergy), tissue rejection, and Hodgkin's disease (HD).
10. Define medical terminology associated with the lymphatic system.

Now study the following parts of words that may help you better understand terminology in this chapter.

Wordbyte	Meaning	Example	Wordbyte	Meaning	Example
anti-	against	*anti*body	chyl-	juice	cisterna *chyli*
axilla-	armpit	*axilla*ry nodes	lact-	milk	*lact*eals

CHECKPOINTS

A. Lymphatic organization (pages 655–658)

■ **A1.** List the components of the lymphatic system.

■ **A2.** Describe the functions of the lymphatic system in this exercise.

a. Lymphatic vessels drain from tissue spaces _____ that are too large to readily pass into blood capillaries. They also transport

_____ from the gastrointestinal (GI) tract to the bloodstream.
b. The defensive functions of the lymphatic system are carried out largely by white

blood cells named _____-cytes. (*B? T?*) lymphocytes destroy foreign substances directly, whereas B lymphocytes lead to the production of

_____.

A3. Contrast in terms of structure and location:
a. Blood capillaries/lymph capillaries

b. Veins/lymphatics

B. Lymphatic tissue (pages 658–660)

■ **B1.** Label structures A–E on Figure LG 22-1. Then color each of those structures according to descriptions below.
(a) ○ Vessels for entrance of lymph into node
(b) ○ Vessel for exit of lymph out of node
(c) ○ Site of lymphocyte production
(d) ○ Channel for flow of lymph in the cortex of the node
(e) ○ Dense strand of lymphocytes, macrophages, and plasma cells.

B2. Describe how lymph is "processed" as it circulates through lymph nodes. How does this function of lymph nodes explain the fact that nodes ("glands") become enlarged and tender during infection?

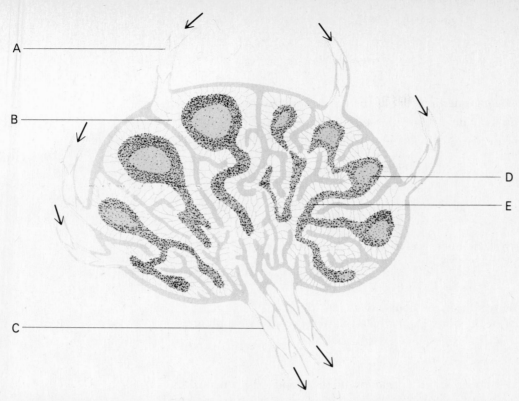

A _____

B _____

C _____

D _____

E _____

Figure LG 22-1 Structure of a lymph node. Color and label as directed in Checkpoint B1.

■ **B3.** Spread, or _____, of cancer may occur via the lymphatic system. Lymph nodes that have become cancerous tend to be enlarged, firm, and (*tender? nontender?*).

■ **B4.** *A clinical challenge.* Which tonsils are most commonly removed in a tonsil-

lectomy? _____ State one disadvantage of loss of tonsillar tissue.

■ **B5.** The spleen is located in the (*upper right? upper left? lower right? lower left?*)

quadrant of the abdomen, immediately inferior to the _____.

■ **B6.** List three functions of the spleen.

_____ _____ _____

■ **B7.** The thymus gland is located in the (*neck? mediastinum?*). Its position is just

posterior to the _____ bone. Its size is relatively (*large? small?*) in children, (*increasing? decreasing?*) in size in adulthood. The thymus helps to produce (*B? T?*)-cells.

C. Lymphatic circulation (pages 661–663)

C1. Describe the general pathway of lymph circulation.

■ **C2.** Answer the following questions about the largest lymphatics.

a. In general, the thoracic duct drains (*three-fourths? one-half? one-fourth?*) of the body's lymph. This vessel starts in the lumbar region as a dilation known as the

_____. This cisterna chyli receives lymph from which areas of the body?

b. In the neck the thoracic duct receives lymph from three trunks. Name these.

c. One-fourth of the lymph of the body drains into the _____ duct. The regions of the body from which this vessel collects lymph are:

d. The thoracic duct empties into the junction of the blood vessels named left

_____ and left _____ veins. The right lymphatic duct empties into veins of the same name on the right side.

e. In summary, lymph fluid starts out originally in plasma or cells and circulates as

_____ fluid. Then as lymph it undergoes extensive cleaning as it passes through nodes. Finally fluid and other substances in lymph are returned

to _____ in veins of the neck.

■ **C3.** The lymphatic system does not have a separate heart for pumping lymph. Describe two principal factors that are responsible for return of lymph from the entire body to major blood vessels in the neck. (Note that these same factors also facilitate venous return.)

■ **C4.** *A clinical challenge.* Explain why a person with a broken left clavicle might exhibit these two symptoms:

a. Edema, especially of the left arm and both legs

b. Excess fat in feces

D. Nonspecific resistance to disease (pages 663–669)

■ **D1.** Identify the types of resistance described in each case as specific (S) or nonspecific (N).

_____ a. Resistance against a single type of pathogen, such as the bacterium causing tuberculosis

_____ b. Specific antibody against a specific antigen (such as a type of bacterium or its toxin)

_____ c. Mechanical barriers such as skin and mucous membranes or chemicals such as acid stomach secretions

_____ d. Inherited resistance

_____ e. The inflammatory process

■ **D2.** *A clinical challenge.* Match the following conditions or factors with the nonspecific resistance that is lost in each case.

Nonspecific resistance lost

_____ a. Cleansing of oral mucosa

_____ b. Cleansing of vaginal mucosa

_____ c. Flushing of microbes from urinary tract

_____ d. Lacrimal fluid with lysozyme

_____ e. Mucociliary escalator

_____ f. Closely packed layers of keratinized cells

_____ g. HCl production; pH 1 or 2

_____ h. Sebum production

Condition/factor

A. Long-term smoking
B. Excessive dryness of skin
C. Use of anticholinergic medications that decrease salivation
D. Vaginal atrophy related to normal aging
E. Dry eyes related to decreased production of tears, as in aging
F. Decreased urinary output, as in prostatic hypertrophy
G. Skin wound
H. Partial gastrectomy (removal of part of stomach)

■ **D3.** Match each of the four steps in phagocytosis with the related description. Note that descriptions are listed in chronological sequence.

| A. Adherence | D. Digestion |
| C. Chemotaxis | I. Ingestion |

_____ a. Phagocytic cells "sniff" the delectable fragrance of invading microbes and especially savor bacteria with the tasty *complement* coating. "Ah, dinner soon!"

_____ b. Phagocytes trap the invaders, possibly wedging them against a blood vessel or a clot.

_____ c. Phagocytic pseudopods like tentacles surround the tasty morsels forming a culinary treat, the *phagocytic vesicle.* "Slurp!"

_____ d. Lysosomal enzymes pounce on the microbial meal, a has-been within a half hour. *Residual bodies,* like the pits of prunes, are discarded. And the phagocytic face smiles in contentment.

D4. Defend or dispute this statement: "Inflammation is a process that can be both helpful and harmful."

■ **D5.** Arrange in correct sequence the four stages of inflammation:

Clot. Fibrin clot formation	Mig. Phagocytic (neutrophil and macrophage) migration
Pus. Pus formation	Dil/perm. Vasodilation and increased capillary permeability

_____ → _____ → _____ → _____

■ **D6.** Now select the phase that best fits each descriptive phrase below.

Clot	Pus	Mig	Dil/perm

_____ a. Brings defensive substances to the injured site and helps remove toxic wastes

_____ b. Histamine, prostaglandins, and kinins enhance this process

_____ c. Causes redness, warmth, and swelling of inflammation

_____ d. Localizes and traps invading organisms

_____ e. Occurs within an hour after initiation of inflammation; involves margination, diapedesis, and chemotaxis

_____ f. White blood cell and debris formation that may lead to abscess or ulcer.

■ **D7.** Identify the chemicals that fit descriptions below.

C. Complement	K. Kinins
D. Defensins	LPF. Leukocytosis-promoting factor
H. Histamine	PG. Prostaglandins
I. Interferon	PR. Properdin

_____ a. Produced by damaged or inflamed tissues (2 answers)

_____ b. Produced by mast cells, basophils, and platelets

_____ c. Stimulates increase in production of neutrophils

_____ d. Induces vasodilation, permeability, chemotaxis, and irritation of nerve endings

_____ e. Neutrophils are attracted by these chemicals present in the inflamed area. (2 answers)

_____ f. Antibiotic chemicals made by neutrophils

_____ g. Group of 20 serum proteins that stimulate release of histamine and also make bacteria "tastier" to phagocytes by opsonization

_____ h. Group of three serum proteins that works with complement in cytolysis of microbes

_____ i. Made by some leucocytes and fibroblasts; stimulates cells near an invading virus to produce antiviral proteins (AVP) that interfere with survival of the virus

402

E. Immunity (specific resistance to disease) (pages 669–678)

■ **E1.** Do this exercise on antigens.

a. An antigen is defined as "any chemical substance which, when introduced into the

body, causes _____."
In general, antigens are (*parts of the body? foreign substances?*).

b. Antigens have two characteristics: _____, the ability to stimu-

late specific antibodies, and *reactivity,* the ability to _____

_____.

c. An antigen with both of these characteristics is called a _____
antigen.

d. Chemically, antigens:

   A. Are all protein
   B. May be proteins, proteins combined with other organic compounds, or polysaccharides

e. Can an entire microbe serve as an antigen? (*Yes? No?*) List the parts of microbes
which may be antigenic.

f. If you are allergic to pollen in the spring or fall or to certain foods, the pollen or
foods serve as (*antigens? antibodies?*) to you.

g. Antibodies form against (*the entire? only a specific region of the?*) antigen. This

region is known as the _____. Most antigens have (*only one?
only two? a number of?*) antigenic determinant sites, and so are called

_____-valent.

h. An individual determinant site has the characteristic of (*reactivity? immuno-

genicity? both?*). By itself, it is called a _____. For individu-
als who are allergic to penicillin, penicillin serves as a (*complete antigen? hapten?*),
which combines with a larger protein in the body to form a compete antigen.

■ **E2.** Chemically, all antibodies are composed of proteins named

_____. Since they are involved with immunity, these are called

immunoglobulins, abbreviated _____. Write the name of the related class of Ig next
to each description.

_____ a. The only type of Ig to cross the placenta, it provides specific resistance of
newborns. Also significantly enhances phagocytosis and neutralization of
toxins in persons of all ages since it is the most abundant type of
antibody.

_____ b. Found in secretions such as mucus, saliva, and tears, so protects against
oral, vaginal, and respiratory infections.

_____ c. Located on mast and basophil cells and involved in allergic reactions, for
example, to certain foods, pollen, or bee venom.

_____ d. Antibodies that can cause agglutination and lysis, and in this way can
destroy invading microbes.

_____ e. Lead to production of antibodies against the antigen.

■ **E3.** Refer to Figure LG 22-2 and check your understanding of antibody structure in this Checkpoint.

a. Color constant and variable portions and crosshatch the heavy chain of the antibody as indicated by color code ovals on the figure.

b. Label *disulfide bonds, antigen bindings sites,* and *hinges.*

c. Label the configuration of the antibody when it is combined with an antigen and when it is in the uncombined state. Note that in the combined state, the constant portions are (*more? less?*) exposed, as for fixing complement.

■ **E4.** Contrast two types of immunity by writing C before descriptions of cellular immunity and H before those describing humoral immunity.

_____ a. Final products are sensitized T lymphocytes

_____ b. Involves formation of antibodies which circulate in the bloodstream

_____ c. Especially effective against bacterial and viral infections

_____ d. Effective against fungi, parasites, certain viruses, cancer, and tissue transplants

_____ e. Involves B cells that develop into plasma cells

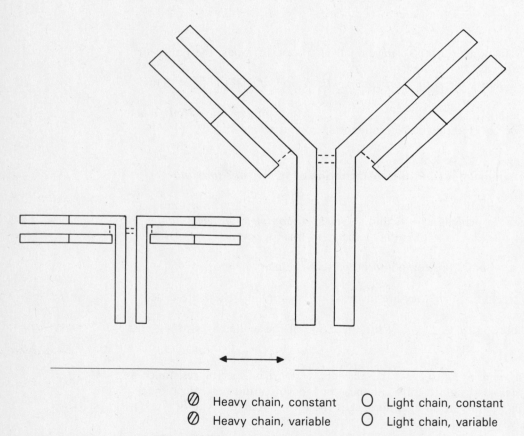

⊘ Heavy chain, constant   ○ Light chain, constant
⊘ Heavy chain, variable   ○ Light chain, variable

Figure LG 22-2 Diagram of an antibody molecule. Color and label as directed in Checkpoint E3.

■ **E5.** Complete this exercise about lymphocytes.

a. Name the two main types of lymphocytes: ____ cells and ____ cells. Both of these

originate in embryonic _____ tissue.

b. Some of these cells migrate to the thymus; there these cells become (*B? T?*) cells. This involves processing of T cells, that is, they develop

_____. Thus, lymphocytes have the ability to differentiate into cells that perform specific immune functions against specific antigens.

c. Immunological competence is conferred (*shortly before and after birth? throughout a lifetime? mostly later in life?*). Therefore removal of the thymus gland (*shortly after birth? at age 30?*) is likely to have more deleterious effects upon immunity.

d. What is the basis for the name B cells?

e. After processing, both T and B cells then move to _____

_____ where they reside until needed.

■ **E6.** Refer to Figure LG 22-3 and complete this Checkpoint about the sequence of events that occur in response to entrance of an antigen into the body.

a. How many different types of T cells are in the body? (*One? Twelve? Thousands?*) How do T cells respond to different antigens?

At any given time (*most? few?*) T cells are active. When one type of antigen enters the body, one type of T cell will be activated.

b. First the invading antigens must be *processed* by cells such as

_____ (*A* in Figure LG 22-3). These cells initiate a series of events that may lead to the destruction of the invader. In the first of these, the macrophages *present* the antigens to T cells (which we will consider in more detail in *C* and *E*). T cells will not recognize an antigen unless macrophages present it

together with _____ antigens.

c. However, in order for T cells to respond to the presented antigens and help eliminate them, T cells must be stimulated by chemicals known as

_____. Name two cells that secrete interleukin-I.

_____ (*A*) and _____ (*O*).

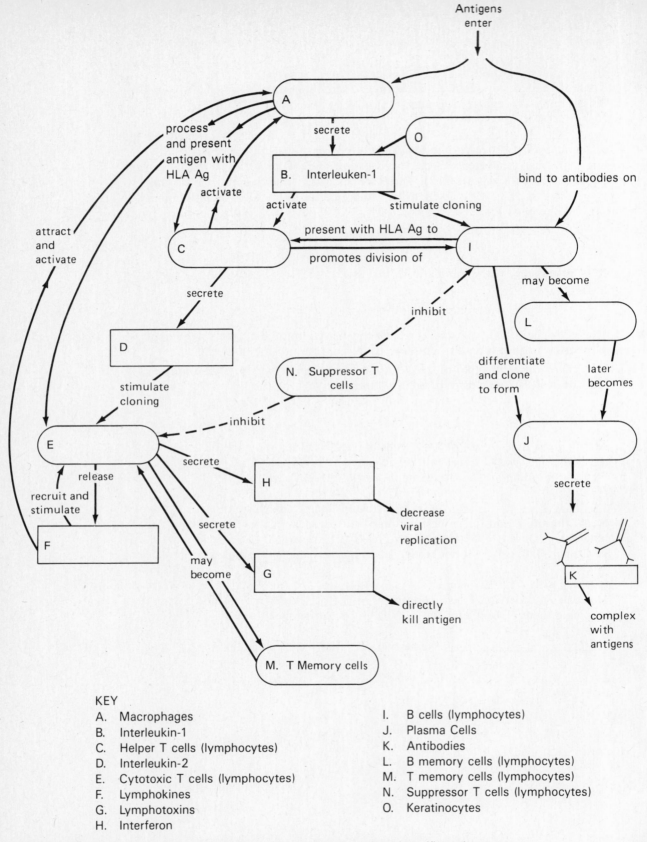

Figure LG 22-3 Diagram of sequence of events in immune response (specific resistance to disease). Ovals contain names of cells; rectangles contain names of chemicals. Solid lines indicate enhancing effects; broken lines show inhibitory relationships. Complete as directed in Checkpoints E6 and E9.

KEY
A. Macrophages
B. Interleukin-1
C. Helper T cells (lymphocytes)
D. Interleukin-2
E. Cytotoxic T cells (lymphocytes)
F. Lymphokines
G. Lymphotoxins
H. Interferon

I. B cells (lymphocytes)
J. Plasma Cells
K. Antibodies
L. B memory cells (lymphocytes)
M. T memory cells (lymphocytes)
N. Suppressor T cells (lymphocytes)
O. Keratinocytes

d. One function of interleukin-1 is stimulation of (*helper? killer?*) T cells (*C*). These cells have several functions as shown on the figure. One is secretion of proteins that enhance the phagocytic effects of macrophages (*A*). Note that action on the figure.

e. A second job description for helper T cells is production of a chemical named

_____ (*D*), which stimulates cloning of T cells. (See *E*.) What is a clone?

f. As a result of presentation of the antigens (by the macrophages) and stimulation by interleukin-2, a large number of cytotoxic (killer) T cells are now said to be

_____ toward the specific invader antigens and ready to combat them.

g. Now consider the array of effects of these cytotoxic T cells. They secrete chemicals

called _____ (*F*). What is the overall effect of lymphokines? (More specifics about these in Checkpoint E7.)

h. Cytotoxic T cells also release chemicals named _____ (*G*). By what means do these directly destroy the antigens?

i. One more function of cytotoxic T cells is secretion of interferon (*H*). This chemical decreases replication of invaders that are (*viral? bacterial?*), and also exhibits a (*positive? negative?*) feedback effect upon killer T cells.

j. In summary, macrophages (*A*) recognize a type of antigen. With the help of interleukins (*B* and *D*), T cells are called to action (*C* and *E*). Both types of T cells release lymphokines (as at *F*) that enhance the phagocytic activities of macrophages (*A*). Cytotoxic (killer) T cells (*E*) also directly kill or decrease replication of antigens (*G* and *H*). All of these activities involve either direct or indirect actions of cells (T cells and macrophages). So the defenses just described are providing (*humoral? cell-mediated?*) immunity.

■ E7. In Checkpoint E6 you considered overall effects of lymphokines. Now match the names of chemicals in this class with their specific actions listed below.

MAF. Macrophage activating factor	MF. Mitogenic factor
MCF. Macrophage chemotactic factor	MIF. Migration inhibitory factor

_____ a. Attracts additional macrophages to the scene of the infection

_____ b. Activates the macrophages

_____ c. Prevents macrophages from moving away from the scene of the infection

_____ d. Stimulates mitosis among lymphocytes so that numbers increase

E8. Contrast natural killer (NK) with cytotoxic (killer) T cells.

■ **E9.** Now that you have considered cell-mediated immune responses, check your understanding of humoral responses in this exercise. Again refer to Figure LG 22-3.

a. In humoral immunity, the final products (antibodies) are made by plasma cells derived from (*T? B?*) lymphocytes. First the antigen binds to specific antibodies on a

(*macrophage? B cell?*). This antigen, along with _____ antigens, is then presented to (*helper? killer?*) T cells, which recognize the antigen and respond by promoting division of (*T? B?*) cells.

b. B cells can then differentiate and divide to form a clone of

_____ cells (*J*). Plasma cells then (*stay in lymphoid tissue?*

circulate in blood?). There they secrete _____ (*K*) that can complex with the specifically related antigen. This complex can fix the nonspecific

proteins called _____ to further enhance the destruction of the antigens. (*For extra review* of complement, refer back to Checkpoints D3 and D7.)

c. Describe a typical rate of antibody production: _____ molecules/sec of antibody may be secreted for a period of several (*minutes? days? year?*) until the plasma cell dies.

d. Some B cells do not differentiate into plasma cells right away. Instead they re-

main as _____ cells (*L*). Once a person is sensitized against a specific antigen, either by infection or by a(n) (*initial? booster dose?*) immunization, the person produces some memory cells. These will provide a (*more? less?*) intense response upon subsequent exposure to the antigen, such as during another

infection or by a _____ dose.

e. T cells (*do? do not?*) produce memory cells. (See *M* on the figure.) Both the B and T memory cells evoke the intense secondary response (as to a booster dose). This is

also called an _____ response to an antigen.

f. Note one other factor that enhances humoral responses. (*Helper? Killer?*) T cells influence B cell differentiation to plasma cells. Identify that relationship on the figure. Now find the type of lymphocyte that can inhibit both plasma cell forma-

tion and killer T functions. This is a _____ cell (____ on the figure).

E10. What is the meaning of the term antibody *titer*?

E11. Define *monoclonal antibodies.*

These antibodies are produced by hybridoma cells. Explain how such a cell can be produced.

E12. Describe three clinical uses of monoclonal antibodies.

E13. Answer these questions about the relationship between immunology and cancer.

a. What are tumor-specific antigens? How are they related to immunologic surveillance?

b. Most researchers believe that tumors are destroyed by (*cellular? humoral?*) immune responses. Therefore (*T? B?*) cells sensitized to the cancer antigens play the major role.

c. State two ways that cancer cells may carry out *immunological escape.*

F. Aging and developmental anatomy of the lymphatic system (pages 678–679)

■ **F1.** The lymphatic system derives from _____-derm, beginning about the (*fifth? seventh? tenth?*) week of gestation.

■ **F2.** Lymphatic vessels arise from lymph sacs that form from early (*arteries? veins?*). Name the embryonic lymph sac that develops vessels to each of the following regions of the body.

a. Pelvic area and lower extremities: _____ lymph sac

b. Upper extremities, thorax, head, and neck: _____ lymph sac.

G. Disorders, medical terminology (pages 679–685)

■ **G1.** Answer these questions about AIDS.

a. AIDS refers to A _____ I _____

D _____ S _____ .

b. Persons with AIDS have blood that is positive for the _____-antibody. Since the immune system is deficient, persons with AIDS are also more susceptible to two

otherwise relatively rare conditions. These are _____ (KS), a

form of skin cancer, and _____ (PCP), a respiratory infection caused by a (*virus? protozoan? fungus?*).

c. Highest risk groups for AIDS include: _____,

_____, and _____. Almost all victims are (*male? female?*). The average time from infection to diagnosis is eight (*days? months? years?*).

d. Transmission of the AIDS virus occurs through (circle all correct answers): (*blood? semen? vaginal secretions? saliva? tears?*) Use of condoms (*dramatically reduces? does not affect?*) risk of AIDS.

e. The AIDS virus carries its genetic code in (*DNA? RNA?*). Via the enzyme

_____ transcriptase, the virus directs synthesis of DNA which takes over host cells. Possibly the first cells affected by the HIV virus are (*macrophage? B? cytotoxic T? helper T?*) cells. A decline in the number of

_____ T cells ultimately inhibits antibody production. (Review Figure LG 22-3C, I–K.)

f. The drug that is most widely tried against the AIDS virus is one that inhibits the

enzyme _____. At this time the primary defense against AIDS is (*a recently developed vaccine? education and safer sexual practices?*)

■ **G2.** Do this exercise on autoimmune disease.

a. Normally the body (*does? does not?*) produce antibodies against its own tissues.

Such self-recognition is called immunological _____. Possible explanations for immunological tolerance are based on actions of

_____ cells (*N* in Figure LG 22-3). These cells normally (*stimulate? inhibit?*) production of antibodies against "self."

b. Such inhibition is lost in _____ diseases, so the body starts attacking itself.

■ **G3.** Based on what you know about antigens and antibodies, complete this exercise about how allergic reaction — anaphylaxis — occurs.

a. In the first stage B cells respond to an allergen (antigen) by producing Ig _____ antibodies. These attach to certain types of body cells, namely,

_____ cells and _____. These cells are

then said to be _____ to the antigen.

b. At a later time exposure to the same antigen causes the IgEs to bind together, leading to release of chemicals from mast cells and basophils. Among these chemicals

is one named _____.

c. Some effects of these chemicals which are associated with anaphylaxis are: (*increased? decreased?*) capillary permeability, leading to extra fluid in tissue spaces

(the condition called _____) and also to lowered blood pressure; (*increased? decreased?*) smooth muscle contraction of bronchial tubes, making breathing (*easier? more difficult?*); (*increased? decreased?*) mucus production, causing a "runny nose."

■ **G4.** Transplantation is likely to lead to tissue rejection because the transplanted organ or tissue serves as an (*antigen? antibody?*). The body tries to reject this foreign tissue

by producing _____ .

■ **G5.** Match types of transplants in the box with descriptions below.

| A. Allograft   I. Isograft   X. Xenograft |

_____ a. Between animals of different species

_____ b. The most successful type of transplant

_____ c. Between individuals of same species, but of different genetic backgrounds, such as a mother's kidney donated to her daughter or a blood transfusion

■ **G6.** Name one drug often used for transplant patients since it is selectively

immunosuppressant. _____

■ **G7.** Match the condition in the box with the description below.

| A. Autoimmune disease   L. Lymphangioma |
| H. Hodgkin's disease   S. Severe combined immunodeficiency (SCID) |

_____ a. A curable malignancy usually arising in lymph nodes

_____ b. Multiple sclerosis (MS), rheumatoid arthritis (RA), and systemic lupus erythematosus (SLE) are examples

_____ c. Benign tumor of lymph vessels

_____ d. Characterized by lack of both B and T cells

ANSWERS TO SELECTED CHECKPOINTS

A1. Lymph, lymph vessels, and lymph organs (such as tonsils, spleen, and lymph nodes).

A2. (a) Proteins, lipids. (b) Lympho, T, antibodies.

B1. (a) A, afferent lymphatic vessel. (b) C, efferent lymphatic vessel. (c) D, germinal center. (d) B, cortical sinus. (e) E, medullary cord.

B3. Metastasis, nontender.

B4. Palatine: as lymphatic tissue does become inflamed at times, it does protect against foreign substances.

B5. Upper left, diaphragm.

B6. B lymphocyte and antibody production; blood reservoir; phagocytosis of microbes, red blood cells, and platelets.

B7. Mediastinum (note that the *thyroid* is in the neck), sternum, large, decreasing, T.

C2. (a) Three-fourths, cisterna chyli; digestive and other abdominal organs and both lower extremities. (b) Left jugular, left subclavian, and left bronchomediastinal. (c) Right lymphatic; right upper extremity and right side of thorax, neck, and head. (d) Internal jugular, subclavian. (e) Interstitial, plasma.

C3. Skeletal muscle contraction squeezing lymphatics which contain valves directing flow of lymph. Respiratory movements.

C4. Fracture of this bone might block the thoracic duct which enters veins close to the left clavicle. Lymph therefore backs up with these results: (a) Extra fluid remains in tissue spaces normally drained by vessels leading to the thoracic duct. (b) Lymph capillaries in the intestine normally absorb fat from foods. Slow lymph flow can decrease such absorption leaving fat in digestive wastes.

D1. (a) S. (b) S. (c) N. (d) N. (e) N.

D2. (a) C. (b) D. (c) F. (d) E. (e) A. (f) G. (g) H. (h) B.

D3. (a) C. (b) A. (c) I. (d) D.

D5. Dil/perm → clot → mig → pus.

D6. (a–c) Dil/perm. (d) Clot. (e) Mig. (f) Pus.

D7. (a) PG, LPF (b) H. (c) LPF. (d) K. (e) K, C. (f) D. (g) C. (h) PR. (i) I.

E1. (a) The body to produce a specific antibody, foreign substances. (b) Immunogenicity, react with the specific antibody. (c) Complete. (d) B. (e) Yes; flagella, capsules, cell walls, and toxins produced by microbes. (f) Antigens. (g) Only a specific region of the, antigenic determinant, a number of, multi-. (h) Reactivity, hapten, hapten.

E2. Globulins, Ig. (a) IgG. (b) IgA. (c) IgE. (d) IgM. (e) IgD.

E3. More.

E4.

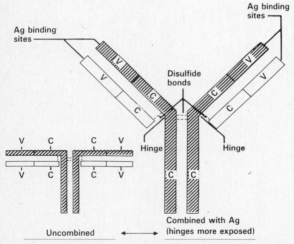

Figure LG 22-2A Diagram of an antibody molecule.

(a) C. (b) H. (c) H. (d) C. (e) H.

E5. (a) B, T, bone marrow (myeloid). (b) T, immunological competence. (c) Shortly before and after birth, shortly after birth. (d) Similar cells in birds were known to be processed in a bursa; the processing site of these cells in humans may be the bone marrow. (e) Lymphoid tissue.

E6. (a) Thousands; have specific sites on them that can react with only a certain type of antigen; few. (b) Macrophages, HLA. (c) Interleukins, macrophages, keratinocytes. (d) Helper. (e) Interleukin-2; a proliferation of cells derived from one cell, in this case, from the one cytotoxic T cell specific for this antigen. (f) Sensitized. (g) Lymphokines, recruit and further enhance activity of both macrophages and cytotoxic T cells. (h) Lymphotoxins such as perforin, lyse invading cells by producing holes in their plasma membranes. (i) Viral, positive. (j) Cell-mediated.

E7. (a) MCF. (b) MAF. (c) MIF. (d) MF.

E9. (a) B, B cell, HLA, helper, B. (b) Plasma, stay in lymphoid tissue, antibodies, complement. (c) 2,000, days, (d) Memory, initial, more, booster. (e) Do, anamnestic. (f) Helper, suppressor T (N).

F1. Meso, fifth.

F2. Veins. (a) Posterior. (b) Jugular.

G1. (a) Acquired immune deficiency syndrome. (b) HIV (human immunodeficiency virus), Kaposi's sarcoma, Pneumocystis carinii pneumonia, fungus. (c) Gay men, IV drug users, and hemophiliacs receiving blood products; male; years. (d) Blood, semen, vaginal secretions, dramatically reduces. (e) RNA, reverse, macrophage, helper. (f) Reverse transcriptase; education and safer sexual practices.

G2. (a) Does not, tolerance, suppressor T, inhibit. (b) Autoimmune.

G3. (a) E, mast, basophils, sensitized. (b) Histamine. (c) Increased, edema, increased, more difficult, increased.

G4. Antigen; antibodies and also T lymphocytes against the tissue.

G5. (a) X. (b) I. (c) A.

G6. Cyclosporine.

G7. (a) H. (b) A. (c) L. (d) S.

Questions 1–3: Arrange the answers in correct sequence.

_____ _____ _____ _____ 1. Flow of lymph through a lymph node:

    A. Afferent lymphatic vessel
    B. Medullary sinus
    C. Cortical sinus
    D. Efferent lymphatic vessel

_____ _____ _____ _____ 2. From first to last in cell-mediated immunity:

    A. Lymphokines and lymphotoxins are released
    B. Interleukin-2 stimulates killer T cells
    C. Helper T cells are activated by interleukin-1 and presentation of the specific antigen
    D. The invading antigen enters macrophage

_____ _____ _____ _____ _____ 3. Activities in humoral immunity, in chronological order

    A. B cells develop in bone marrow or other part of the body
    B. B cells differentiate and divide (clone), forming plasma cells
    C. B cells migrate to lymphoid tissue
    D. B cells are activated by specific antigen
    E. Antibodies are released and are specific against the antigen that activated the B cell

Questions 4–18: Circle the letter preceding the one best answer to each question. Take particular note of underlined terms.

4. Both the thoracic duct and the right lymphatic duct empty directly into:

    A. Axillary lymph nodes
    B. Superior vena cava
    C. Cisterna chyli
    D. Subclavian arteries
    E. Junction of internal jugular and subclavian veins

5. Which cells are the most dominant microphages?

    A. Histiocytes   D. Neutrophils
    B. Basophils   E. Eosinophils
    C. Lymphocytes

6. All of these are examples of nonspecific defenses EXCEPT:

    A. Antigens and antibodies
    B. Saliva
    C. Complement
    D. Interferon
    E. Skin
    F. Phagocytes

7. All of the following correctly match lymphocytes with their functions EXCEPT:

    A. Helper T cells: stimulate B cells to divide and differentiate
    B. Cytotoxic (killer) T cells: produce lymphokines that attract and activate macrophages and lymphotoxins that directly destroy antigens
    C. Natural killer (NK) cells: suppress action of cytotoxic T cells and B cells
    D. B cells: become plasma cells that secrete antibodies

8. All of the following match types of allergic reactions with correct descriptions EXCEPT:

    A. Type I: anaphylactic shock, for example from exposure to iodine or bee venom
    B. Type II: transfusion reaction in which recipient's IgG or IgM antibodies attack donor's red blood cells
    C. Type III: Involves Ag-Ab-complement complexes that cause inflammations, as in lupus (SLE) or rheumatoid arthritis (RA)
    D. Type IV: cell-mediated reactions that involve immediate responses by B cells, as in the TB skin test.

9. Choose the FALSE statement about lymphatic vessels.

 A. Lymph capillaries are more permeable than blood capillaries.
 B. Lymphatics have thinner walls than veins.
 C. Like arteries, lymphatics contain no valves.
 D. Lymph vessels are blind-ended.

10. Choose the FALSE statement about non-specific defenses.

 A. Complement functions protectively by being converted into histamine.
 B. Histamine increases permeability of capillaries so leukocytes can more readily reach the infection site.
 C. Complement and properdin are both enzymes found in serum.
 D. Opsonization enhances phagocytosis.

11. Choose the FALSE statement about T cells.

 A. Some are called cytotoxic cells.
 B. They are called T cells because they are processed in the thymus.
 C. They are involved primarily in humoral immunity.
 D. Like B cells, they originate from stem cells in bone marrow.

12. Choose the one TRUE statement about AIDS.

 A. It appears to be caused by a virus that carries its genetic code in DNA.
 B. Persons at highest risk are homosexual men and women.
 C. Persons with AIDS experience a decline in helper T cells and in antibody production.
 D. AIDS is transmitted primarily through semen and saliva.

13. Choose the FALSE statement about lymph nodes.

 A. Lymphocytes are produced here.
 B. Lymph nodes are distributed evenly throughout the body, with equal numbers in all tissue.
 C. Lymph may pass through several lymph nodes in a number of regions before returning to blood.
 D. Lymph nodes are shaped roughly like kidney (or lima) beans.

14. Choose the FALSE statement about lymphatic organs.

 A. The palatine tonsils are the ones most often removed in a tonsillectomy.
 B. The spleen is the largest lymphatic organ in the body.
 C. The thymus reaches its maximum size at age 40.
 D. The spleen is located in the upper left quadrant of the abdomen.

15. Choose the FALSE statement.

 A. Skeletal muscle contraction aids lymph flow.
 B. Mucous membranes are less effective than skin in preventing entrance of microbes into the body.
 C. Interferon is produced by viruses.
 D. An allergen is an antigen, not an antibody.

16. Choose the TRUE statement about antigens.

 A. Antigens are usually made by the body
 B. All antigens are proteins.
 C. Microbes serve as antigens, not antibodies.
 D. Most antigens are univalent.

17. Choose the TRUE statement about antibodies.

 A. Most antibodies are foreign substances, not made by the body.
 B. Antibodies are usually composed of one light and one heavy polypeptide chain.
 C. In the uncombined state antibodies are in the Y shape.
 D. All antibodies are protein.

18. All of the following are chemicals released by cytotoxic (killer) T cells EXCEPT:

 A. Interleukin-1
 B. Lymphotoxins
 C. Interferon
 D. Macrophage activating factor (MAF)

Questions 19–20: Circle T (true) or F (false). If the statement is false, change the underlined word or phrase so that the statement is correct.

T F 19. Immunogenicity means the ability of an antigen to <u>react with</u> a specific antibody.

T F 20. A person with autoimmune disease produces <u>fewer than normal antibodies</u>.

Questions 21–25: fill-ins. Complete each sentence with the word or phrase that best fits.

_____ 21. ____ Lymphocytes provide humoral immunity, and ____ lymphocytes and macrophages offer cellular immunity.

_____ 22. ____ are a group of chemicals secreted by killer T cells that attract and activate macrophages and recruit even more killer T cells.

_____ 23. List the four fundamental signs or symptoms of inflammation.

_____ 24. The class of immunoglobins most associated with allergy are Ig ____ .

_____ 25. Three examples of mechanical factors that provide nonspecific resistance to disease are ____ .

ANSWERS TO MASTERY TEST: ■ Chapter 22

Arrange

1. A C B D   3. A C D B E
2. D C B A

Multiple Choice

4. E	12. C
5. D	13. B
6. A	14. C
7. C	15. C
8. D	16. C
9. C	17. D
10. A	18. A
11. C	

True-False

19. F. Stimulate formation of
20. F. Antibodies against the individual's own tissues

Fill-ins

21. B, T
22. Lymphokines
23. Redness, warmth, pain, and swelling
24. E
25. Skin, mucosa, cilia, epiglottis. Also flushing by tears, saliva, and urine.

FRAMEWORK 23

Respiratory System

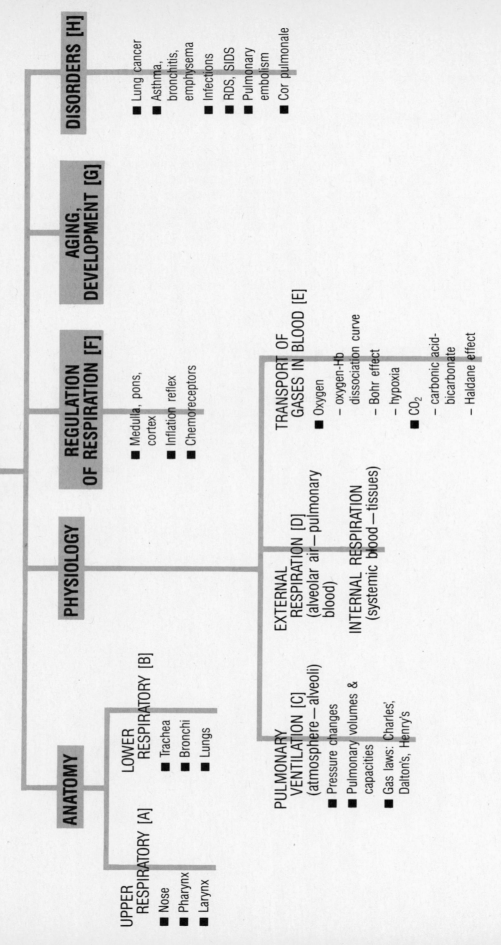

ANATOMY

UPPER RESPIRATORY [A]
- Nose
- Pharynx
- Larynx

LOWER RESPIRATORY [B]
- Trachea
- Bronchi
- Lungs

PHYSIOLOGY

PULMONARY VENTILATION [C]
(atmosphere—alveoli)
- Pressure changes
- Pulmonary volumes & capacities
- Gas laws: Charles', Dalton's, Henry's

EXTERNAL RESPIRATION [D]
(alveolar air—pulmonary blood)

INTERNAL RESPIRATION (systemic blood—tissues)

TRANSPORT OF GASES IN BLOOD [E]
- Oxygen
  – oxygen-Hb dissociation curve
  – Bohr effect
  – hypoxia
- CO_2
  – carbonic acid-bicarbonate
  – Haldane effect

REGULATION OF RESPIRATION [F]
- Medulla, pons, cortex
- Inflation reflex
- Chemoreceptors

AGING, DEVELOPMENT [G]

DISORDERS [H]
- Lung cancer
- Asthma, bronchitis, emphysema
- Infections
- RDS, SIDS
- Pulmonary embolism
- Cor pulmonale

The Respiratory System

Oxygen is available all around us. But to get an oxygen molecule to a muscle cell in the stomach or a neuron of the brain requires coordinated efforts of the respiratory system with its transport adjunct, the cardiovascular system. Oxygen traverses a system of ever-narrowing and diverging airways to reach millions of air sacs (alveoli). Each alveolus is surrounded by a meshwork of pulmonary capillaries, much like a balloon encased in nylon stocking. Here oxygen changes places with the carbon dioxide wastes in blood, and oxygen travels through the blood to reach the distant stomach or brain cells. Interference via weakened respiratory muscles (as in normal aging) or extremely narrowed airways (as in asthma or bronchitis) can limit oxygen delivery to lungs.

Decreased diffusion of gases from alveoli to blood (as in emphysema, pneumonia, or pulmonary edema) can profoundly affect total body function and potentially lead to death.

Take a couple of big breaths, and look at the Chapter 23 Framework and the key terms for each section.

TOPIC OUTLINE AND OBJECTIVES

A. Upper respiratory passageways

1. Identify the organs of the respiratory system and describe their functions.

B. Lower respiratory passageways, lungs

2. Explain the structure of the alveolar-capillary (respiratory) membrane and its function in the diffusion of respiratory gases.

C. Respiration: ventilation

3. Describe the events involved in inspiration and expiration.

D. Respiration: exchange of respiratory gases

E. Transport of respiratory gases

4. Explain how respiratory gases are carried by blood.

F. Control of respiration

5. Describe the various factors that control the rate of respiration.

G. Aging and development of the respiratory system

6. Describe the effects of aging on the respiratory system.
7. Describe the development of the respiratory system.

H. Disorders, medical terminology

8. Define bronchogenic carcinoma (lung cancer), bronchial asthma, bronchitis, emphysema, pneumonia, tuberculosis (TB), respiratory distress syndrome (RDS) of the newborn, respiratory failure, sudden infant death syndrome (SIDS), coryza (common cold), influenza (flu), pulmonary embolism, pulmonary edema, and smoke inhalation injury as disorders of the respiratory system.
9. Define medical terminology associated with the respiratory system.

WORDBYTES

Now study the following parts of words that may help you better understand terminology in this chapter.

Wordbyte	Meaning	Example	Wordbyte	Meaning	Example
ex-	out	*ex*piration	pneumo-	air	*pneumo*thorax
in-	in	*in*spiration	pulmo-	lung	*pulmo*nary
intra-	within	*intra*alveolar	spir-	breathe	*spir*ometer
-pnea	breathe	a*pnea*			

CHECKPOINTS

A. Upper respiratory passageways (pages 690–696)

A1. Explain how the respiratory and cardiovascular systems work together to accomplish gaseous exchange among the atmosphere, blood, and cells.

■ **A2.** Name the structures of the nose that are designed to carry out each of the following functions.

a. Warm, moisten, and filter air

b. Sense smell

c. Assist in speech

■ **A3.** Write *LP* (laryngopharynx), *NP* (nasopharynx), or *OP* (oropharynx) to indicate locations of each of the following structures.

_____ a. Adenoids

_____ b. Palatine tonsils

_____ c. Lingual tonsils

_____ d. Opening (fauces) from oral cavity

_____ e. Opening into larynx and esophagus

A4. On Figure LG 23-1, cover the key and identify all structures associated with the nose, palate, pharynx, and larynx. Color structures with color code ovals.

A5. Explain how the larynx prevents food from entering the trachea. Hint: notice the arrow on Figure LG 23-1.

■ **A6.** The _____ is a space between the true vocal cords. The

_____ cartilages are pyramid-shaped cartilages of the larynx attached to the vocal cords.

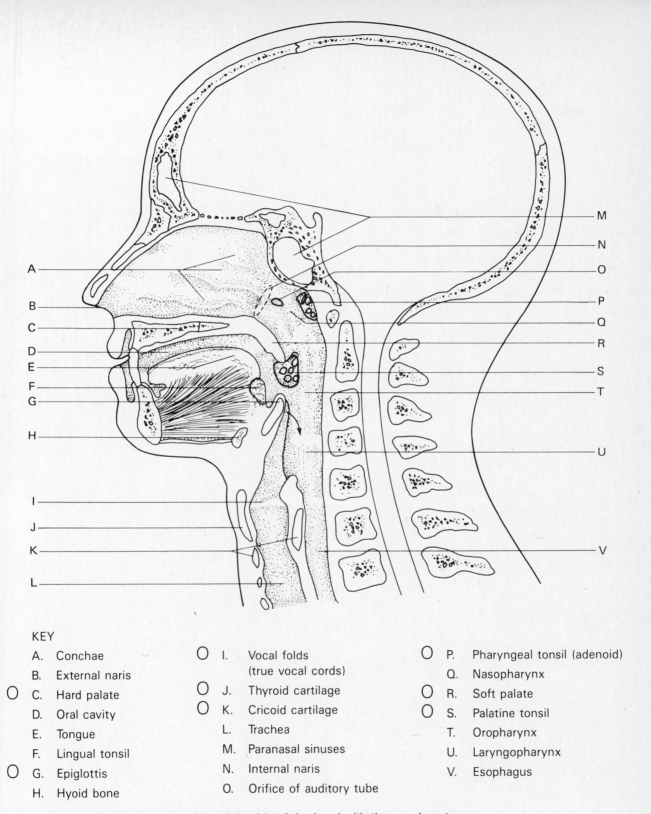

A.
B.
C.
D.
E.
F.
G.
H.
I.
J.
K.
L.

M
N
O
P
Q
R
S
T
U
V

KEY

A. Conchae
B. External naris
C. Hard palate
D. Oral cavity
E. Tongue
F. Lingual tonsil
G. Epiglottis
H. Hyoid bone

I. Vocal folds
    (true vocal cords)
J. Thyroid cartilage
K. Cricoid cartilage
L. Trachea
M. Paranasal sinuses
N. Internal naris
O. Orifice of auditory tube

P. Pharyngeal tonsil (adenoid)
Q. Nasopharynx
R. Soft palate
S. Palatine tonsil
T. Oropharynx
U. Laryngopharynx
V. Esophagus

Figure LG 23-1 Sagittal section of the right side of the head with the nasal septum removed. Color and identify labeled structures as directed in Checkpoint A4. Arrow refers to Checkpoint A5.

A7. Tell how the larynx produces sound. Explain how pitch is controlled and what causes male pitch usually to be lower than female pitch.

■ **A8.** *A clinical challenge.* After a larynx is removed (for example, due to cancer), what other structures help the laryngectomee to speak?

B. Lower respiratory passageways, lungs (pages 696-702)

■ **B1.** Complete this Checkpoint about the trachea.

a. The trachea is commonly known as the _____. It is located

(*anterior? posterior?*) to the esophagus. About _____ cm (_____ in.) long, the

trachea terminates at the _____, which is a Y-shaped intersection with the primary bronchi.

b. The tracheal wall is lined with a _____ membrane, and

strengthened by 16–18 C-shaped rings composed of _____.

■ **B2.** *A clinical challenge.* Anna has aspirated a small piece of candy. Dr. Lennon expects to find it in the right bronchus rather than in the left. Why?

B3. Identify tracheobronchial tree structures *I–N* on Figure LG 23-2, and color each of those structures indicated with color code ovals.

■ **B4.** Bronchioles (*do? do not?*) have cartilage rings. How is this fact significant during an asthma attack?

B5. On Figure LG 23-2, color the two layers of the pleura and the diaphragm according to color code ovals. Also cover the key and identify all other parts of the diagram from *A–H*.

■ **B6.** Write the correct term for each of these conditions.

a. Inflammation of the pleura: _____

b. Air in the pleural cavity: _____

c. Blood in the pleural cavity: _____

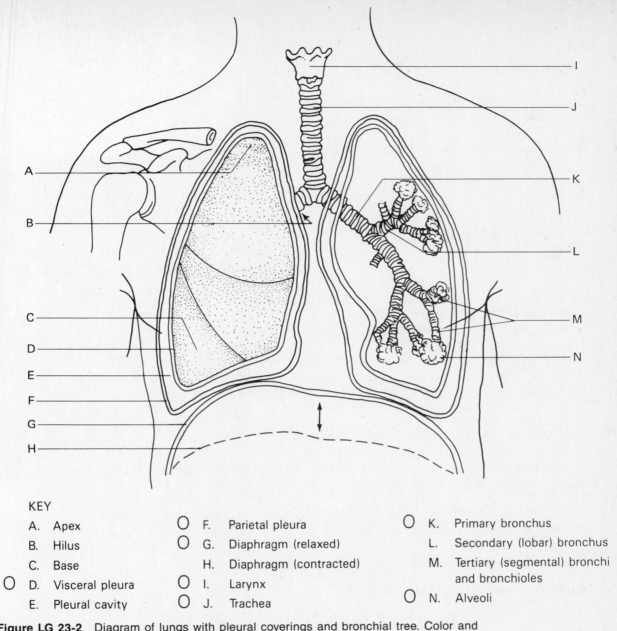

A. ━━━━━━━━

B. ━━━━━━━━

C. ━━━━━━━━

D. ━━━━━━━━

E. ━━━━━━━━

F. ━━━━━━━━

G. ━━━━━━━━

H. ━━━━━━━━

I ━━━━━━━━

J ━━━━━━━━

K ━━━━━━━━

L ━━━━━━━━

M ━━━━━━━━

N ━━━━━━━━

KEY

A. Apex

B. Hilus

C. Base

○ D. Visceral pleura

E. Pleural cavity

○ F. Parietal pleura

○ G. Diaphragm (relaxed)

H. Diaphragm (contracted)

○ I. Larynx

○ J. Trachea

○ K. Primary bronchus

L. Secondary (lobar) bronchus

M. Tertiary (segmental) bronchi
      and bronchioles

○ N. Alveoli

Figure LG 23-2 Diagram of lungs with pleural coverings and bronchial tree. Color and identify labeled structures as directed in Checkpoints B3 and B5.

■ **B7.** Answer these questions about the lungs. (As you do the exercise, locate the parts of the lung on Figure LG 23-2.)

a. The broad, inferior portion of the lung that sits on the diaphragm is called the

_____. The upper narrow apex of each lung extends just su-

perior to the _____. The costal surfaces lie against the

_____.

b. Along the mediastinal surface is the _____ where the root of

the lung is located. This root consists of _____.

c. Answer these questions with *right* or *left*. In which lung is the cardiac notch

located? _____ Which lung has just two lobes and so only

two lobar bronchi? _____Which lung has a horizontal fissure?

_____ Write *S* (superior), *M* (middle), and *I* (inferior) on the
three lobes of the right lung.

■ **B8.** Describe the structures of a lobule in this exercise.

a. Arrange in order the structures through which air passes as it enters a lobule en

route to alveoli. ____ ____ ____
   A. Alveolar ducts
   B. Respiratory bronchiole
   T. Terminal bronchiole

b. In order for air to pass from alveoli to blood in pulmonary capillaries, it must pass

through the _____ (respiratory) membrane. Identify structures
in this pathway: *A–E* on Figure LG 23-3.

c. *For extra review.* Now label layers (*1–6*) of the alveolar-capillary membrane in the
insert in Figure LG 23-3.

d. *A clinical challenge.* Normally the respiratory membrane is extremely (*thick? thin?*).
Suppose pulmonary capillary blood pressure rises dramatically, pushing extra fluid
out of blood. The presence of excess fluid in interstitial areas and ultimately in

alveoli is known as pulmonary _____. Diffusion occurs
(*more? less?*) readily through the fluid-filled membrane.

e. The alveolar epithelial layer of the respiratory membrane contains three types of
cells. Forming most of the layer are flat (*squamous? septal? macrophage?*) cells
which are ideal for diffusion of gases. Septal cells produce an important phos-

pholipid substance called _____, which reduces tendency for
collapse of alveoli due to surface tension. Macrophage cells help to remove

_____ from lungs.

C. Respiration: ventilation (pages 702–708)

■ **C1.** Match each of the three phases of respiration with the correct description.

| E. External respiration   I. Internal respiration   V. Ventilation |

_____ a. Exchange of air between atmosphere   _____ c. Exchange of gas between blood and
             and alveoli   cells
_____ b. Exchange of gas between alveoli and
             blood

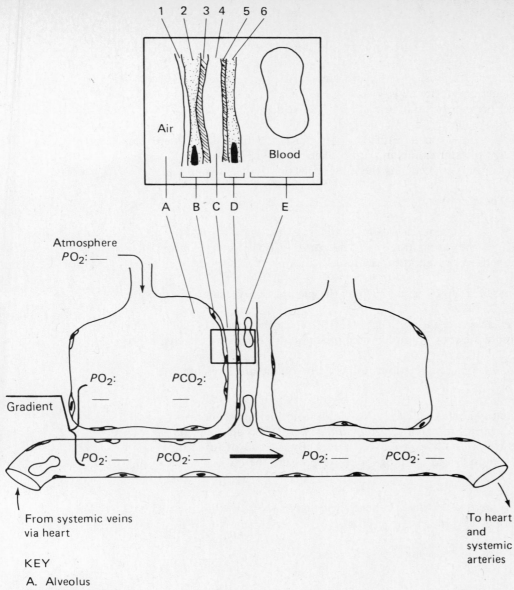

Atmosphere
PO_2: —

Gradient

PO_2: — PCO_2: —

PO_2: — PCO_2: — PO_2: — PCO_2: —

From systemic veins
via heart

To heart
and
systemic
arteries

KEY

A. Alveolus
B. Alveolar wall
C. interstitial space
D. Capillary wall
E. Blood plasma and blood cells

Figure LG 23-3   Diagram of alveoli and pulmonary capillary. Insert enlarges alveolar-capillary membrane. Numbers refer to Checkpoint B8. Complete partial pressures (in mm Hg) as directed in Checkpoint D3.

■ **C2.** Refer to Figure 23-12e (page 705) in the text and describe the process of ventilation in this exercise.

a. In diagram (a), just before the start of inspiration, pressure within the lungs (called

_____) is _____ mm Hg. This is (*more than? less than? the same as?*) atmospheric pressure.

b. At the same time, pressure within the pleural cavity (called

_____ pressure) is _____ mm Hg. This is (*more than? less than? the same as?*) intrapulmonic and atmospheric pressure.

c. The first step in inspiration occurs as the muscles in the floor and walls of the

thorax contract. These are the _____ and

_____ muscles. Note in diagram (b) (and also in Figure LG 23-2) that the size of the thorax (*increases? decreases?*). Since the two layers of pleura tend to adhere to one another, the lungs will (*increase? decrease?*) in size also.

d. Increase in volume of a closed space such as the pleural cavity causes the pressure

there to (*increase? decrease?*) to _____ mm Hg. Since the lungs also increase in size (due to pleural cohesion), intrapulmonic pressure also (*increases? decreases?*) to

_____ mm Hg. This inverse relationship between volume and pressure is a

statement of _____'s law.

e. A pressure gradient is now established. Air flows from high pressure area (*alveoli? atmosphere?*) to low pressure area (*alveoli? atmosphere?*). So air flows (*into? out of?*) lungs. Thus (*inspiration? expiration?*) occurs. By the end of inspiration, sufficient air will have moved into the lungs to make pressure there equal to atmospher-

ic pressure, that is, _____ mm Hg.

f. Inspiration is a(n) (*active? passive?*) process, whereas expiration is normally a(n) (*active? passive?*) process. In certain respiratory disorders, such as COPD, accessory

muscles help to force air out. Name two of these: _____

_____

■ **C3.** Answer these questions about compliance.

a. Imagine trying to blow up a new balloon. Initially, the balloon resists your efforts. Compliance is the ability of a substance to yield elastically to a force; in this case it is the ease with which a balloon can be inflated. So a new balloon has a (*high? low?*) level of compliance, whereas a balloon that has been inflated many times has (*high? low?*) compliance.

b. Similarly, alveoli that inflate easily have (*high? low?*) compliance. The presence of a

coating called _____ lining the inside of alveoli prevents alveolar walls from sticking together during ventilation, and so (*increases? decreases?*) compliance.

c. Surfactant production is especially developed during the final weeks before birth. A premature infant may lack adequate surfactant; this disorder is known as

_____.

d. Collapse of all or part of a lung may occur as a result of lack of surfactant or

other factors. This lung collapse is known as _____.

C4. Refer to Exhibit 23-1 (page 707) in your text. Read carefully descriptions of modified respiratory movements such as sobbing, yawning, sighing, or laughing while demonstrating those actions yourself.

C5. Each minute the average adult takes _____ breaths (respirations). Check your

own respiratory rate and write it here: _____ breaths per minute.

■ **C6.** Maureen is breathing at the rate of 15 breaths per minute. She has a tidal volume

of 480 ml per breath. Her *minute volume of respiration* is _____.

■ **C7.** Of the total amount of air that enters the lungs with each breath, about (*99%? 70%? 30%? 5%?*) actually enters the alveoli. The remaining amount of air is much like the last portion of a crowd trying to rush into a store: it does not succeed in entering the alveoli during an inspiration, but just reaches airways and then is quickly ushered out during the next expiration. Such air is known as anatomic

_____ and constitutes about _____ ml of a typical breath.

■ **C8.** Match the lung volumes and capacities with the descriptions given. You may find it helpful to refer to Figure 23-14 (page 708) in the text.

ERV. Expiratory reserve volume	TLC. Total lung capacity
FRC. Functional residual capacity	TV. Tidal volume
IRV. Inspiratory reserve volume	VC. Vital capacity
RV. Residual volume	

_____ a. The amount of air taken in with each inspiration during normal breathing is called ____.

_____ b. At the end of a normal expiration the volume of air left in the lungs is called ____. Emphysemics who have lost elasticity of their lungs cannot exhale adequately, so this volume will be large.

_____ c. Forced exhalation can remove some of the air in FRC. The maximum volume of air that can be expired beyond normal expiration is called ____. This volume will be small in emphysema patients.

_____ d. Even after the most strenuous expiratory effort, some air still remains in the lungs; this amount, which cannot be removed vountarily, is called ____.

_____ e. The volume of air that represents a person's maximum breathing ability is called ____. This is the sum of ERV, TV, and IRV.

_____ f. Adding RV to VC gives ____.

_____ g. The excess air a person can take in after a normal inhalation is called ____.

■ **C9.** Indicate normal volumes for each of the following.

a. TV = _____ ml

b. TLC = _____ ml (_____ liters)

c. VC = _____ ml

D. Respiration: exchange of respiratory gases (pages 708–712)

■ **D1.** Check your understanding of gas laws by matching the correct law with the condition that it explains.

| B. Boyle's law | D. Dalton's law | H. Henry's law |

_____ a. If a patient breathes air highly concentrated in oxygen (as in a hyperbaric chamber), a higher percentage of oxygen will dissolve in the blood and tissues.

_____ b. The total atmospheric pressure (760 mm Hg) is due mostly to pressure caused by nitrogen, partly to PO_2, and slightly to PCO_2.

_____ c. Under high PN_2 more nitrogen dissolves in blood. The *bends* occurs when pressure decreases and nitrogen forms bubbles in tissue as it comes out of solution.

_____ d. As the size of the thorax increases, the pressure within it decreases.

■ **D2.** Refer to Figure LG 23-4 and do this exercise.

a. The atmosphere contains enough gaseous molecules to exert pressure upon a

column of mercury to make it rise about ____ mm. The atmosphere is said to have a pressure of 760 mm Hg.

b. Air is about ____ percent nitrogen and ____ percent oxygen. To show this, color the nitrogen molecules *green* and oxygen molecules *red* in Figure LG 23-4b. Note that

carbon dioxide is not represented in the figure since only about ____ percent of air is CO_2.

c. Dalton's law explains that of the total 760 mm of atmospheric pressure, a certain amount is due to each type of gas. Determine the portion of the total pressure due to nitrogen molecules:

79% × 760 mm Hg = _____ mm Hg

This is the partial pressure of nitrogen, or _____.

d. Calculate the partial pressure of oxygen.

e. In Figure LG 23-4a, color the part of the Hg column due to N_2 pressure (up to the 600 mm Hg mark) *green*. Then color the portion of the Hg column due to O_2 pressure *red*.

■ **D3.** Answer these questions about external and internal respiration.

a. A primary factor in the diffusion of gas across a membrane is the difference in

concentration of the gas (reflected by _____ pressures) on the two sides of the membrane. On Figure LG 23-3 write values for PO_2 (in mm Hg) in each of the following areas. (Refer to text Figure 23-16, page 711 for help.)
    Atmospheric air (Recall this value from Figure LG 23-4.)
    Alveolar air (Note that this value is lower than that for atmospheric PO_2 since some alveolar O_2 enters blood)
    Blood entering lungs

b. Calculate the PO_2 difference (gradient) between alveolar air and blood entering lungs.

____ mm Hg – ____ mm Hg = ____ mm Hg

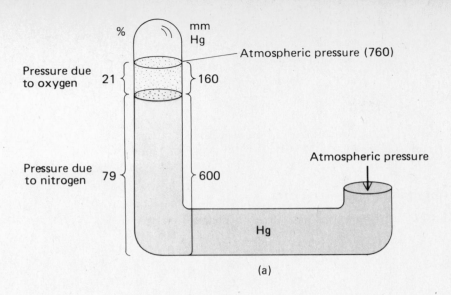

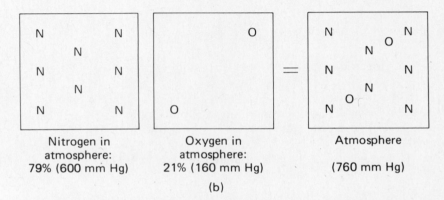

Figure LG 23-4 Atmospheric pressure. (a) Effect of gas molecules on column of mercury (Hg). (b) Partial pressures contributing to total atmospheric pressure. Color according to directions in Checkpoint D2.

c. Three other factors that increase exchange of gases between alveoli and blood are: (*large? small?*) surface area of lungs; (*thick? thin?*) respiratory membrane; and (*increased? decreased?*) blood flow through lungs (as in exercise).

d. By the time blood leaves lungs to return to heart and systemic arteries, its PO_2 is normally (*greater than? the same as? less than?*) PO_2 of alveoli. Write the correct value on Figure LG 23-3.

e. Now fill in all three PCO_2 values on Figure LG 23-2.

f. *A clinical challenge.* Erin has blood drawn from her brachial artery to determine her arterial blood gases. Her PO_2 is 56 and her PCO_2 is 48. Are these typical

values for a healthy adult? _____

■ **D4.** With increasing altitude the air is "thinner," that is, gas molecules are farther apart, so atmospheric pressure is lower. Atop a 25,000 foot mountain, this pressure is only 282 mm Hg. Oxygen still accounts for 21 percent of the pressure. What is PO_2 at that level?

From this calculation you can see limitations of life (or modifications that must be made) at high altitudes. If atmospheric PO_2 is 59.2, neither alveolar nor blood PO_2 could surpass that level.

E. Transport of respiratory gases (pages 712–716)

■ **E1.** Answer these questions about oxygen transport. Refer to Figure 23-17 (page 713) in your text.

a. One hundred ml of blood contains about _____ ml of oxygen. Of this, over 19.5

ml is carried as _____. Only a small amount of oxygen is carried in the dissolved state since oxygen has a (*high? low?*) solubility in blood or water.

b. Oxygen is attached to the _____ atoms in hemoglobin. The

chemical formula for oxyhemoglobin is _____. When hemoglobin carries all of

the oxygen it can hold, it is said to be fully _____. High PO_2 in alveoli will tend to (*increase? decrease?*) oxygen saturation of hemoglobin.

c. Refer to Figure 23-18 (page 714) in your text. Note that arterial blood, with a PO_2

of about 105 mm Hg, has its hemoglobin _____% saturated with oxygen. (This

may be expressed as SO_2 = _____%.)

d. List four factors that will enhance the dissociation of oxygen from hemoglobin so that oxygen can enter tissues. (*Hint:* Think of conditions within active muscle tissue.)

e. *For extra review.* Demonstrate the effect of temperature on the oxygen–hemoglobin dissociation curve. Draw a vertical line on Figure 23-20 (page 715) of the text at PO_2 = 40 (the value for venous blood). Compare the percent saturation of hemoglobin (SO_2) at these two body temperatures:

38°C (100.4°F) ____% SO_2

43°C (109.4°F) ____% SO_2

In other words, at a higher body temperature, (*more? less?*) oxygen will be attached to hemoglobin, while (*more? less?*) oxygen will enter tissues to fuel metabolism.

f. By the time blood enters veins to return to the heart, its oxygen saturation of

hemoglobin (SO_2) is about ____%. Note on Figure 23-18 (page 714) of your text that although PO_2 drops from 105 in arterial blood to 40 in venous blood, oxygen saturation drops (*more? less?*) dramatically, that is, from 97% to 75%. Of what significance is this?

■ **E2.** Carbon monoxide has about _____ times the affinity that oxygen has for hemoglobin. State the significance of this fact.

E3. State one example of use of 100% oxygen and discuss potential problems of such practice.

E4. Contrast *anemic hypoxia* with *stagnant hypoxia*.

■ **E5.** Complete this Checkpoint about transport of carbon dioxide.

a. Write the percentage of CO_2 normally carried in each of these forms: _____ per-

cent is present in bicarbonate ion (HCO_3^-); _____ percent is bound to the globin

portion of hemoglobin; _____ percent is dissolved in plasma.

b. Carbon dioxide (CO_2) produced by cells of your body diffuses into red blood cells

(RBCs) and combines with water to form _____

c. Carbonic acid tends to dissociate into two products. One is H^+ which binds to

_____. The other product is _____

(bicarbonate), which is carried in (*RBCs? plasma?*) in exchange for a (*K^+? Cl^-?*)
ion that shifts into the RBC.

d. Now write the entire sequence of reactions described in (*a–c*). Be sure to include the enzyme that catalyzes the first reaction. Notice that the reactions show that increase in CO_2 in the body (as when respiratory rate is slow) will tend to cause a build-up of acid (H^+) in the body.

$$CO_2 + \underline{\hspace{2cm}} \xrightarrow[\text{(enzyme)}]{} \underline{\hspace{3cm}} \rightarrow$$

$$H^+ + \underline{\hspace{3cm}}$$
   ↓   ↓
Binds Shifts to plasma in
to Hb exchange for Cl^-

E6. Note that as the red blood cells reach lung capillaries, the same reactions you just studied occur, but in reverse. Study Figure 23-22b (page 717) in the text carefully. Then list the major steps that occur in the lungs so that CO_2 can be exhaled.

F. Control of respiration (pages 717–721)

■ **F1.** Complete the table about respiratory control areas. Indicate whether the area is located in the medulla (*M*) or pons (*P*).

Name	M/P	Function
a.	M	Controls rhythm; consists of inspiratory and expiratory areas
b. Pneumotaxic		
c.		Prolongs inspiration and inhibits expiration

■ **F2.** Answer these questions about respiratory control.

a. The main chemical change that stimulates respiration is increase in blood level of

____, which is directly related to (*decrease in PO₂? increase in PCO₂?*) of blood.

b. Cells most sensitive to changes in blood CO_2 are located in the (*medulla? pons? aorta and carotid arteries?*).

c. An increase in arterial blood PCO_2 is called _____. Write an

arterial PCO_2 value that is hypercapnic. ____ mm Hg (*Even slight? Only severe?*) hypercapnia will stimulate the respiratory system, leading to (*hyper? hypo?*)-ventilation.

d. State two locations of chemoreceptors sensitive to changes in PO_2.

(*Even slight? Only severe?*) decreases in PO_2 level of blood will stimulate these chemoreceptors and lead to hyperventilation. Give an example of a PO_2 low

enough to evoke such a response. ____ mm Hg

e. Increase in body temperature (as in fever), as well as stretching of the anal sphincter, will cause ____-crease in the respiratory rate.

f. Take a deep breath. Imagine the _____ receptors in your airways being stimulated. These will cause (*excitation? inhibition?*) of the inspiratory and apneustic areas, resulting in expiration. This reflex, known as the

_____ reflex, prevents overinflation of lungs.

G. Aging and development of the respiratory system (pages 721–722)

■ **G1.** Describe possible effects of these age-related changes.

a. Pulmonary blood vessels become sclerosed, so are more rigid and resistant to blood flow.

b. Chest wall becomes more rigid as bones and cartilage lose flexibility.

c. Decreased macrophage and ciliary action of lining of respiratory tract.

■ **G2.** Describe development of the respiratory system in this exercise.

a. The laryngotracheal bud is derived from _____-derm. List structures formed from this bud.

b. Identify portions of the respiratory system derived from mesoderm.

H. Disorders, medical terminology (pages 722–727)

H1. A host of effects upon the respiratory system are associated with smoking. Describe these two.

a. Bronchogenic carcinoma (Include the roles of basal cells and excessive mucus production.)

b. Emphysema (Note that as walls of alveoli break down, surface area of the

respiratory membrane ____-creases, so amount of gas diffused ____-creases also.)

■ **H2.** Match the condition with the correct description.

BA. Bronchial asthma	P. Pneumonia
CB. Chronic bronchitis	RF. Respiratory failure
D. Dyspnea	TB. Tuberculosis
E. Emphysema	

_____ a. Permanent inflation of lungs due to loss of elasticity; rupture and merging of alveoli, followed by their replacement by fibrous tissue

_____ b. Inflammation of bronchi with excessive mucus production for at least three months per year for at least two consecutive years

_____ c. Acute infection or inflammation of alveoli which fill with fluid

_____ d. Spasms of small passageways with wheezing and dyspnea

_____ e. Caused by a species of Mycobacterium; lung tissue is destroyed and replaced with inelastic connective tissue

_____ f. Difficult, painful breathing; shortness of breath

_____ g. Occurs when PO_2 drops below 50 mm Hg and PCO_2 rises above 50 mm Hg

_____ h. A group of conditions known as chronic obstructive pulmonary disease (COPD) (3 answers)

■ **H3.** Discuss respiratory distress syndrome (RDS) in this exercise.

a. Before birth fetal lungs are filled with (*air? fluid?*). Inflation of lungs after birth depends largely on the presence of a chemical known as

_____. This chemical is a (*lipoprotein? phospholipid?*) produced by (*basal cells of bronchi? septal cells of alveoli?*).

b. Surfactant (*raises? lowers?*) surface tension so that alveolar walls are less likely to stick together. Thus surfactant (*facilitates? inhibits?*) lung inflation.

c. RDS especially targets (*premature? full term?*) infants. Explain why.

H4. To what does SIDS refer? _____
What is the mortality rate of SIDS?

Discuss possible causes of SIDS.

■ **H5.** Contrast pulmonary embolism (*PE*) with pulmonary edema (*P edema*) by identifying the related descriptions.

_____ a. Defined as abnormal accumulation of fluid in interstitial areas and/or alveoli of lungs

_____ b. Likely to occur if infection such as pneumonia increases permeability of pulmonary capillaries

_____ c. Likely to occur if left ventricle fails so blood backlogs and increases pressure in pulmonary capillaries

_____ d. Defined as presence of blood clot or other foreign material in pulmonary vessels, thus obstructing blood flow

_____ e. Likely to result from deep vein thrombosis (DVT), especially in bedridden persons, or from release of fat from marrow of fractured bones

H6. Explain how the abdominal thrust (Heimlich maneuver) helps to remove food that might otherwise cause death by choking.

■ **H7.** Write the *ABC*'s of CPR.

ANSWERS TO SELECTED CHECKPOINTS

A2. (a) Mucosa lining nose, septum, conchae, meati and sinuses, lacrimal drainage, and coarse hairs in vestibule. (b) Olfactory region lies superior to superior nasal conchae. (c) Sounds resonate in sinuses surrounding nose.
A3. (a) NP. (b–d) OP. (e) LP.
A6. Glottis, arytenoid.
A8. Oral and nasal cavities, pharynx and paranasal sinuses continue to act as resonating chambers. Muscles of face, tongue, lips, pharynx, and esophagus also help in forming words.
B1. (a) Windpipe, anterior, 12 (4.5), carina. (b) Mucous, cartilage.
B2. The right bronchus is more vertical and slightly wider than the left.

B4. Do not; spasms can collapse airways.
B6. (a) Pleurisy. (b) Pneumothorax. (c) Hemothorax.
B7. (a) Base, clavicle, ribs. (b) Hilus; bronchi, pulmonary vessels and nerves. (c) Left, left, right.
B8. (a) T R A. (b) Alveolar–capillary. See key to Figure LG 23-3. (c) 1, Surfactant; 2, alveolar epithelium; 3, epithelial basement membrane; 4, interstitial space; 5, capillary basement membrane; 6, capillary endothelium. (d) Thin, edema, less. (e) Squamous, surfactant, debris.
C1. (a) V. (b) E. (c) I.

C2. (a) Intraalveolar or intrapulmonic, 760, the same as. (b) Intrathoracic or intrapleural, 756; less than. (c) Diaphragm, external intercostal, increases, increase. (d) Decrease, 754, decreases, 758, Boyle. (e) Atmosphere, alveoli, into, inspiration, 760. (f) Active, passive; abdominal and internal intercostals.

C3. (a) Low, high. (b) High, surfactant, increases. (c) Respiratory distress syndrome (or hyaline membrane disease). (d) Atelectasis.

C6. 7200 ml/min (= 15 breaths/min × 480 ml/breath).

C7. 70%, dead space, 150.

C8. (a) TV. (b) FRC. (c) ERV. (d) RV. (e) VC. (f) TLC. (g) IRV.

C9. (a) 500. (b) 6,000, 6. (c) 4,800.

D1. (a) H. (b) D. (c) H. (d) B.

D2. (a) 760. (b) 79, 21, 0.04. (c) 600.4, PN_2. (d) 21% × 760 mm Hg = about 160 mm Hg = PO_2.

D3. (a) Partial; see Figure LG 23-3A. (b) 105 − 40 =65. (c) Large, thin, increased. (d) The same as, 105. (e) See Figure LG 23-3A. (f) No. Typical brachial arterial values are same as for alveoli or blood leaving lungs: PO_2 = 105, PCO_2 = 40. These values indicate inadequate gas exchange.

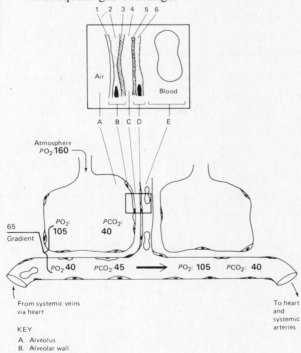

Figure LG 23-3A Diagram of alveoli and pulmonary capillary. Insert enlarges alveolar-capillary membrane.

D4. 282 × 21% = 59.2 mm Hg.

E1. (a) 20, oxyhemoglobin, low. (b) Iron, HbO_2, saturated, increase. (c) 97, 97. (d) Increase in temperature, PCO_2, acidity, and DPG. (e) 63, 36, less, more. (f) 75, less. In the event that respiration is temporarily halted, even venous blood has much oxygen attached to hemoglobin and available to tissues.

E2. 200, oxygen-carrying capacity is drastically reduced.

E5. (a) 70, 23, 7. (b) H_2CO_3 (carbonic acid). (c) Hemoglobin (as H•Hb); HCO_3^-, plasma, Cl^-. (d) $CO_2 + H_2O \xrightarrow{\text{carbonic anhydrase}} H_2CO_3$ (carbonic acid) → $H^+ + HCO_3^-$.

F1.

Name	M/P	Function
a. **Medullary rhythmicity**	M	Controls rhythm; consists of inspiratory and expiratory areas
b. Pneumotaxic	P	**Limits inspiration and facilitates expiration**
c. **Apneustic**	P	Prolongs inspiration and inhibits expiration

F2. (a) H^+, increase in PCO_2. (b) Medulla. (c) Hypercapnia, any value higher than 40, Even slight, hyper. (d) Aortic and carotid bodies, only severe, usually below 60. (e) In. (f) Stretch, inhibition, inflation (Hering-Breuer).

G1. (a) Blood backs up into right ventricle, causing it to pump extra hard and hypertrophy. This is cor pulmonale. (b) Decreased vital capacity and so decreased arterial PO_2. (c) At risk for pneumonia.

G2. (a) Endo; lining of larynx, trachea, bronchial tree, and alveoli. (b) Smooth muscle, cartilage, and other connective tissues of airways.

H2. (a) E. (b) CB. (c) P. (d) BA. (e) TB. (f) D. (g) RF. (h) BA, CB, E.

H3. (a) Fluid, surfactant, phospholipid, septal cells of alveoli. (b) Lowers, facilitates. (c) Premature. Septal cells do not produce adequate amounts of surfactant until between weeks 28–32 of the 39-week human gestational period. A baby born at 7 months (30 weeks), for example, would be at high risk for RDS.

H5. (a) P edema. (b) P edema. (c) P edema. (d) PE. (e) PE.

H7. Establish Airway, ventilate (Breathing), and reestablish Circulation.

Questions 1–4: Arrange the answers in correct sequence.

_____ _____ _____ 1. From first to last, the steps involved in inspiration:

    A. Diaphragm and intercostal muscles contract
    B. Thoracic cavity and lungs increase in size
    C. Intrapulmonic pressure decreases to 758 mm Hg

_____ _____ _____ 2. From most superficial to deepest:

    A. Parietal pleura
    B. Visceral pleura
    C. Pleural cavity

_____ _____ _____ _____ _____ 3. From superior to inferior:

    A. Bronchioles
    B. Bronchi
    C. Larynx
    D. Pharynx
    E. Trachea

_____ _____ _____ _____ _____ _____ 4. Pathway of inspired air:

    A. Alveolar ducts
    B. Bronchioles
    C. Lobar bronchi
    D. Primary bronchi
    E. Segmental bronchi
    F. Alveoli

Questions 5–9: Circle the letter preceding the one best answer to each question.

5. All of the following terms are matched correctly with descriptions EXCEPT:

    A. Internal nares: choanae
    B. External nares: nostrils
    C. Posterior of nose: vestibule
    D. Pharyngeal tonsils: adenoids

6. Which of these values (in mm Hg) would be most likely for PO_2 of blood in the femoral artery?

    A. 40   D. 160
    B. 45   E. 760
    C. 100   F. 0

7. Pressure and volume in a closed space are inversely related, as described by ___ law.

    A. Boyle's   C. Dalton's
    B. Starling's   D. Henry's

8. Choose the correct formula for carbonic acid:

    A. HCO_3^- D. HO_3C_2
    B. H_3CO_2 E. H_2C_3O
    C. H_2CO_3

9. A procedure in which an incision is made in the trachea and tube inserted into the trachea is known as a:

    A. Tracheostomy   C. Intubation
    B. Bronchogram   D. Pneumothorax

Questions 10–20: Circle T (true) or F (false). If the statement is false, change the underlined word or phrase so that the statement is correct.

T   F   10. In the chloride shift Cl⁻ moves into red blood cells in exchange for $\underline{H^+}$.

T   F   11. When chemoreceptors sense increase in PCO_2 or increase in acidity of blood (H^+), respiratory rate will be stimulated.

T   F   12. Both increased temperature and increased acid content tend to cause oxygen to bind more tightly to hemoglobin.

T   F   13. Under normal circumstances intrapleural pressure is always less than atmospheric.

T   F   14. The pneumotaxic and apneustic areas controlling respiration are located in the pons.

T   F   15. Fetal hemoglobin has a lower affinity for oxygen than maternal hemoglobin does.

T   F   16. Most CO_2 is carried in the blood in the form of bicarbonate.

T   F   17. The alveolar wall does contain macrophages that remove debris from the area.

T   F   18. Intrapulmonic pressure means the same thing as intrapleural pressure.

T   F   19. Inspiratory reserve volume is normally larger than expiratory reserve volume.

T   F   20. The PO_2 and PCO_2 of blood leaving the lungs are about the same as PO_2 and PCO_2 of alveolar air.

Questions 21–25: fill-ins. Complete each sentence with the word or phrase that best fits.

_____ 21. The process of exchange of gases between alveolar air and blood in pulmonary capillaries is known as ____.

_____ 22. Take a normal breath and then let it out. The amount of air left in your lungs is the capacity called ____ and it usually measures about ____ ml.

_____ 23. The Bohr effect states that when more H^+ ions are bound to hemoglobin, less ____ can be carried by hemoglobin.

_____ 24. The epiglottis, thyroid, and cricoid cartilages are all parts of the ____.

_____ 25. ____ is a chemical that lowers surface tension and therefore increases inflatability (compliance) of lungs.

ANSWERS TO MASTERY TEST: ■ Chapter 23

Arrange

1. A B C
2. A C B
3. D C E B A
4. D C E B A F

Multiple Choice

5. C   8. C
6. C   9. A
7. A

True-False

10. F. HCO₃⁻
11. T
12. F. Dissociate from
13. T
14. T
15. F. Higher
16. T
17. T
18. F. Intraalveolar
19. T
20. T

Fill-ins

21. External respiration (or diffusion)
22. Functional residual capacity (FRC), 2400
23. Oxygen
24. Larynx
25. Surfactant

FRAMEWORK 24

Digestive System

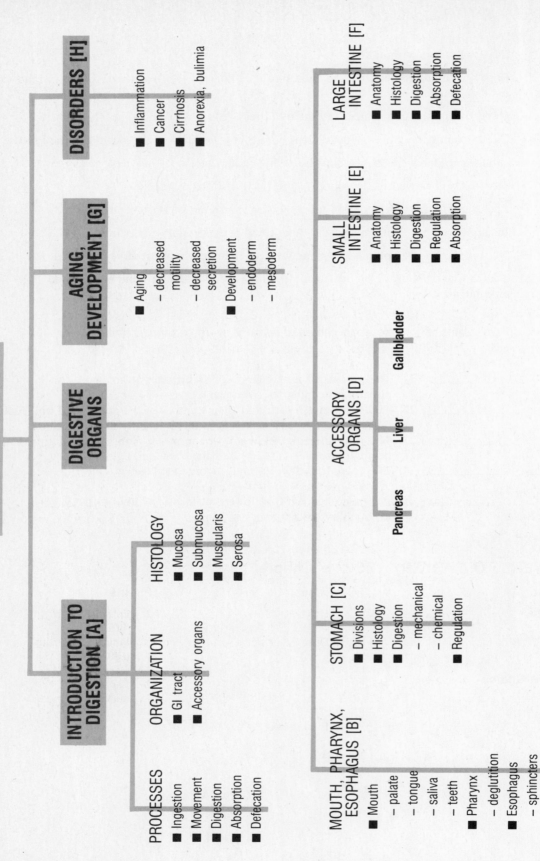

INTRODUCTION TO DIGESTION [A]

PROCESSES
- Ingestion
- Movement
- Digestion
- Absorption
- Defecation

ORGANIZATION
- GI tract
- Accessory organs

HISTOLOGY
- Mucosa
- Submucosa
- Muscularis
- Serosa

DIGESTIVE ORGANS

MOUTH, PHARYNX, ESOPHAGUS [B]
- Mouth
  – palate
  – tongue
  – saliva
  – teeth
- Pharynx
  – deglutition
- Esophagus
  – sphincters
  – peristalsis

STOMACH [C]
- Divisions
- Histology
- Digestion
  – mechanical
  – chemical
- Regulation

ACCESSORY ORGANS [D]

Pancreas   **Liver**   **Gallbladder**

SMALL INTESTINE [E]
- Anatomy
- Histology
- Digestion
- Regulation
- Absorption

LARGE INTESTINE [F]
- Anatomy
- Histology
- Digestion
- Absorption
- Defecation

AGING, DEVELOPMENT [G]
- Aging
  – decreased motility
  – decreased secretion
- Development
  – endoderm
  – mesoderm

DISORDERS [H]
- Inflammation
- Cancer
- Cirrhosis
- Anorexia, bulimia

The Digestive System

CHAPTER 24

Food comes in very large pieces (like whole oranges, stalks of broccoli, and slices of bread) that must fit into very small spaces in the human body (like liver cells or brain cells). The digestive system makes such change possible. Foods are minced and enzymatically degraded so that absorption into blood en route to cells becomes a reality. The gastrointestinal (GI) tract provides a passageway complete with mucous glands to help food slide along, muscles to propel food forward and others (sphincters and valves) that regulate flow, and secretions that break apart foods or modify pH to suit local enzymes. Accessory structures such as teeth, tongue, pancreas, and liver lie outside the GI tract, yet each contributes to the mechanical and chemical dismantling of food from bite-sized to right-sized pieces. The ultimate fate of the absorbed products of digestion is the story line of metabolism (in Chapter 25).

You will find food for thought in the Chapter 24 Framework and its terminology.

TOPIC OUTLINE AND OBJECTIVES

A. Introduction to digestion

1. Describe the mechanism that regulates food intake.
2. Identify the organs of the gastrointestinal (GI) tract and the accessory organs of digestion and their functions in digestion.

B. Mouth, pharynx and esophagus

C. Stomach

D. Accessory organs: pancreas, liver, gallbladder

E. Small intestine

3. Describe the mechanical movements of the gastrointestinal tract.
4. Explain how salivary secretion, gastric secretion, gastric emptying, pancreatic secretion, bile secretion, and small intestinal secretion are regulated.
5. Define absorption and explain how the end products of digestion are absorbed.

F. Large intestine

6. Define the processes involved in the formation of feces and defecation.

G. Aging and developmental anatomy

7. Describe the effects of aging on the digestive system.
8. Describe the development of the digestive system.

H. Disorders and medical terminology

9. Describe the clinical symptoms of the following disorders: dental caries, periodontal disease, peritonitis, peptic ulcers, appendicitis, gastrointestinal tumors, diverticulitis, cirrhosis, hepatitis, gallstones, anorexia nervosa, and bulimia.
10. Define medical terminology associated with the digestive system.

WORDBYTES

Now study the following parts of words that may help you better understand terminology in this chapter.

Wordbyte	Meaning	Example	Wordbyte	Meaning	Example
amyl-	starch	*amyl*ase	gluco-, glyco-	sugar, sweet	*gluco*se, hyper-*glyc*emia
-ase	enzyme	malt*ase*	hex-	six	*hex*ose
caec-, cec-	blind	*cec*um	lith-	stone	chole*lith*iasis
chole-, cholecyst-	gallbladder	*chole*cystectomy	-ose	sugar	lact*ose*
chym-	juice	*chym*otrypsin	-rhea	to flow	diar*rhea*
crypt-	hidden	*crypt*s of Lieberkuhn	stoma-	mouth, opening	colo*stom*y
dent-	tooth	*dent*ures	taen-	ribbon	*taen*ia coli
-ectomy	removal of	append*ectomy*	vermi-	worm	*vermi*form appendix
entero-	intestine	*entero*gastrone			
-gen	to produce	glyco*gen*			

CHECKPOINTS

A. Introduction to digestion (pages 733–736)

A1. Explain why food is vital to life. Give three specific examples of uses of foods in the body.

■ **A2.** List the five basic activities of the digestive system.

_____   _____   _____

_____   _____

■ **A3.** _____ digestion occurs by action of enzymes (such as those

in saliva) and intestinal secretion, whereas _____ digestion involves action of the teeth and muscles of the stomach and intestinal wall.

■ **A4.** Cover the key and identify all digestive organs on Figure LG 24-1. Then color the structures with color code ovals; these five structures are all (*parts of the gastrointestinal tract? accessory organs?*).

■ **A5.** Color layers of the GI wall indicated by color code ovals on Figure LG 24-2.

■ **A6.** Now match the names of layers of the GI wall with the correct description.

Muc. Mucosa	Ser. Serosa
Mus. Muscularis	Sub. Submucosa

_____ a. Also known as the peritoneum, it forms mesentery and omentum

_____ b. Consists of epithelium, lamina propria, and muscularis mucosa

_____ c. Connective tissue containing glands, nerves, blood and lymph vessels

_____ d. Consists of an inner circular layer and an outer longitudinal layer

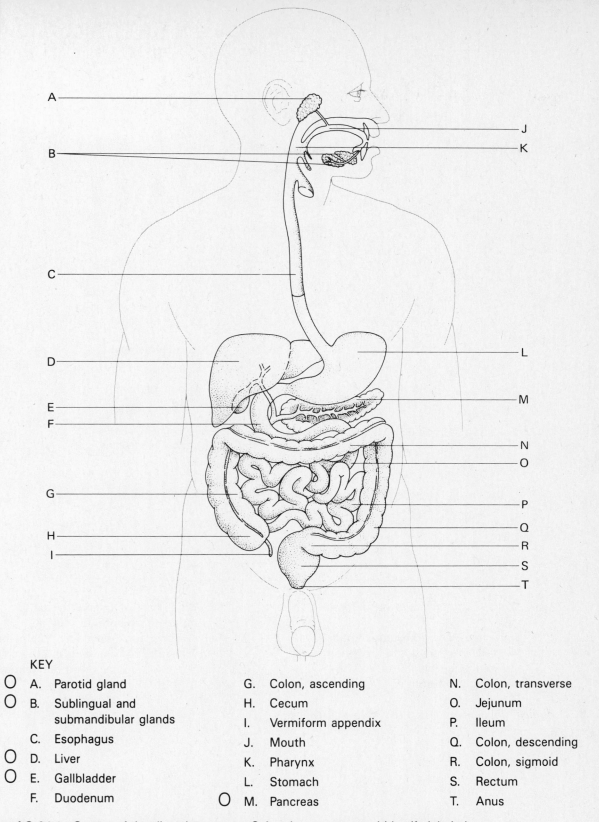

A

B

C

J

K

D

E

F

G

H

I

L

M

N

O

P

Q

R

S

T

KEY

○ A. Parotid gland

○ B. Sublingual and
       submandibular glands

C. Esophagus

○ D. Liver

○ E. Gallbladder

F. Duodenum

G. Colon, ascending

H. Cecum

I. Vermiform appendix

J. Mouth

K. Pharynx

L. Stomach

○ M. Pancreas

N. Colon, transverse

O. Jejunum

P. Ileum

Q. Colon, descending

R. Colon, sigmoid

S. Rectum

T. Anus

Figure LG 24-1 Organs of the digestive system. Color, draw arrows, and identify labeled
structures as directed in Checkpoints A4, F1, and F5.

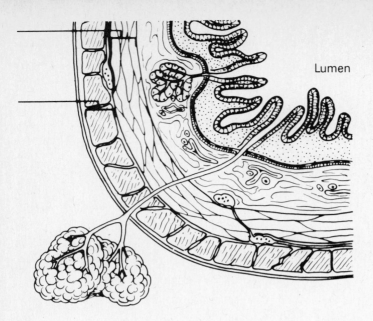

Lumen

KEY
O Mucous membrane
O Muscularis
O Serous membrane
O Submucosa layer

Figure LG 24-2 Gastrointestinal (GI) tract and related gland seen in cross section. Refer to Checkpoint A5 and color layers of the wall according to color code ovals. Label muscularis layers.

■ **A7.** Match the names of these peritoneal extensions with the correct descriptions.

F. Falciform ligament	M. Mesentery
G. Greater omentum	Meso. Mesocolon
L. Lesser omentum	

_____ a. Attaches liver to anterior abdominal wall

_____ b. Binds intestine to posterior abdominal wall; provides route for blood and lymph vessels and nerves to reach small intestine

_____ c. Binds part of large intestine to posterior abdominal wall

_____ d. "Fatty apron"; covers and helps prevent infection in small intestine

_____ e. Suspends stomach and duodenum from liver

A8. *A clinical challenge.* Define the following terms:

a. Ascites

b. Peritonitis

B. Mouth, pharynx, and esophagus (pages 736–745)

■ **B1.** Refer to Figure LG 24-3 and do this activity.

a. Label the following structures:

Hard palate	Palatine tonsils	Uvula of soft palate
Lingual frenulum	Palatoglossal arch	Vestibule

b. Color teeth in lower right of figure as indicated by color code ovals.

c. Number teeth (1–8) on the upper right according to times of eruptions. One (the first one to erupt) is done for you.

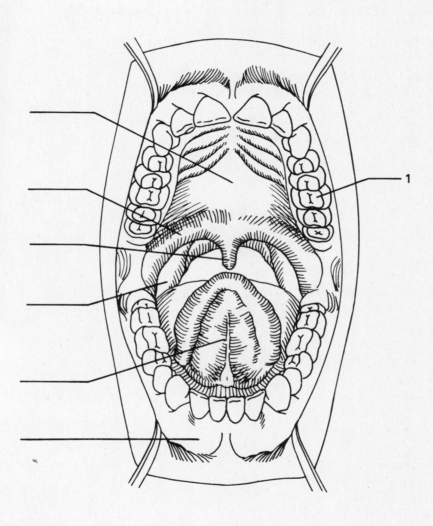

○ Cuspids   ○ Molars

○ Incisors   ○ Premolars (bicuspids)

Figure LG 24-3 Mouth (oral cavity). Complete as directed in Checkpoint B1.

B2. Contrast terms in each pair:

a. Intrinsic muscles of the tongue—extrinsic muscles of the tongue

b. Fungiform papillae— filiform papillae

B3. Identify the three salivary glands on Figure LG 24–1 and visulize their locations on yourself.

■ **B4.** Complete this exercise about salivary glands.

a. Which glands are largest? (*Parotid? Sublingual? Submandibular?*).
b. Which secrete the thickest secretion due to presence of much mucus?

_____

c. About 1 to 1½ (*tablespoons? cups? liters?*) of saliva is secreted daily.
d. State three functions of saliva.

e. The pH of the mouth is appropriate for action of salivary amylase. This is about pH (*2? 6.5? 9?*).

■ **B5.** Arrange the following in correct sequence:

_____ _____ _____ a. From most superficial to deepest:
 A. Root
 B. Neck
 C. Crown
_____ _____ _____ b. From most superficial to deepest within a tooth:
 A. Enamel
 B. Dentin
 C. Pulp cavity
_____ _____ _____ c. From hardest to softest:
 A. Enamel
 B. Dentin
 C. Pulp cavity

■ **B6.** What is the function of the periodontal ligament?

■ **B7.** Salivary amylase is an enzyme that digests _____. Most of the starch ingested (*is? is not?*) broken down by the time food leaves the mouth. What inactivates amylase in the stomach?

■ **B8.** Match each stage of deglutition with the correct description.

> E. Esophageal   V. Voluntary
> P. Pharyngeal

_____ a. Soft palate and epiglottis close off respiratory passageways.

_____ b. Tongue pushes food back into oropharynx.

_____ c. Peristaltic contractions push bolus from pharynx to stomach.

■ **B9.** Failure of the lower esophageal sphincter to close results in the sensation of

_____. Consequently, the esophageal lining may be irritated by (*acidic? basic?*) contents of the stomach that enter the esophagus. Failure of this

sphincter to relax is a condition known as _____. Resulting distention of the esophagus causes pain that may be confused with

_____ pain.

B10. Summarize digestion in the mouth, pharynx, and esophagus by completing parts a and b of Table LG 24-1.

C. Stomach (pages 745–752)

■ **C1.** Refer to Figure LG 24-1 and complete this Checkpoint.

a. Arrange these regions of the stomach according to the pathway of food from first to last:

> body   fundus
> cardia pylorus

_____ → _____ → _____ → _____

b. Choose from these answers for the following questions

> GC. Greater curvature   LC. Lesser curvature

Which is more lateral and inferior in location? _____

To which curvature is the greater omentum attached? _____

The lesser omentum? _____

Table LG 24-1 Summary of Design

Digestive Organs	Carbohydrate	Protein
a. Mouth, salivary glands	Salivary amylase: digests starch to maltose	
b. Pharynx, esophagus		
c. Stomach		
d. Pancreas		
e. Intestinal juices		
f. Liver	No enzymes for digestion of carbohydrates	
g. Large intestine	No enzymes for digestion of carbohydrates	No enzymes for digestion of proteins

Lipid	Mechanical	Other Functions
	Deglutition, peristalsis	
		1. Secretes intrinsic factor 2. Produces hormone stomach gastrin
Pancreatic lipase: digests about 80% of fats		
No enzymes for diges-tion of lipids		

■ **C2.** Complete this table about gastric secretions.

Name of Cell	Type of Secretion	Function of Secretion
a. Zygomatic (chief)		
b. Mucous		
c.	HCl	
d.	Gastrin	

■ **C3.** How does the structure of the stomach wall differ from that in other parts of the GI tract?

a. Mucosa

b. Muscularis

■ **C4.** Answer these questions about chemical digestion in the stomach.

a. Pepsin is most active at very (*acid? alkaline?*) pH.

b. State two factors that enable the stomach to digest protein without digesting its own cells (which are composed largely of protein).

c. If mucus fails to protect the gastric lining, the condition known as

_____ may result.

d. Another enzyme produced by the stomach is _____ which

digests _____. In adults it is quite (*effective? ineffective?*). Why?

■ **C5.** Answer these questions about control of gastric secretion.

a. Name the three phases of gastric secretion: _____,

_____, and _____. Which of these
causes gastric secretion to begin when you smell or taste food?

_____.

b. Gastric glands are stimulated mainly by the _____ nerves.
Their fibers are (*sympathetic? parasympathetic?*). Two stimuli which cause vagal

impulses to stimulate gastric activity are _____ and

_____.

c. The hormone _____ is released when foods, especially
(*carbohydrate? protein? lipid?*) and alcohol, reach the (*fundic? pyloric?*) region of
the stomach. This hormone travels through the blood to all parts of the body; its
target areas are gastric glands which are (*stimulated? inhibited?*) and the pyloric
sphincter which is (*contracted? relaxed?*).

d. When food (chyme) reaches the intestine, nerves initiate the

_____ reflex which (*stimultes? inhibits?*) further gastric secretion. Three hormones released by the intestine also inhibit gastric secretion as well

as gastric motility. Name these three hormones: _____,

_____, and _____.

■ **C6.** Food stays in the stomach for about ____ hours. Which food type leaves the

stomach most quickly? _____ Which type stays in the stomach

longest? _____

■ **C7.** Answer these questions about vomiting (emesis).
a. Identify the two strongest stimuli for vomiting

b. Such stimuli are transmitted to the _____ which is the site of
the vomiting center. Nerve impulses then convey instructions to (*contract? relax?*)
abdominal muscles and (*contract? relax?*) esophageal sphincters.
c. Explain how prolonged vomiting can lead to serious disturbances in homeostasis.

■ **C8.** The stomach is responsible for (*much? little?*) absorption of foods. What types of
substances are absorbed by the stomach?

C9. Complete part c of Table LG 24-1, describing the role of the stomach in digestion.

D. Accessory organs: pancreas, liver, gallbladder (pages 752–759)

D1. Refer to Figure LG 24-4 and color all of the structures with color code ovals according to colors indicated.

■ **D2.** Complete these statements about the pancreas.

a. The pancreas lies posterior to the _____.
b. The pancreas is shaped roughly like a fish, with its head in the curve of the

_____ and its tail nudging up next to the

_____.

c. The pancreas contains two kinds of glands. Ninety-nine percent of its cells produce (*endocrine? exocrine?*) secretion. One type of these secretions is (*acid? alkaline?*) fluid to neutralize the chyme entering from the stomach.
d. One enzyme in the pancreatic secretions is trypsin; it digests (*fats? carbohydrates? proteins?*). Trypsin is formed initially in the inactive form (trypsinogen) and is activated by (*HCl? NaHCO$_3$? enterokinase?*).
e. Most of the amylase and lipase produced in the body is secreted by the pancreas. Describe the functions of these two enzymes.

f. All exocrine secretions of the pancreas empty into ducts (____ and ____ on Figure

LG 24-4). These empty into the _____.
g. *A clinical challenge.* In most persons the pancreatic duct also receives bile flowing through the _____ (*F* on the figure). State one possible complication that may occur if a gallstone blocks the pancreatic duct (as at point *H* on the figure).

h. The endocrine portions of the pancreas are known as the _____

_____. These cells secrete two hormones,

_____ and _____. Typical of all hormones, these pass into (*ducts? blood vessels?*), specifically into vessels that empty

into the _____ vein.

D3. Fill in part d of Table LG 24-1, describing the role of the pancreas in digestion.

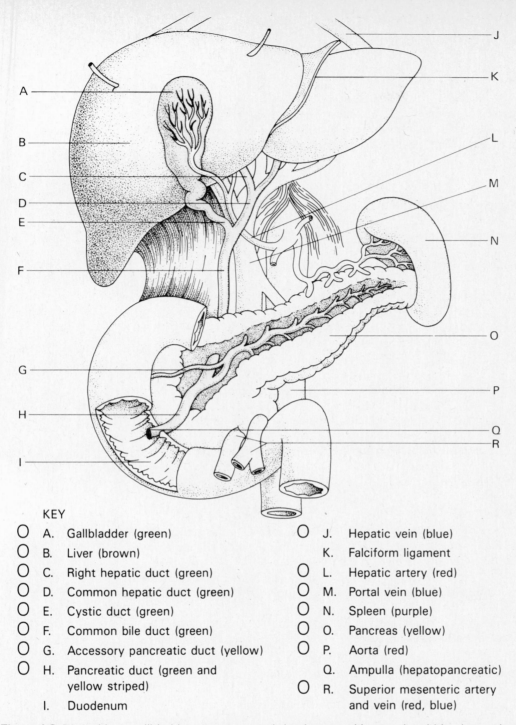

KEY

○ A.	Gallbladder (green)
○ B.	Liver (brown)
○ C.	Right hepatic duct (green)
○ D.	Common hepatic duct (green)
○ E.	Cystic duct (green)
○ F.	Common bile duct (green)
○ G.	Accessory pancreatic duct (yellow)
○ H.	Pancreatic duct (green and yellow striped)
I.	Duodenum

○ J.	Hepatic vein (blue)
K.	Falciform ligament
○ L.	Hepatic artery (red)
○ M.	Portal vein (blue)
○ N.	Spleen (purple)
○ O.	Pancreas (yellow)
○ P.	Aorta (red)
Q.	Ampulla (hepatopancreatic)
○ R.	Superior mesenteric artery and vein (red, blue)

Figure LG 24-4 Liver, gallbladder, pancreas, and duodenum, with associated blood vessels and ducts. (Stomach has been removed). Refer to Checkpoints D1, D2, D6, and D9; color and draw arrows as directed.

■ **D4.** Describe regulation of the pancreas in this exercise.

a. The pancreas is stimulated by (*sympathetic? parasympathetic?*) nerves, specifically

the _____ nerves, and also by hormones produced by cells

located in the wall of the _____ .

b. The hormone cholecystokinin (CCK) activates the pancreas to secrete fluid rich in (HCO_3^- ? *digestive enzymes such as trypsin, amylase, and lipase?*). What is the nature of the pancreatic fluid produced under the influence of secretin?

■ **D5.** Answer these questions about the liver.

a. This organ weighs about ____ kg (____ lb). It lies in the _____

quadrant of the abdomen. Of its _____ lobes, the _____ is the largest.

b. The _____ ligament separates the right and left lobes. In the edge of this ligament is the ligamentum teres (round ligament of the liver), which

is the obliterated _____ vein.

c. Blood enters the liver via vessels named _____ and

_____ . In liver lobules, blood mixes in channels called

_____ before leaving the liver via vessels named

_____ .

■ **D6.** Complete Figure LG 24-4 as directed.

a. Identify the pathway of bile from liver to intestine by following the structures you colored green in Figure LG 24-4: *C D F H*. The general direction of bile is from (*superior to inferior? inferior to superior?*).

b. Now draw arrows to show direction of blood through the liver. In order to reach the inferior vena cava, blood must flow (*superiorly? inferiorly?*) through the liver.

■ **D7.** While blood is in liver sinusoids, hepatic cells have ample opportunity to act on this blood, modify it, and add new substances to it. Check your understanding of important functions of the liver in this Checkpoint.

a. Bile secreted from the liver is composed largely of the pigment named

_____ , which is a breakdown product of

_____ cells. Excessive amounts of this pigment give skin a yel-

lowish color, a condition known as _____ . Two functions of

bile are _____ of fats and _____ of fats (and fat-soluble vitamins).

b. The liver synthesizes the anticoagulant named _____ and also most other plasma proteins. List three:

c. Name three types of cells phagocytized by the liver:

d. Identify two types of "poisons" detoxified by the liver.

e. The liver can convert excess glucose to _____ and

_____, and also reverse that process. (More on this in Chapter 25.)

f. The liver stores the four fat-soluble vitamins named ____, ____, ____, and ____. It

also stores a vitamin necessary for erythropoiesis, namely vitamin ____. And the

liver cooperates with kidneys and skin to activate vitamin ____.

g. Note that the liver (*does? does not?*) secrete digestive enzymes.

D8. Complete part f of Table LG 24-1.

■ **D9.** Imagine a terrific one-act play that you might attend on an upcoming Saturday night. The cast of characters includes all chemicals and organs identified by CAPITAL letters. The stage is set with organs/characters in position much as in Figure LG 24-4.

a. CHYME strolls in from the stomach (located offstage) and moves into the duodenum. Note that the doorway through which CHYME entered must therefore be the

_____ sphincter. CHYME today is particularly rich in food

types such as _____ (which are plastered all over CHYME's costume).

b. The presence of CHYME bearing these foods serves as a stimulus to the walls of

the _____ to secrete the hormone CCK, whose full name is

_____ (CCK enters and waves.)

c. CCK then exits from the intestinal area by way of (*common bile duct? portal vein?*). CCK (*goes directly to the gallbladder without passing any other organ? meanders in the bloodstream through all parts of the body?*). (CCK now takes a scenic trip through the blood vessels of arms, legs, ears, heart, as desired.)

d. At last CCK arrives at the GALLBLADDER, "knowing" to attach there since GALLBLADDER has specific receptors for CCK. CCK directs GALLBLADDER to (*contract? relax?*).

e. GALLBLADDER harbors many molecules of BILE SALTS. Contraction of GALLBLADDER expels BILE SALTS. These pass into the (*portal vein? cystic duct?*) and via the common bile duct to the duodenum.

f. A GALLSTONE (biliary calculus) becomes lodged in the bile pathway, such as the common bile duct. If scenes a–d of this play are repeated, how will scene e differ?

g. A cholecystectomy may need to be performed. If so, which player must terminate

his/her role in the cast? _____

h. In addition to the bile-related activities already performed by one of our stars, CCK, what other functions might CCK also portray in upcoming productions?

E. Small intestine (pages 759–768)

■ **E1.** Describe the structure of the intestine in this Checkpoint.

a. The average diameter of the small intestine is:

A. 2.5 cm (1 in.)   B. 5.0 cm (2 in.)

b. Its average length is about:

A. 3 m (10 ft)   B. 6 m (20 ft)

c. Name its three main parts (in sequence from first to last):

_____ → _____ → _____

■ **E2.** Match the correct term related to the intestine to the description that fits.

DBG. Duodenal (Brunner's) glands	MV. Microvilli
GC. Goblet cells	PC. Plicae circulares
IG. Intestinal glands	V. Villi

_____ a. Fingerlike projections up to 1 mm high that give the intestinal lining a velvety appearance and increase absorptive surface

_____ b. Fingerlike projections of plasma membrane

_____ c. Circular folds that further increase intestinal surface area

_____ d. Crypts of Lieberkuhn that secrete digestive enzymes

_____ e. One-celled glands that secrete mucus

_____ f. Glands in the submucosa; secrete alkaline substances that protect mucosa

E3. Contrast *segmentation* and *peristalsis,* the two types of movement of the small intestine.

■ **E4.** Write the main steps in the digestion of each of the three major food types.

a. *Carbohydrates*

Polysaccharides → _____ → _____ → monosaccharides

b. *Proteins*

Proteins → _____ → _____ → _____

c. *Lipids*

Neutral fats → emulsified fats → _____ _____

E5. Complete the description of functions of intestinal juices by filling in part e of Table LG 24-1. Then review digestion of each of the three major food groups by reading your columns vertically.

■ **E6.** *For extra review* of roles of GI organs in digestion, identify which chemicals in the list in the box are made by each of the following organs.

```
┌──────────────────────────────────────────────────────────┐
│   A.  Amylase              MLS. Maltase, lactase, sucrose │
│   B.  Bile                  P.  Pepsin                    │
│ CCK.  Cholecystokinin      Pd.  Peptidases                │
│  En.  Enterokinase          S.  Secretin                  │
│  EG.  Enteric gastrin      SG.  Stomach gastrin           │
│ HCl.  HCl                  TCC. Trypsin, chymotrypsin,    │
│   L.  Lipase                      carboxypeptidase        │
└──────────────────────────────────────────────────────────┘
```

_____ a. Stomach   _____ c. Small intestine

_____ b. Pancreas   _____ d. Liver

■ **E7.** *For extra review* of secretions involved with digestion, match names of secretions listed for Checkpoint E6 with descriptions below. Blank lines follow some descriptions. On these indicate whether the secretion is classified as an enzyme (*E*), hormone (*H*), or neither of these (*N*). The first one is done for you.

_____ a. Causes contraction of gallbladder and relaxation of the sphincter of the

hepatopancreatic ampulla so that bile enters duodenum __**H**__

_____ b. Stimulates bile production in the liver ____

_____ c. Stimulates pancreas to produce secretions rich in enzymes such as amylase and lipase ____

_____ d. Stimulates pancreas to secrete alkaline fluid to neutralize stomach acid ____

_____ e. Stimulates intestinal secretions and inhibits gastric secretions and motility (2 answers)

_____ f. Stimulate gastric secretions and motility (2 answers) _____ _____

_____ g. Active form of a proteolytic enzyme formed in the stomach

_____ h. Activates pepsinogen ____

_____ i. Starts protein digestion in the GI tract

_____ j. Protein-digesting enzymes produced by the pancreas

_____ k. Activates trypsinogen to trypsin ____

_____ l. Intestinal enzymes that complete the breakdown of protein to amino acids

_____ m. Most effective of this type of enzyme produced by the pancreas; digests fats

_____ n. Emulsifies fats before they can be digested effectively ____

_____ o. Starch-digesting enzymes secreted by salivary glands and pancreas

_____ p. Intestinal enzymes that complete carbohydrate breakdown, digesting disaccharides to simple sugars

■ **E8.** Describe the absorption of end products of digestion in this exercise.

a. Almost all absorption takes place in the (*large? small?*) intestine. Glucose and amino acids move into the epithelial cells lining the intestine by (*active transport? diffusion?*).

b. Simple sugars, amino acids, and short-chain fatty acids are absorbed into (*blood?*

lymph?*) capillaries which lead to the _____ vein. These sub-

stances can then be stored or metabolized in the _____.

c. Long-chain fatty acids and monoglycerides first combine with

_____ salts to form (*micelles? chylomicrons?*). This enables fatty acids and monoglycerides to enter epithelial cells in the intestinal lining and soon enter lacteals leading to the (*portal vein? thoracic duct?*). Most bile salts are ultimately (*eliminated in feces? recycled to the liver?*).

d. *A clinical challenge.* Which vitamins are absorbed with the help of bile? (*Water-soluble? Fat-soluble?*) Circle the fat-soluble vitamins: A B_{12} C D E K. Obstruction of bile pathways may lead to symptoms related to vitamin deficiency. Name two.

e. About ____ liters (or quarts) of fluids are ingested or secreted into the GI tract each day. Of this, all but about 0.5 to 1.0 liter is reabsorbed into blood capillaries in the walls of the (*small? large?*) intestine. Several hundred milliliters of fluid are also reabsorbed each day into the (*small? large?*) intestine.

f. When inadequate water reabsorption occurs, as in (*constipation? diarrhea?*), then

_____ such as Na^+ and Cl^- are also lost.

F. Large intestine (pages 768–773)

F1. Identify the regions of the large intestine in Figure LG 24-1. Draw arrows to indicate direction of movement of intestinal contents.

■ **F2.** The total length of the large intestine is about ____ m (____ ft). More than 90 per-

cent of its length consists of the part known as the _____.

■ **F3.** Match each structure in the large intestine with the related description.

AC. Anal canal	IS. Ileocecal sphincter (valve)
H. Haustra	VA. Vermiform appendix
HF. Hepatic flexure	

_____ a. Valve between small and large intestine

_____ b. Blind-ended tube attached to cecum

_____ c. Portion of colon located between ascending and transverse colons

_____ d. Pouches that give large intestine its puckered appearance

_____ e. Terminal 3 cm (1 inch) of rectum

F4. Inflammation and enlargement of the rectal veins is a condition called

_____. These may occur as a result of repeated episodes of constipation and are related to (*high? low?*)-fiber diet. Contrast first- and third-degree hemorrhoids.

■ **F5.** Taeniae coli are parts of the (*submucosa? muscularis?*) wall of the large intestine. These fibers run (*circularly? longitudinally?*); label them on Figure LG 24-1.

■ **F6.** *A clinical challenge.* Certain groups of individuals lack the normal flora of bacteria in the intestine. Name two such groups.

Name one sign or symptom resulting from inadequate bacterial synthesis of vitamin K.

F7. List the chemical components of feces.

F8. Describe the process of defecation. Include these terms: *rectal receptors, rectal muscles, sphincters, diaphragm,* and *abdominal muscles.*

F9. Describe how these two techniques can be used to help diagnose cancer of the GI tract.
a. Fecal occult blood testing

b. Colonoscopy

F10. Complete part g of Table LG 24-1, describing the role of the large intestine in digestion.

G. Aging and developmental anatomy of the digestive system (pages 773–775)

■ **G1.** *A clinical challenge.* List changes with aging that may lead to a decreased desire to eat among the elderly population.
a. Related to the upper GI tract (to stomach)

b. Related to the lower GI tract (stomach and beyond)

■ **G2.** Indicate which germ layer, endoderm (E) or mesoderm (M) gives rise to each of these structures.

_____ a. Epithelial lining and digestive glands of the GI tract

_____ b. Liver, gallbladder, and pancreas

_____ c. Muscularis layer and connective tissue of submucosa

G3. List GI structures derived from each portion of the primitive gut.

a. Foregut

b. Midgut

c. Hindgut

H. Disorders, medical terminology (pages 775–779)

H1. Explain how tooth decay occurs. Include the roles of bacteria, plaque, and acid. List the most effective known measures for preventing dental caries.

H2. Briefly describe these disorders, stating possible causes of each.

a. Periodontal disease

b. Appendicitis

c. Cirrhosis

■ **H3.** Explain why a patient may be given large doses of antibiotics prior to abdominal surgery.

■ **H4.** Which of the following is an inflammation? (_Diverticulosis? Diverticulitis?_)

456

■ **H5.** Match the terms with the descriptions.

A. Anorexia nervosa	D. Diarrhea
B. Bulimia	F. Flatus
Cho. Cholecystitis	Htb. Heartburn
Colit. Colitis	Hem. Hemorrhoids
Colos. Colostomy	Hep. Hepatitis
Con. Constipation	P. Peptic ulcer

_____ a. Incision of the colon, creating artificial anus

_____ b. Inflammation of the liver

_____ c. Inflammation of the colon

_____ d. Burning sensation in region of esophagus and stomach; probably due to gastric contents in lower esophagus

_____ e. Frequent defecation of liquid feces

_____ f. Inflammation of the gallbladder

_____ g. Infrequent or difficult defecation

_____ h. Craterlike lesion in the GI tract due to acidic gastric juices

_____ i. Excess air (gas) in stomach or intestine, usually expelled through anus

_____ j. Binge-purge syndrome

_____ k. Loss of appetite and self-imposed starvation

ANSWERS TO SELECTED CHECKPOINTS

A2. Ingestion, movement, digestion, absorption, and defecation.

A3. Chemical, mechanical.

A4. Accessory organs.

A5.

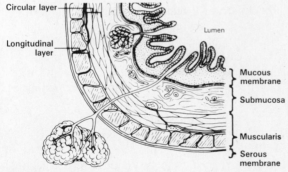

Figure LG 24-2A Gastrointestinal (GI) tract and related gland seen in cross section.

A6. (a) Ser. (b) Muc. (c) Sub. (d) Mus.

A7. (a) F. (b) M. (c) Meso. (d) G. (e) L.

B1.

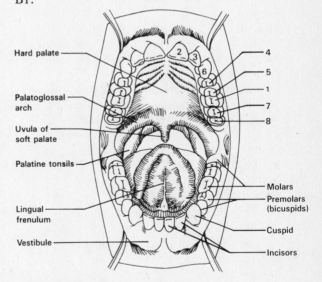

Figure LG 24-3A Mouth (oral cavity).

B4. (a) Parotid. (b) Sublingual. (c) Liters. (d) Dissolving medium for foods, lubrication, source of lysozyme and salivary amylase. (e) 6.5.

B5. (a) C B A. (b) A B C. (c) A B C.

B6. Anchors tooth to bone.

B7. Starch, is not, acidic pH of stomach.

B8. (a) P. (b) V. (c) E.

B9. Heartburn, acidic, achalasia, heart.

C1. (a) Cardia, fundus, body, pylorus. (b) GC, GC, LC.

C2.

Name of Cell	Type of Secretion	Function of Secretion
a. Zygomatic (chief)	Pepsinogen	Precursor of pepsin
b. Mucous	Mucus	Protects gastric lining from acid and pepsin
c. Parietal	HCl	Activates pepsinogen to pepsin.
	Intrinsic factor	Facilitates absorption of vitamin B_{12}
d. Enteroendocrine	Gastrin	Stimulates secretion of HCl and pepsinogen; contracts lower esophageal sphincter and relaxes pyloric sphincter

C3. (a) Arranged in rugae. (b) It has three, rather than two, layers of smooth muscle; the extra one is an oblique layer located inside the circular layer.

C4. (a) Acid. (b) Pepsin is released in the inactive state (pepsinogen); mucus protects the stomach lining from pepsin. (c) Ulcer. (d) Gastric lipase; emulsified fats, as in butter; ineffective; its optimum pH is 5 or 6 and most fats have yet to be emulsified (by bile from liver).

C5. (a) Cephalic, gastric, intestinal; cephalic. (b) Vagus; parasympathetic; sight, smell, or thought of food; presence of food in the stomach. (c) Stomach gastrin, protein, pyloric, stimulated, relaxed. (d) Enterogastric, inhibits, CCK, GIP, secretin.

C6. 2-6, carbohydrate, fat.

C7. (a) Irritation and distension of the stomach. (b) Medulla, contract, relax. (c) Loss of electrolytes (such as HCl) as well as fluids can lead to fluid/electrolyte imbalances.

C8. Little; water, electrolytes, certain drugs (such as aspirin), and alcohol.

D2. (a) Stomach. (b) Duodenum, spleen. (c) Exocrine, alkaline. (d) Proteins, enterokinase. (e) Amylase digest carbohydrates, including starch; lipase digests lipids. (f) G, H, duodenum. (g) Common bile duct; pancreatic proteases such as trypsin, chymotrypsin, and carboxypeptidase may digest tissue proteins of the pancreas itself. (h) Pancreatic islets (of Langerhans); insulin, glucagon; blood vessels, portal.

D4. (a) Parasympathetic, vagus, small intestine. (b) Digestive enzymes such as trypsin, amylase, and lipase; alkaline (HCO_3^-) fluid.

D5. (a) 1.4, 3.0, upper right, 4, right. (b) Falciform, umbilical. (c) Hepatic artery, portal vein, sinusoids, hepatic veins.

D6. (a) Superior to inferior. (b) Draw arrows from L and M towards B and then J; superiorly.

D7. (a) Bilirubin, red blood, jaundice; emulsification, absorption. (b) Heparin; prothrombin, fibrinogen and albumin. (c) RBCs, WBCs, and some bacteria. (d) Products of protein digestion such as ammonia, and substances such as medications and alcohol. (e) Glycogen and fat. (f) A, D, E, K; B_{12}, D. (g) Does not.

D9. (a) Pyloric, fats and partially digested proteins. (b) Duodenum, cholecystokinin. (c) Portal vein, meanders in the bloodstream through all parts of the body. (d) Contract. (e) Cystic duct. (f) Obstruction by the stone backs up bile, leading to spasms of smooth muscle (biliary colic) in bile pathways and entrance of new player: INTENSE PAIN. (g) GALLBLADDER. (h) Stimulation of pancreatic and intestinal secretions, inhibition of gastric secretion and motility, and opening of hepatopancreatic sphincter.

E1. (a) A. (b) B. (c) Duodenum, jejunum, ileum.

E2. (a) V. (b) MV. (c) PC. (d) IG. (e) GC. (f) DBG.

E4. (a) Shorter chain polysaccharides, disaccharides. (b) Shorter chain polypeptides, dipeptides, amino acids. (c) Fatty acids and glycerol.

E6. (a) HCl, L, P, SG. (b) A, L, TCC. (c) CCK, En, EG, MLS, Pd, S. (d) B.

E7. (a) CCK. (b) S, H. (c) CCK, H. (d) S, H. (e) CCK, S. (f) EG, SG, both H. (g) P. (h) HCl, N. (i) P. (j) TCC. (k) En, E. (l) Pd. (m) L. (n) B, N. (o) A. (p) MLS.

E8. (a) Small, active transport. (b) Blood, portal, liver. (c) Bile, micelles, thoracic duct, recycled to the liver. (d) Fat-soluble, A D E K; examples, A: night blindness; D: rickets or osteomalacia due to decreased calcium absorption; K: excessive bleeding. See Exhibit 25-5 (pages 810–812) in the text. (e) 9, small, large. (f) Diarrhea, electrolytes.

F2. 1.5, 5, colon.

F3. (a) IS. (b) VA. (c) HF. (d) H. (e) AC.

F5. Muscularis, longitudinally. See Figure 24-22 (page 769) of the text.

F6. Newborns, persons who have been on long-term antibiotic therapy; excessive bleeding.

G1. (a) Decrease taste sensations, gum inflammation (pyorrhea) leading to loss of teeth and loose-fitting dentures, and difficulty swallowing (dysphagia). (b) Decreased muscle tone and neuromuscular feedback lead especially to constipation.

G2. (a) E. (b) E. (c) M.

H3. To kill intestinal bacteria and reduce the risk of peritonitis.

H4. Diverticulitis.

H5. (a) Colos. (b) Hep. (c) Colit. (d) Htb. (e) D. (f) Cho. (g) Con. (h) P. (i) F. (j) B. (k) A.

MASTERY TEST: Chapter 24

Questions 1–10: Circle the letter preceding the one best answer to each question.

1. Which of these organs is not part of the GI tract, but is an accessory organ?

   A. Mouth   D. Small intestine
   B. Pancreas   E. Esophagus
   C. Stomach

2. The main function of salivary and pancreatic amylase is to:

   A. Lubricate foods
   B. Help absorb fats
   C. Digest polysaccharides to smaller carbohydrates
   D. Digest disaccharides to monosaccharides
   E. Digest polypeptides to amino acids

3. Which enzyme is most effective at pH 1 or 2?

   A. Gastric lipase   D. Salivary amylase
   B. Maltase   E. Pancreatic
   C. Pepsin   amylase

4. Choose the TRUE statement about fats.

   A. They are digested mostly in the stomach.
   B. They are the type of food that stays in the stomach the shortest length of time.
   C. They stimulate release of gastrin.
   D. They are emulsified and absorbed with the help of bile.

5. Each of the following hormones or enzymes is correctly matched with its function EXCEPT:

   A. Stomach gastrin—relaxes pyloric sphincter
   B. Enteric gastrin—relaxes pyloric sphincter
   C. Secretin—stimulates pancreas to secrete pancreatic juices rich in digestive enzymes and stimulates gallbladder to release bile
   D. CCK—stimulates secretion of intestinal juice and inhibits gastric secretions

6. Which of the following is under only nervous (not hormonal) control?

   A. Salivation
   B. Gastric secretion
   C. Intestinal secretion
   D. Pancreatic secretion

7. All of the following are enzymes involved in protein digestion EXCEPT:

   A. Amylase
   B. Trypsin
   C. Carboxypeptidase
   D. Pepsin
   E. Chymotrypsin

8. All of the following chemicals are produced by the walls of the small intestine EXCEPT:

   A. Lactase
   B. Secretin
   C. CCK
   D. Trypsin
   E. Peptidases

9. Which type of movement is used primarily to propel chyme through the intestinal tract, rather than to mix chyme and enzymes.

   A. Peristalsis
   B. Rhythmic segmentation
   C. Haustral churning

10. Choose the FALSE statement about layers of the wall of the GI tract.

    A. Most large blood and lymph vessels are located in the submucosa.
    B. The myenteric plexus is part of the muscularis layer.
    C. Most glandular tissue is located in the layer known as the mucosa.
    D. The mucosa layer forms the peritoneum.

Questions 11–15: Arrange the answers in correct sequence.

_____ _____ _____ 11. From anterior to posterior:

   A. Glossopalatine arch
   B. Pharyngopalatine arch
   C. Palatine tonsils

_____ _____ _____ _____ 12. GI tract wall, from deepest to most superficial:

   A. Mucosa
   B. Muscularis
   C. Serosa
   D. Submucosa

_____ _____ _____ _____ _____ 13. Pathway of chyme:

   A. Ileum
   B. Jejunum
   C. Cecum
   D. Duodenum
   E. Pylorus

_____ _____ _____ _____ _____ 14. Pathway of bile:

   A. Bile canaliculi
   B. Common bile duct
   C. Common hepatic duct
   D. Right and left hepatic ducts
   E. Hepatopancreatic ampulla and duodenum

_____ _____ _____ _____ _____ 15. Pathway of wastes:

   A. Ascending colon
   B. Transverse colon
   C. Sigmoid colon
   D. Descending colon
   E. Rectum

Questions 16–20: Circle T (true) or F (false). If the statement is false, change the underlined word or phrase so that the statement is correct.

T F 16. In general, the sympathetic nervous system <u>stimulates</u> salivation and secretions of the gastric and intestinal glands.

T F 17. The principal chemical activity of the stomach is to begin digestion of <u>protein</u>.

T F 18. The esophagus produces <u>no digestive enzymes or mucus</u>.

T F 19. <u>Villi, microvilli, goblet cells, and rugae</u> are all located in the walls of the small intestine.

T F 20. Cirrhosis and hepatitis are diseases of the <u>liver</u>.

Questions 21–25: fill-ins. Complete each sentence with the word or phrase that best fits.

_____ 21. Stomach motility is ____-creased by the enterogastric reflex, CCK and GIP, and ____-creased by stomach gastrin and parasympathetic nerves.

_____ 22. Mumps involves inflammation of the ____ salivary glands.

_____ 23. Most absorption takes place in the ____, although some substances, such as ____, are absorbed in the stomach.

_____ 24. Mastication is a term that means ____.

_____ 25. Epithelial lining of the GI tract, as well as the liver and pancreas, are derived from ____-derm.

ANSWERS TO MASTERY TEST: ■ Chapter 24

Multiple Choice

1. B
2. C
3. C
4. D
5. C
6. A
7. A
8. D
9. A
10. D

Arrange

11. A C B
12. A D B C
13. E D B A C
14. A D C B E
15. A B D C E

True-False

16. F. Inhibits
17. T
18. F. No digestive enzymes, but it does produce mucus
19. F. Villi, microvilli, and goblet cells (Rugae are in wall of the stomach.)
20. T

Fill-ins

21. De, in
22. Parotid
23. Small intestine; water, electrolytes, alcohol, and some drugs such as aspirin
24. Chewing
25. Endo

FRAMEWORK 25

Metabolism

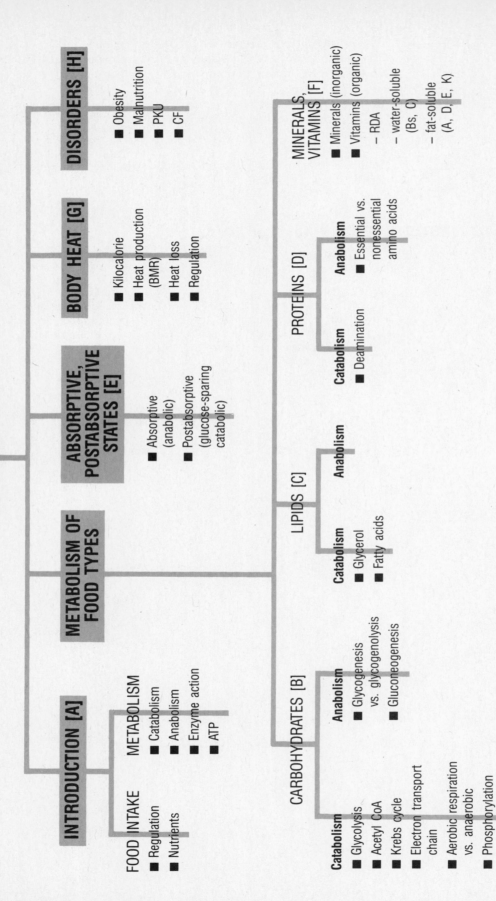

INTRODUCTION [A]

FOOD INTAKE
- Regulation
- Nutrients

METABOLISM
- Catabolism
- Anabolism
- Enzyme action
- ATP

METABOLISM OF FOOD TYPES

CARBOHYDRATES [B]

Catabolism
- Glycolysis
- Acetyl CoA
- Krebs cycle
- Electron transport chain
- Aerobic respiration vs. anaerobic
- Phosphorylation

Anabolism
- Glycogenesis vs. glycogenolysis
- Gluconeogenesis

LIPIDS [C]

Catabolism
- Glycerol
- Fatty acids

Anabolism

PROTEINS [D]

Catabolism
- Deamination

Anabolism
- Essential vs. nonessential amino acids

ABSORPTIVE, POSTABSORPTIVE STATES [E]
- Absorptive (anabolic)
- Postabsorptive (glucose-sparing catabolic)

MINERALS, VITAMINS [F]
- Minerals (inorganic)
- Vitamins (organic)
  - RDA
  - water-soluble (Bs, C)
  - fat-soluble (A, D, E, K)

BODY HEAT [G]
- Kilocalorie
- Heat production (BMR)
- Heat loss
- Regulation

DISORDERS [H]
- Obesity
- Malnutrition
- PKU
- CF

Metabolism

Ingested foods, once digested, absorbed, and delivered, are used by body cells. This array of biochemical reactions within cells is known as metabolism. Products of digestion may reassembled, as in formation of human protein from the amino acids of meat or beans, or the synthesis of fats from ingested oils. These reactions are examples of anabolism. The flip side of metabolic currency is catabolism, as in the breakdown of complex carbohydrates stored in liver or muscle to provide simple sugars for quick energy. Catabolism releases energy that maintains body heat and provides ATP to fuel activity. Vitamins and minerals play key roles in metabolism — for example, in enzyme synthesis and function.

The Chapter 25 Framework provides an organizational preview of the metabolism of the major food groups. Study the key terms and concepts it presents.

TOPIC OUTLINE AND OBJECTIVES

A. Introduction to metabolism

1. Define a nutrient and list the functions of the six principal classes of nutrients.
2. Define metabolism and explain the role of ATP in anabolism and catabolism.
3. Describe the characteristics and importance of enzymes in metabolism.

B. Carbohydrate metabolism

C. Lipid metabolism

D. Protein metabolism

4. Describe the metabolism of carbohydrates, lipids, and proteins.

E. Absorptive and postabsorptive states; regulation of metabolism

5. Distinguish between the absorptive and postabsorptive (fasting) states.

F. Minerals, vitamins

6. Compare the sources, functions, and importance of minerals in metabolism.

G. Metabolism and heat

7. Describe the various mechanisms involved in heat production and heat loss.
8. Define basal metabolic rate (BMR) and explain several factors that affect it.
9. Explain how normal body temperature is maintained and describe fever, heat cramp, heatstroke, heat exhaustion, and hypothermia as abnormalities of temperature regulation.

H. Disorders

10. Define obesity, malnutrition, phenylketonuria (PKU), cystic fibrosis (CF), and celiac disease.

WORDBYTES

Now study the following parts of words that may help you better understand terminology in this chapter.

Wordbyte	Meaning	Example	Wordbyte	Meaning	Example
ana-	up	*ana*bolism	de-	remove, from	*de*amination
calor-	heat	*Calor*ie	-lysis	destruction	hydro*lysis*
cata-	down	*cata*bolism			

CHECKPOINTS

A. Introduction to metabolism (pages 785–790)

■ **A1.** Complete this Checkpoint about regulation of food intake.

a. The site within the brain that is the location of feeding and satiety centers is the (*medulla? hypothalamus?*). When the satiety center is active, an individual will feel (*hungry? satiated or full?*).

b. The feeding center is constantly active except when it is inhibited by the

_____ center. Circle all of the answers that will keep the feeding center active and enhance the desire to eat.

↓ Glucose in blood
↓ Amino acids in blood
↓ Fats entering intestine (with related ↓ in CCK release)
↓ Environmental temperature
↓ Stretching of stomach

■ **A2.** List the six classes of nutrients on lines below.

_____   _____   _____

_____   _____   _____

■ **A3.** Explain why metabolism might be thought of as an "energy-balancing act."

A4. Complete this table comparing catabolism with anabolism.

Process	Definition	Releases or Uses Energy	Examples
a. Catabolism			
b. Anabolism			

464

■ **A5.** In order for molecules to react with one another, they must have a certain

amount of energy, known as _____ energy. Answer the following questions about this energy.

a. One way to attain such energy is to heat the reacting molecules. Why would this be a problem in living systems?

b. How can enzymes increase chemical reaction rates without requiring an increase in body temperature?

■ **A6.** Do this exercise about characteristics of enzymes.

a. Chemically, all enzymes consist (either entirely or mostly) of

_____. The molecular weight of most enzymes is (*100–2,000? 10,000–millions?*).

b. Since enzymes speed chemical reaction rates, they are known as

_____. They react with specific molecules called

_____.

c. The rate of production of enzymes depends on the cells' needs at a given moment and (*is? is not?*) controlled by genes.

d. Names of most enzymes end in the letters _____. Enzyme names also give clues to their functions. For example, enzymes that remove hydrogen are called

_____, whereas those such as amylase or protease that split

apart substrates by adding water are known as _____.

e. Whole enzymes are sometimes known as (*apo? holo?*)-enzymes. These consist of a protein portion called the (*apo? holo?*)-enzyme, as well as a cofactor.

f. Cofactors are of two types. Some are metal ions, such as

_____. So, for example, a protein (apoenzyme) plus calcium ion may serve as a holoenzyme.

g. The other type of cofactor is known as a ____-enzyme. Most coenzymes are derived from (*minerals? vitamins?*). List three examples of coenzymes:

_____. So, for example, NAD may function as a coenzyme by

picking up _____ atoms from the substrate.

h. The substrate interacts with a specific region on the enzyme called the

_____. An intermediate known as an

_____ complex is formed. Very quickly the (*enzyme? substrate?*) is transformed in some way (such as broken down or transferred to another substrate), while the (*enzyme? substrate?*) is recycled for further catalytic action.

■ **A7.** Consider catabolism (energy production) in this Checkpoint.

a. Organic nutrients such as glucose are rich in hydrogen; energy is contained within the C–H bonds. In catabolism, much of the energy in glucose is ultimately released

and stored in the high-energy molecule named ____. Write the chemical formulas to show this overall catabolic conversion:

_____	→	_____	+	ATP
Glucose	→	*Carbon dioxide*	+	*adenosine triphosphate*

Note that the inorganic compound carbon dioxide is energy-(*rich? poor?*) since it lacks C–H bonds.

b. The reaction shown above is an oversimplification of the entire catabolic process, which actually involves many steps. The hydrogens (H) removed from glucose must

first be trapped by H-carrying coenzymes, such as _____ or _____, and ulti-

mately the H's combine with oxygen to form _____. In other words, glucose is dehydrogenated (loses H) while oxygen is reduced (gains H).

c. Refer to Figure LG 25-1, which shows just two steps of many oxidation-reduction reactions in catabolism. Visualize a football game with the ball marked *H* (for hydrogen). Each player serves as a receiver of the ball, and then passes the ball on down the field. As a tribute to the fact that oxygen is the most common ultimate receiver of H (down at the goalpost), any player (such as #3 in the figure, oxygen) who is a receiver of H is said to be a(n) (*oxidizer? reducer?*). And that player (#3) is him/herself (*reduced? oxidized?*) upon receiving the H.

d. Note in the figure that as individuals pass the H/ball, they experience dehydrogena-

tion, also known as _____ since oxygen will ultimately receive that ball. Notice that as one player is oxidized (passes H), another is simultaneous-

ly reduced (receives the ball). This is known as a _____ oxidation-reduction reaction.

e. Catabolism involves a complex series of stepwise reactions in which hydrogens are transferred in the form of (*H? H⁺ + e⁻?*). Since oxygen is the final molecule that serves as an oxidizer, catabolism is often described as biological (*reduction? oxidation?*).

f. Along the way, energy from the original C–H bonds of glucose (and present in the

electrons of H) is trapped in _____. This process is known as

_____ since it involves the addition of a phosphate to ADP. A more complete expression of the total catabolism of glucose (shown briefly in *a* above) is included below. Write the correct labels under each chemical: *oxidized, oxidizer, reduced, reducer:*

Glucose	+	O₂	+	ADP	→	CO₂	+	H₂O	+	ATP
(_____)		(_____)				(_____)		(_____)		

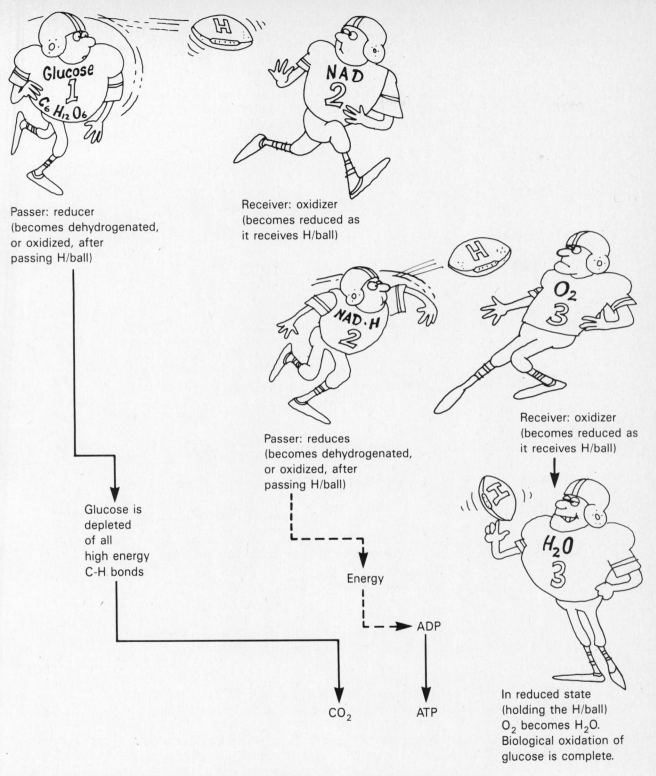

Figure LG 25-1 Representation of biological oxidation as a football game with three players (glucose, NAD, and oxygen) passing and receiving the ball (hydrogen). Refer to Checkpoint A7. Note that in actuality many more players are on the "team," but are not shown in this abbreviated "game."

B. Carboydrate metabolism (pages 790–801)

■ **B1.** Answer these questions about carbohydrate metabolism.

a. The story of carbohydrate metabolism is really the story of

_____ metabolism, since this is the most common carbohydrate (and in fact the most common energy source) in the human diet.

b. What other carbohydrates besides glucose are ingested? How are these converted to glucose?

c. Just after a meal, the level of glucose in the blood (*increases? decreases?*). Cells

use some of this glucose; by _____ glucose, they release
energy.

d. List three mechanisms by which excess glucose is used or discarded.

e. Increased glucose level of blood is known as _____. This can
occur after a meal containing concentrated carbohydrate or in the absence of the

hormone _____, since this hormone (*facilitates? inhibits?*)
entrance of glucose into cells. Lack of insulin occurs in the condition known as

_____.

f. As soon as glucose enters cells, glucose combines with _____

to form glucose-6-phosphate. This process is known as _____.

■ **B2.** Describe the process of glycolysis in this exercise.

a. Glucose is a _____-carbon molecule (which also has hydro-
gens and oxygens on it). During glycolysis each glucose molecule is converted to

two _____-carbon molecules named

_____.

b. (*A lot? A little?*) energy is released from glucose during glycolysis. This process
requires (*one? many?*) step(s) and (*does? does not?*) require oxygen.

c. The fate of pyruvic acid depends on whether sufficient _____
is available. If it is, pyruvic acid undergoes chemical change in phases 2 and 3 of
glucose catabolism (see below). These processes are (*aerobic? anaerobic?*), that is,
they require oxygen.

d. If the respiratory rate cannot keep pace with glycolysis, insufficient oxygen is available to break down pyruvic acid. In the absence of sufficient oxygen, pyruvic acid

is temporarily converted to _____. This is likely to occur dur-

ing active _____.

e. Several mechanisms prevent the accumulation of an amount of lactic acid that

might be harmful. The liver changes some back to _____

and _____. Excess PCO_2 caused by exercise (*stimulates? inhibits?*) respiratory rate, so more oxygen is available for breakdown of the pyruvic

acid. Lactic acid contributes to muscle _____, which will also
cause the person to want to rest. When exercise has slowed or stopped, sufficient

oxygen is inspired so that the oxygen _____ is paid.

■ **B3.** Review glucose catabolism up to this point and summarize the remaining phases
by doing this exercise. Refer to Figure LG 25-2. A six-carbon compound named

(a) _____ is converted to two molecules of

(b) _____ via a number of steps, together called

(c) _____, occurring in the (d) _____ of
the cell.
    In order for pyruvic acid to be further broken down, it must undergo a

(e) _____ step. This involves removal of the one-carbon

molecule, (f) _____. The remaining two-carbon

(g) _____ group is attached to a carrier called

(h) _____, forming (i) _____.
    This compound hooks on to a four-carbon compound called

(j) _____. The result is a six-carbon molecule of

(k) _____. The name of this compound is given to the cycle of

reactions called the (l) _____ cycle, occurring in the

(m) _____ of the cell.
    Two main types of reactions occur in the Krebs cycle. One is the decarboxylation

reaction, in which (n) _____ molecules are removed. (Notice
these locations on Figure 25-6, page 794 of your text.) (What happens to the CO_2
molecules that are removed?) As a result the six-carbon citric acid is eventually
shortened to regenerate oxaloacetic, giving the cyclic nature to this process.
    The other major type of reaction involves removal of (o)

_____ atoms during oxidation of compounds such as isocitric
and α-ketoglutaric acids. (See Figure 25-6, page 794 in your text.) These hydrogen

atoms are carried off by two coenzymes named (p) _____ and
"trapped" for use in phase 3.

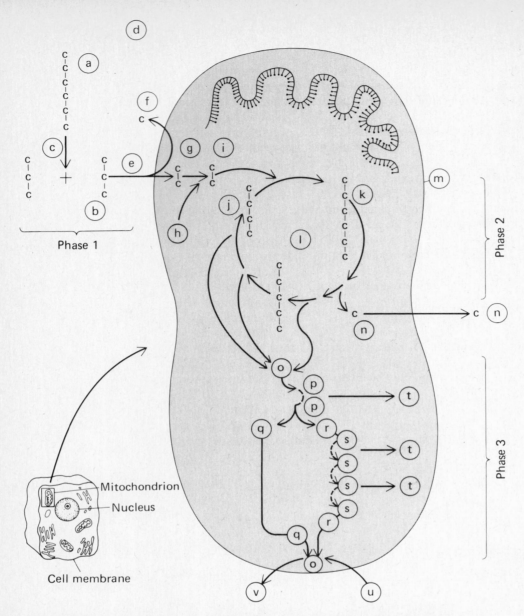

Figure LG 25-2 Summary of glucose catabolism. Letters refer to Checkpoint B3.

In phase 3, hydrogen atoms on NAD and FAD (and later coenzyme Q) are

ionized to (q) _____ and (r) _____. Electrons are shuttled along a chain of

(s) _____. During electron transport energy in these electrons
(derived from hydrogen atoms in ingested foods) is "tapped" and stored in

(t) _____.
Finally, the electrons, depleted of some of their energy, are reunited with

hydrogen ions and with (u) _____ to form

(v) _____. Notice why your body requires oxygen—to keep
drawing hydrogen atoms off of nutrients (in biological oxidation) so that energy from
them can be stored in ATP for use in cellular activities.

■ **B4.** Write a chemical equation showing the total catabolism of glucose.

■ **B5.** Most of the 38 ATPs generated from the total oxidation of glucose derive from (*glycolysis? the Krebs cycle?*). Of those generated by the Krebs cycle, most are formed from the oxidation of hydrogens carried by (*NAD? FAD?*). About what percent of the energy originally in glucose is stored in ATP after glucose is completely catabolized? (*10%? 43%? 60%? 99%?*)

■ **B6.** Complete the exercise about the reaction shown below.

            (1)

Glucose ⇌ Glycogen

            (2)

a. You learned earlier that excess glucose may be stored as glycogen. In other words,

glycogen consists of large branching chains of _____. Name the process of glycogen formation by labeling 1 in the chemical reaction above.

b. Between meals, when glucose is needed, glycogen can be broken down again to release glucose. Label 2 above with the name of this process.

c. Which of these two reactions is anabolic? _____

d. Where is most (80 percent) of glycogen in the body stored?

e. Identify a hormone that stimulates reaction (1). Write its name on line (1). As more glucose is stored in the form of glycogen, the blood level of glucose

_____-creases. Now write below line (2) two hormones that stimulate glycogenolysis.

f. In order for glycogenolysis to occur, a _____ group must be added to glucose as it breaks away from glycogen. The enzyme catalyzing this reac-

tion is known as _____. Does the reverse reaction (shown in 2) require phosphorylation also? (*Yes? No?*)

B7. Define *gluconeogenesis* and briefly discuss how it is related to other metabolic reactions.

■ **B8.** Circle the processes at left and the hormones at right that lead to increased blood glucose level.

Processes

Glycogenesis
Glycogenolysis
Glycolysis
Gluconeogenesis

Hormones

Insulin
Glucagon
Epinephrine
Cortisol
Thyroxine
Growth hormone

For extra review, refer to Figures LG 18-2 and 18-7, pages LG 315 and 324.

■ **B9.** The two molecules that appear to be most involved in interconversions among

amino acids, fatty acids, and carbohydrates are _____ and

_____ .

C. Lipid metabolism (pages 801–803)

C1. List six examples of structural or functional roles of fats in the body.

_____   _____   _____

_____   _____   _____

C2. Is the fat stored in your subcutaneous tissue "the same fat" that was in that location two years ago? Explain.

■ **C3.** Complete the exercise about fat catabolism.

a. Fats must first be broken into _____ and

   _____ . Glycerol is converted into _____
   and enters glycolysis pathways.

b. Recall that fatty acids are long chains of carbons, with attached hydrogens and a few oxygens. Two-carbon pieces are "snipped off" of fatty acids by a process called

   _____ , occurring in the _____ .

c. Some of these two-carbon pieces, in _____ , can enter
   Krebs cycle reactions; in this way fats can release energy just as carbohydrates do.

d. When large numbers of acetyl CoA's form, they tend to pair chemically:

   Acetic acid + acetic acid → _____
   (2C)   (2C)   (4C)

   The presence of excessive amounts of these acids in blood (*raises? lowers?*) blood pH. Since these acids are called "keto" acids, this condition is known as

   _____ .

e. A slight alteration of acetoacetic acid (removal of CH_3) forms acetone. Collectively,

   acetoacetic acid and acetone are known as _____ bodies. Formation of ketone bodies occurs in the liver; the process is known as

   _____ .

f. Ketogenesis occurs when cells are forced to turn to fat catabolism. State two reasons why cells might carry out excessive fat catabolism leading to ketogenesis and ketosis.

g. Explain why a diabetic might have sweet breath. Tell how this relates to fat metabolism.

C4. Briefly describe how a diet that is excessive in carbohydrates or proteins can lead to formation of fat deposits in the body.

D. Protein metabolism   (page 803)

■ **D1.** Which of the three major nutrients (carbohydrates, lipids, and proteins) fulfills each function?

a. Most direct source of energy; stored in body least: _____
b. Second as source of energy; used more for body-building:

_____

c. Used least for energy; used most for body-building: _____

■ **D2.** Contrast the caloric values of the three major food types.

a. Carbohydrates and proteins each produce _____ Cal/g or (since a pound con-

tains 454 g) _____ Cal/lb.

b. Fats produce _____ Cal/g or _____ Cal/lb.

D3. Throughout your study of systems of the body so far, you have learned about a variety of roles of proteins. In one or two words each, list at least six functions of proteins in the body. Include some structural and some regulatory roles.

_____   _____

_____   _____

_____   _____

■ **D4.** Name a hormone that enhances transport of amino acids into cells.

■ **D5.** Refer to Figure LG 25-3 and describe the uses of protein.

Two sources of proteins are shown: (a) _____ and

(b) _____. Amino acids from these sources may undergo a

process known as (c) _____ to form new body proteins. If other
energy sources are used up, amino acids may undergo catabolism. The first step is

(d) _____, in which an amino group (NH$_2$) is removed and

converted (in the liver) to (e) _____, a component of

(f) _____, which exits in urine. The remaining portion of the

amino acid may enter (g) _____ pathways at a number of points
(Figure 25-12, page 804 in your text). In this way, proteins can lead to formation of

(h) _____ .

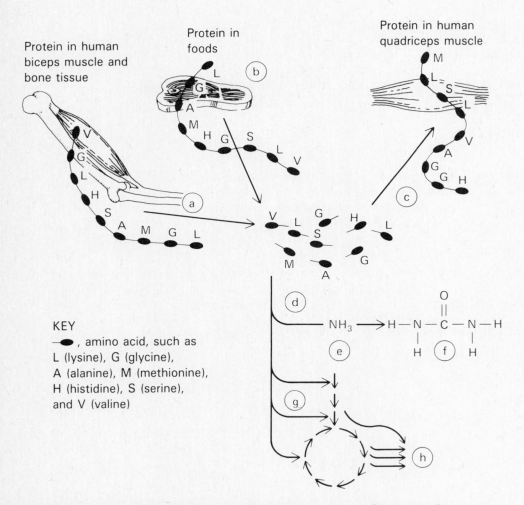

Figure LG 25-3   Metabolism of protein. Lowercase letters refer to Checkpoint D5.

D6. Contrast *essential amino acids* and *nonessential amino acids.*

E. Absorptive and postabsorptive states, control of metabolism
(pages 803–806)

■ **E1.** Contrast these two states by writing A if the description refers to the absorptive state and P if it refers to the postabsorptive state.

_____ a. Period when the body is fasting

_____ b. Time when the body is absorbing nutrients from the GI tract

_____ c. State during which the principal concern is formation of stores of glucose (as glycogen) and fat

_____ d. State during which glucose (stored in glycogen in liver and muscle) is released

_____ e. State during which most systems (excluding nervous system) switch over to fat as energy source

■ **E2.** Do this exercise about maintenance of glucose level between meals.

a. The normal blood glucose level in the postabsorptive state is about (*30–50? 70–110? 150–170? 600–700?*) mg glucose/100 ml blood.

b. Now list four sources of glucose that may be called upon during the postabsorptive state.

_____   _____

_____   _____

c. Can fatty acids serve as a source of glucose? (*Yes? No?*) Explain.

d. Fats can be used to generate ATP by being broken down into

_____ which can then enter the Krebs cycle. Therefore fatty acids are said to provide a glucose (*sparing? utilizing?*) mechanism between meals.

E3. Complete the table describing how each hormone assists in regulation of metabolism. (Refer to Exhibit 25-3, page 807, of your text.)

Hormone	Source	Action
a.		Stimulates glycogenesis in liver; stimulates glucose uptake in other cells
b.	Alpha cells of islets of Langerhans in pancreas	
c. Growth hormone (GH)		
d.		Stimulates adrenal cortex to produce glucocorticoids
e. Glucocorticoids		
f.	Thyroid gland	
g.	Adrenal medulla	
h. Testosterone		

■ **E4.** Circle the hormones in Checkpoint E3 above that are hyperglycemic, that is, tend to *increase* blood glucose.

F. Minerals, vitamins (pages 806–813)

■ **F1.** Define *minerals*.

Minerals make up about _____ percent of body weight and are concentrated in the

_____ .

■ **F2.** Study Exhibit 25-4 (pages 808–809) in your text. Then check your understanding of minerals by doing this matching exercise.

Ca. Calcium	Fe. Iron	Na. Sodium
Cl. Chlorine	I. Iodine	P. Phosphorus
Co. Cobalt	K. Potassium	S. Sulfur
F. Fluorine	Mg. Magnesium	

_____ a. Main anion in extracellular fluid, part of HCl in stomach; component of table salt

_____ b. Involved in generation of nerve impulse, helps to regulate osmosis, acts in buffer systems

_____ c. Most abundant cation in the body, found mostly in bones and teeth; necessary for normal muscle contraction and for blood clotting

_____ d. Important component of hemoglobin and cytochromes

_____ e. Main cation inside of cells; used in nerve transmission

_____ f. Essential component of thyroxin

_____ g. Constituent of vitamin B_{12}, so necessary for red blood cell formation

_____ h. Important component of amino acids, vitamins, and hormones

_____ i. Improves tooth structure

_____ j. Found mostly in bones and teeth; important in buffer system and in ATP processes; component of DNA and RNA.

■ **F3.** Vitamins are (*organic? inorganic?*). Most vitamins (*can? cannot?*) be synthesized in the body. In general, what are the functions of vitamins?

■ **F4.** Contrast the two principal groups of vitamins, and list the main vitamins in each group.

a. Fat-soluble _____

b. Water-soluble _____

■ **F5.** The intitials RDA found on vitamins and minerals refer to

R _____ D _____ A _____ .

■ **F6.** Select the vitamin that fits each description.

A	B₁	B₂	B₁₂	C	D	E	K

_____ a. This serves as a coenzyme that is essential for blood clotting, so it is called the antihemorrhagic vitamin; synthesized by intestinal bacteria.

_____ b. Its formation depends upon sunlight on skin and also on kidney and liver activation; necessary for calcium absorption.

_____ c. Riboflavin is another name for it; a component of FAD; necessary for normal integrity of skin, mucosa and eye.

_____ d. This vitamin acts as an important coenzyme in carbohydrate metabolism; deficiency leads to beriberi.

_____ e. Formed from carotene, it is necessary for normal bones and teeth; it prevents night blindness.

_____ f. This substance is also called ascorbic acid; deficiency causes anemia, poor wound healing, and scurvy.

_____ g. This coenzyme, the only B vitamin not found in vegetables, is necessary for normal erythropoiesis; absorption from GI tract depends on intrinsic factor.

_____ h. Also known as tocopherol, it is necessary for normal red blood cell membranes; deficiency is associated with sterility in some animals.

G. Metabolism and heat (pages 813–818)

G1. Define these terms:

a. Heat

b. Calorimeter

■ **G2.** Contrast *calorie* and *kilocalorie.*

If you ingest, digest, absorb, and metabolize a slice of bread, the energy released from the bread equals about 80–100 (*cal? kcal?*).

■ **G3.** Answer these questions about catabolism.
Catabolism of foods (*uses? releases?*) energy. Most of the heat produced by the body comes from catabolism by (*oxidation? reduction?*) of nutrients.

G4. Discuss factors that affect how rapidly your body catabolizes, that is, your metabolic rate. Include: *exercise, nervous system, hormones,* and *body temperature, specific dynamic action (SDA), age,* and *pregnancy.*

G5. Define *basal metabolic rate* (*BMR*).

List four conditions that are necessary in order for the body to be in a basal state.

_____   _____

_____   _____

■ **G6.** Calculate a typical BMR in this exercise.

a. BMR may be determined by measuring the amount of _____
consumed within a period of time, since oxygen is necessary for the metabolism of
foods. Normally for each liter of oxygen consumed, the body metabolizes enough

food to release about _____ kilocalories.

b. Suppose you consume 15 liters of oxygen in an hour. Your body would release

_____ kCal of heat in that hour.

c. BMR also factors in differences in body size. A typical body surface (as if you
could lay carpet over all of your skin is about (*0.5? 2? 5? 10?*) square meters. If
you used 71.7 kCal in an hour, and your body surface is 2 m^2, your BMR

is _____ kCal/m^2/hr.

d. If a normal value for BMR for a person your age is 39.5 kCal/m^2/hr, and your
BMR is 35.9 kCal/m^2/hr, then your BMR is about 10 percent (*above? below?*) normal. This would be expressed as a BMR of (*+10? −10?*).

e. A BMR of −20 is more likely to be due to (*hyper? hypo?*)-thyroidism.

■ **G7.** Your body temperature usually (*does? does not?*) rise much as a result of
catabolism. Why?

G8. Write the percent of heat loss by each of the following routes (at room temperature). Then write an example of each of these types of heat loss. One is done for you.

a. Radiation ____

c. Convection __15__ Cooling by draft while taking
a shower

b. Conduction ____

d. Evaporation ____

■ **G9.** During an hour of active exercise Bill produces 1 liter of sweat. The amount of

heat loss (cooling) that accompanies this much evaporation is _____ Cal. On a hu-
mid day, (*more? less?*) evaporation of Bill's sweat occurs, so Bill would be cooled
(*more? less?*) on such a day.

G10. Lil is outside on a snowy day without a warm coat. Her skin is pale and chilled and she begins to shiver. Explain how these responses are attempts of her body to maintain homeostasis of temperature. What other responses might raise her body temperature? Include these words in your answers: *sympathetic, metabolism, adrenal medulla, thyroid,* and *coat.*

■ **G11.** Arrange in order the events believed to occur in the production of fever of

39.4°C (103°F) and recovery from this state. ___ ___ ___ ___ ___ ___

A. 39.4°C (103°F) temperature is reached and maintained.

B. PGs cause the hypothalamic "thermostat" to be reset from normal 37.5°C (98.6°F) to a higher temperature such as 39.4°C (103°F).

C. Vasoconstriction and shivering occur as heat is conserved in order to raise body temperature to 39.4°C (103°F) level directed by the hypothalamus.

D. Source of pyrogens is removed (for example, bacteria are killed by antibiotics), lowering hypothalamic "thermostat" to 37.5°C (98.6°F).

E. "Crisis" occurs. Obeying hypothalamic orders, the body shifts to heat loss mechanisms such as sweating and vasodilation, returning body temperature to normal.

F. Infectious organisms cause phagocytes to release interleukin-1, which then causes hypothalamic release of prostaglandins (PGs).

G12. In what ways is a fever beneficial?

G13. Describe each of these temperature abnormalities.

a. Heat cramp

b. Heat exhaustion

H. Disorders (pages 818–821)

■ **H1.** Circle the correct answer to complete each statement.

a. Obesity is defined as a body weight ___ percent above desirable weight.

10–20   100
30–50

b. Increase in the total number of adipocytes, as well as the amount of fat in them occurs in:

HP. Hyperplastic (lifelong) obesity
HT. Hypertropic (adult-onset) obesity

c. A surgery in which the upper jejunum is anastomosed to the lower ileum is known as the:

Intestinal bypass   Gastric balloon
Gastroplasty

d. Protein deficiency despite adequate caloric intake is a description of:

Marasmus   Kwashiorkor

■ **H2.** *A clinical challenge.* Explain why an individual with kwashiorkor may have a large abdomen even though the person's diet is inadequate.

■ **H3.** Individuals with PKU are unable to convert _____ to

_____. What are results of toxic levels of phenylalanine?

■ **H4.** Cystic fibrosis (*is? is not?*) an inherited disorder. It affects (*endo? exo?*)-crine glands. List three regions of the body in which glands are affected.

■ **H5.** *A clinical challenge.* Explain why the child with cystic fibrosis may require a low fat diet and is at risk for developing tetany.

■ **H6.** Persons with celiac disease require a diet that is lacking in most (*meats? vegetables? grains?*). The problem stems from a water-insoluble protein named

_____ which causes destruction of the intestinal lining. Name two grains that are acceptable in the diet of a celiac disease patient.

_____   _____

ANSWERS TO SELECTED CHECKPOINTS

A1. (a) Hypothalamus, satiated or full. (b) Satiety, all answers.

A2. Carbohydrates, proteins, lipids, minerals, vitamins, and water.

A3. All chemical reactions either require energy (as in anabolism) or release energy (as in catabolism).

A5. Activation. (a) Temperature increases could kill living cells. (b) Increase frequency of collisions, orient molecules so that they are more likely to react, lower activation energy.

A6. (a) Protein, 10,000–millions. (b) Catalysts, substrates. (c) Is. (d) -ase, dehydrogenases, hydrolases. (e) Holo, apo. (f) Magnesium, zinc, calcium. (g) Co; vitamins; NAD, NADP and FAD; hydrogen. (h) Active site, enzyme-substrate, substrate, enzyme.

A7. (a) ATP; $C_6H_{12}O_6 \rightarrow CO_2 + $ ATP; poor. (b) NAD or FAD, H_2O. (c) Oxidizer, reduced. (d) Oxidation, coupled. (e) $H^+ + e^-$, oxidation. (f) ATP, phosphorylation;

Glucose + O_2 + ADP →
(reducer) (oxidizer)

CO_2 + H_2O + ATP
(oxidized) (reduced)

B1. (a) Glucose. (b) Starch, sucrose, lactose; by enzymes located mostly in liver cells. (c) Increases, oxidizing. (d) Stored as glycogen, converted to fat and stored, excreted in urine. (e) Hyperglycemia, insulin, facilitates, diabetes mellitus. (f) Phosphate, phosphorylation.

B2. (a) Six, three, pyruvic acid. (b) A little, many, does not. (c) Oxygen, aerobic. (d) Lactic acid, exercise. (e) Pyruvic acid, glucose, stimulates, fatigue, debt.

B3. (a) Glucose. (b) Pyruvic acid. (c) Glycolysis. (d) Cytoplasm. (e) Transition. (f) CO_2. (g) Acetyl. (h) Coenzyme A. (i) Acetyl coenzyme A. (j) Oxaloacetic acid. (k) Citric acid. (l) Citric acid (or Krebs). (m) Mitochondria. (n) CO_2 which are exhaled. (o) Hydrogen. (p) NAD and FAD. (q) H^+. (r) Electrons (e^-). (s) Cytochromes or electron transfer system. (t) ATP. (u) Oxygen. (v) Water.

B4. $C_6H_{12}O_6 + 6 O_2 \rightarrow 38$ ATP $+ 6 CO_2 + 6 H_2O$

B5. The Krebs cycles, NAD, 43.

B6. (a) Glucose; 1: glycogenesis. (b) 2: Glycogenolysis. (c) Glycogenesis. (d) Muscles. (e) Insulin; de; glucagon and epinephrine. (f) Phosphate, phosphorylase, yes.

B8. Processes: glycogenolysis and gluconeogenesis; hormones: all except insulin.

B9. Pyruvic acid and acetyl coenzyme A (see Figure 25-10, page 800, of the text).

C3. (a) Fatty acids, glycerol, glyceraldehyde-3-phosphate. (b) Beta oxidation, liver. (c) Acetyl coenzyme A. (d) Acetoacetic acid, lowers, ketosis. (e) Ketone, ketogenesis. (f) Examples of reasons: starvation, fasting diet, lack of glucose in cells due to lack of insulin (diabetes mellitus), excess growth hormone (GH) which stimulates fat catabolism. (g) As acetone (formed in ketogenesis) passes through blood in pulmonary vessels, some is exhaled and detected by its sweet aroma.

D1. (a) Carbohydrates. (b) Lipids. (c) Proteins.

D2. (a) 4, about 1,800. (b) 9, about 4000.

D4. Growth hormone (GH) and insulin.

D5. (a) Worn-out cells as in bone or muscle. (b) Ingested foods. (c) Anabolism. (d) Deamination. (e) Ammonia. (f) Urea. (g) Glycolytic or Krebs cycle. (h) CO_2 + H_2O + ATP.

E1. (a) P. (b) A. (c) A. (d) P. (e) P.

E2. (a) 70–110. (b) Liver glycogen (4-hour supply), glycerol from fats, muscle glycogen and lactic acid during vigorous exercise, amino acids from tissue proteins. (c) No, since acetyl CoA cannot be readily converted into pyruvic acid and then to glucose. (d) Acetyl CoA, sparing.

E4. (b) Glucagon. (c) GH. (d) ACTH. (e) Glucocorticoids. (f) Thyroxine. (g) Epinephrine.

F1. Inorganic substances; 4, skeleton.

F2. (a) Cl. (b) Na. (c) Ca. (d) Fe. (e) K. (f) I. (g) Co. (h) S. (i) F. (j) P.

F3. Organic; cannot; most serve as coenzymes, maintaining growth and metabolism.

F4. (a) A D E K. (b) B complex and C.

F5. Recommended dietary allowances.

F6. (a) K. (b) D. (c) B_2. (d) B_1. (e) A. (f) C. (g) B_{12} (h) E.

G2. A kilocalorie is the amount of heat required to raise 1000 ml (1 liter) of water 1°C; a kilocalorie (kCal) = 1000 calories (Cal); kCal.

G3. Releases, oxidation.

G6. (a) Oxygen, 4.783. (b) 71.7 (= 4.783 kCal/l O_2 × 15 l O_2/hr). (c) 2, 35.9 (= 71.7/2). (d) Below, −10. (e) Hypo.

G7. Does not. About half of the energy released by catabolism is stored in ATP and used for body activities. The remaining energy is released as heat and regulated by heat loss mechanisms.

G9. 580 (= 0.58 Cal/ml water × 1000 ml/liter), less, less.

G11. F B C A D E.

H1. (a) 10–20. (b) HP. (c) Intestinal bypass. (d) Kwashiorkor.

H2. Protein deficiency (due to lack of essential amino acids) decreases plasma proteins. Blood has less osmotic pressure, so fluids exit from blood. Movement of these into the abdomen (ascites) increases the size of the abdomen. Fatty infiltration of the liver also adds to the abdominal girth.

H3. Phenylalanine, tyrosine; toxicity to the brain with possible mental retardation.

H4. Is: exo; respiratory, pancreas, salivary, and sweat glands.

H5. Pancreatic ducts are blocked, so lipases do not reach the intestine to digest fats. So fats and fat-soluble vitamins (including D) are not absorbed. Vitamin D and fat are necessary for calcium absorption; hypocalcemia can cause overstimulation of nerves to muscles (tetany).

H6. Grains, gluten, rice, and corn.

MASTERY TEST: Chapter 25

Questions 1–14: Circle the letter preceding the one best answer to each question.

1. Which of these processes is anabolic?

    A. Pyruvic acid → CO_2 + H_2O + ATP
    B. Glucose → pyruvic acid + ATP
    C. Protein synthesis
    D. Digestion of starch to maltose
    E. Glycogenolysis

2. All of these statements are true of coenzyme A EXCEPT:

    A. It acts as a carrier.
    B. It is nonprotein.
    C. It functions in the transition step between glycolysis and the Krebs cycle.
    D. It carries a two-carbon acetyl group.
    E. It is used in glycolysis.

3. Which hormone is said to be hypoglycemic since it tends to lower blood sugar?

    A. Glucagon
    B. Glucocorticoids
    C. Growth hormone
    D. Insulin
    E. Epinephrine

4. The complete oxidation of glucose yields all of these products EXCEPT:

    A. ATP
    B. Oxygen
    C. Carbon dioxide
    D. Water

5. All of these processes occur exclusively or primarily in liver cells except one, which occurs in virtually all body cells. This one is:

    A. Gluconeogenesis
    B. Beta oxidation of fats
    C. Ketogenesis
    D. Deamination
    E. Krebs cycle

6. The process of forming glucose from fats or proteins is called:

    A. Gluconeogenesis
    B. Glycogenesis
    C. Ketogenesis
    D. Deamination
    E. Glycogenolysis

7. All of the following are parts of fat absorption and fat metabolism EXCEPT:

    A. Lipogenesis
    B. Beta oxidation
    C. Chylomicron formation
    D. Ketogenesis
    E. Glycogenesis

8. Which of the following vitamins is water-soluble? A, C, D, E, K?

9. At room temperature most body heat is lost by:

    A. Evaporation
    B. Convection
    C. Conduction
    D. Radiation

10. Choose the FALSE statement about vitamins.

    A. They are organic compounds
    B. They regulate physiological processes.
    C. Most are synthesized by the body.
    D. Many act as parts of enzymes or coenzymes.

11. Choose the FALSE statement about temperature regulation.

    A. Some aspects of fever are beneficial.
    B. Fever is believed to be due to a "resetting of the body's thermostat."
    C. Vasoconstriction of blood vessels in skin will tend to conserve heat.
    D. Heat-producing mechanisms which occur when you are in a cold environment are primarily parasympathetic.

12. Which of the following processes involves a cytochrome chain?

    A. Glycolysis
    B. Transition step between glycolysis and Krebs cycle.
    C. Krebs cycle
    D. Electron transport system

13. Which of the following can be represented by the equation: Glucose → pyruvic acids + small amount ATP?

    A. Glycolysis
    B. Transition step between glycolysis and Krebs cycle
    C. Krebs cycle
    D. Electron transport system

14. Choose the one FALSE statement about anabolism and catabolism.

    A. Anabolic reactions are more likely (than catabolic reactions) to involve dehydration synthesis reactions.
    B. Anabolic reactions are energy-requiring.
    C. Catabolic reactions are usually hydrolysis reactions.
    D. Anabolic reactions result in production of large quantities of ATP.

Question 15: Arrange the answers in correct sequence.

_____ _____ _____ _____ _____ _____ 15. Steps in complete oxidation of glucose:
  A. Glycolysis takes place.
  B. Pyruvic acid is converted to acetyl coenzyme A.
  C. Hydrogens are picked up by NAD and FAD; hydrogens ionize.
  D. Krebs cycle releases CO_2 and hydrogens
  E. Oxygen combines with hydrogen ions and electrons to form water.
  F. Electrons are transported along cytochromes and energy from electrons is stored in ATP.

Questions 16–20: Circle T (true) or F (false). If the statement is false, change the underlined word or phrase so that the statement is correct.

T F 16. Catabolism of carbohydrates involves <u>oxidation,</u> which is a process of <u>addition of hydrogens.</u>

T F 17. Nonessential amino acids are those that <u>are not used in the synthesis of human protein.</u>

T F 18. The complete oxidation of glucose to CO_2 and H_2O yields <u>4</u> ATPs.

T F 19. Anabolic reactions are <u>synthetic reactions that release energy.</u>

T F 20. <u>Carbohydrates, proteins, and vitamins</u> provide energy and serve as building materials.

Questions 21–25: fill-ins. Complete each sentence with the word or phrase that best fits.

_____ 21. Catabolism of each gram of carbohydrate or protein results in release of about ___ kCal, whereas each gram of fat leads to about ___ kCal.

_____ 22. The body's "favorite" source of energy (since it is most easily catabolized) is ___, while the least favorite is ___.

_____ 23. ___ is the mineral that is most common in extracellular fluid (ECF); it is also important in osmosis, buffer systems, and in nerve impulse conduction.

_____ 24. Aspirin and acetominophen (Tylenol) reduce fever by inhibiting synthesis of ___ so that the body's "thermostat" located in the ___ is reset to a lower temperature.

_____ 25. In uncontrolled diabetes mellitus, the person may go into a state of acidosis due to the production of ___ acids resulting from excessive breakdown of ___.

Multiple Choice

1. C	8. C
2. E	9. D
3. D	10. C
4. B	11. D
5. E	12. D
6. A	13. A
7. E	14. D

Arrange

15. A B D C F E

True-False

16. F. Oxidation, removal of hydrogens
17. F. Can be synthesized by the body
18. F. 38
19. F. Synthetic reactions that require energy
20. F. Carbohydrates and proteins but not vitamins

Fill-ins

21. 4, 9
22. Carbohydrate, protein
23. Sodium (Na^+)
24. Prostaglandins (PG), hypothalamus
25. Acetoacetic (keto-), fats

FRAMEWORK 26

Urinary System

KIDNEYS

ANATOMY [A]

External

Internal

- Cortex, medulla, pelvis
- Nephron
- Blood supply
- JGA

PHYSIOLOGY [B]

Glomerular Filtration

- Passive
- *P*eff
- GFR
- Hemodialysis

Tubular Reabsorption

- Passive & active
- Renin-angiotensin
- Water reabsorption
- Countercurrent mechanism

Tubular Secretion

- Active & passive
- Effects of ADH and aldosterone
- Effects on pH

HOMEOSTASIS [C]

- Kidneys, lungs, skin, GI

PATHWAY OF URINE [D]

- Ureters
- Bladder
- Urethra

URINE [E]

- Urinalysis
- Renal function

AGING, DEVELOPMENT [F]

DISORDERS [G]

The Urinary System

CHAPTER 26

Wastes like urea or hydrogen ions (H^+) are products of metabolism that must be removed or they will lead to deleterious effects. The urinary system works with lungs, skin, and the GI tract to eliminate wastes. Kidneys are designed to filter blood and selectively determine which chemicals stay in the blood or exit in the urine. The process of urine formation in the kidney requires an intricate balance of factors such as blood pressure and hormones (ADH and aldosterone). Examination of urine (urinalysis) can provide information about the status of the blood and possible kidney malfunction. A healthy urinary system requires an exit route for urine through ureters, bladder, and urethra.

Start your study of the urinary system with a look at the Chapter 26 Framework and familiarize yourself with the key concepts and terms.

TOPIC OUTLINE AND OBJECTIVES

A. Kidneys: anatomy

1. Identify the external and internal gross anatomical features of the kidneys.
2. Define the structural adaptations of a nephron for urine formation.

B. Kidneys: physiology

3. Discuss the process of urine formation through glomerular filtration, tubular reabsorption, and tubular secretion.

C. Homeostasis: other excretory organs

D. Ureters, urinary bladder, urethra

4. Discuss the structure and physiology of the ureters, urinary bladder, and urethra.

E. Urine

5. List and describe the physical characteristics and normal chemical constituents of urine.

6. Define albuminuria, glycosuria, hematuria, pyuria, ketosis, bilirubinuria, urobilinogenuria, casts, and renal calculi.

F. Aging and development of the urinary system

7. Describe the effects of aging on the urinary system.
8. Describe the development of the urinary system.

G. Disorders, medical terminology

9. Discuss the causes of gout, glomerulonephritis, pyelitis, pyelonephritis, cystitis, nephrosis, polycystic disease, renal failure, and urinary tract infections (UTIs).
10. Define medical terminology associated with the urinary system.

Now study the following parts of words that may help you better understand terminology in this chapter.

Wordbyte	Meaning	Example	Wordbyte	Meaning	Example
anti-	against	*anti*diuretic	ren-	kidney	*ren*al vein
calyx	cup	*calyc*es	retro-	behind, backward	*retro*peritoneal
juxta-	near	*juxta*glomerular			
nephr-	kidney	*nephr*itis	urin-, uro-	urinary	*uro*logist

CHECKPOINTS

A. Kidneys: anatomy (pages 826–832)

■ **A1.** Describe functions of the urinary system in this exercise.

a. List waste products eliminated through urine.

b. Kidneys produce _____, which regulates blood pressure, and

_____, which is vital to normal red blood cell formation.

Kidneys also activate vitamin ____ .

■ **A2.** Identify the organs that make up the urinary system on Figure LG 26-1 and answer the following questions about them.

a. The kidneys are located at about (*waist? hip?*) level, between

_____ and _____ vertebrae. Each kidney is about ____ cm (____ in.) long. Visualize kidney location and size on yourself.

b. The kidneys are in an extreme (*anterior? posterior?*) position in the abdomen. They

are described as _____ since they are posterior to the peritoneum.

c. Identify the parts of the internal structure of the kidney on Figure LG 26-1. Then color all structures indicated by color code ovals.

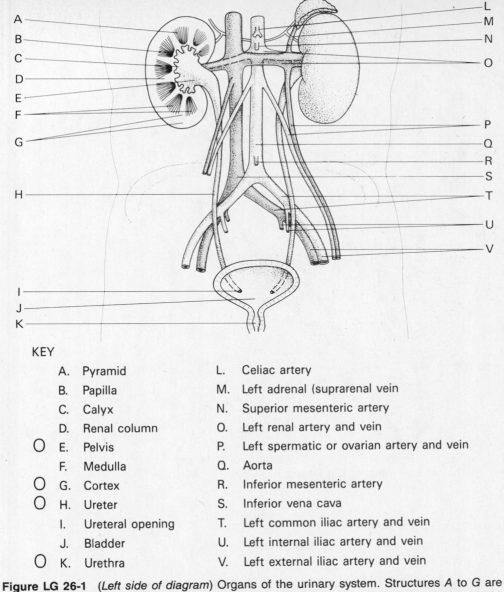

A
B
C
D
E
F
G

H

I
J
K

L
M
N

O

P
Q
R
S

T

U

V

KEY

A. Pyramid	L. Celiac artery
B. Papilla	M. Left adrenal (suprarenal vein
C. Calyx	N. Superior mesenteric artery
D. Renal column	O. Left renal artery and vein
○ E. Pelvis	P. Left spermatic or ovarian artery and vein
F. Medulla	Q. Aorta
○ G. Cortex	R. Inferior mesenteric artery
○ H. Ureter	S. Inferior vena cava
I. Ureteral opening	T. Left common iliac artery and vein
J. Bladder	U. Left internal iliac artery and vein
○ K. Urethra	V. Left external iliac artery and vein

Figure LG 26-1 (*Left side of diagram*) Organs of the urinary system. Structures *A* to *G* are parts of the kidney. Identify and color as directed in Checkpoints A2 and D1. (*Right side of diagram*) Blood vessels of the abdomen. Color as directed in Chapter 21, Checkpoint E8, page LG 387.

■ **A3.** Do this exercise about the functional unit of the kidney, a nephron.

a. On Figure LG 26-2, label parts A–F of the nephron, with letters arranged according to the flow of urine.

b. Which one of those structures forms part of the renal corpuscle? The other part of

the renal corpuscle is cluster of capillaries known as a _____
and numbered (*4? 6B?*) in the figure.

c. Identify the part of the renal tubule on the figure that consists of cuboidal epithelium with many microvilli that increase absorptive surface. _____ Which part consists of simple squamous epithelium? _____

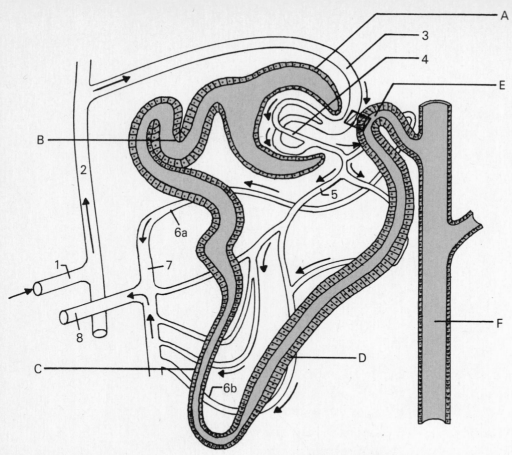

Figure LG 26-2 Diagram of a nephron. Letters refer to Checkpoint A3. Numbers refer to Checkpoints A3 and A7. The boxed area refers to Checkpoint A8.

■ **A4.** Describe the endothelial-capsular membrane.

a. Arrange in order the layers that fluid or solutes pass through as they move from blood into the forming urine.

    A. Filtration slits between podocytes of visceral layer of glomerular capsule
    B. Basement membrane of the glomerulus
    C. Endothelial pores of capillary membrane

b. List the functional advantages of these structures: *endothelial pores* and *filtration slits*.

A5. Contrast the locations of the *cortical nephron* and the *juxtaglomerular nephron*. Note which parts of the nephrons lie in the cortex and which parts are in the medulla.

■ **A6.** About what percent of cardiac output passes through the kidneys each

minute? _____ %

■ **A7.** Follow the pathway that blood takes through kidneys by naming vessels 1–8 on Figure LG 26-2. Some arrows are shown; add more to reinforce your understanding of the pathway of blood.

1. _____   3. _____   5. _____   7. _____

2. _____   4. _____   6. _____   8. _____

■ **A8.** The boxed structure on Figure LG 26-2 is known as the

_____. It consists of muscle cells of the _____-ferent arteriole as well as macula densa cells of the (*proximal? distal?*) convoluted tubule. The JGA cells secrete renin that helps control (*red blood cell formation? blood pressure?*).

B. Kidneys: physiology (pages 832–845)

■ **B1.** In what main ways is blood modified as it passes through the kidneys?

■ **B2.** List the three steps in urine production:

_____   _____   _____

■ **B3.** Now describe the first step in urine production in this Checkpoint.

a. Glomerular filtration is a process of (*pushing? pulling?*) fluids and solutes out of

_____ and into the fluid known as

_____.

b. (*All types? All small substances? Only selected small substances?*) are forced out of blood. Refer to Figure LG 26-3, area A (plasma in glomeruli) and area B (filtrate just beyond the capsule). Compare composition of these two fluids (described in the first two columns of Exhibit 26-1, page 834, in your textbook). Note that during filtration all solutes are freely filtered from blood except

_____. Why is so little protein filtered?

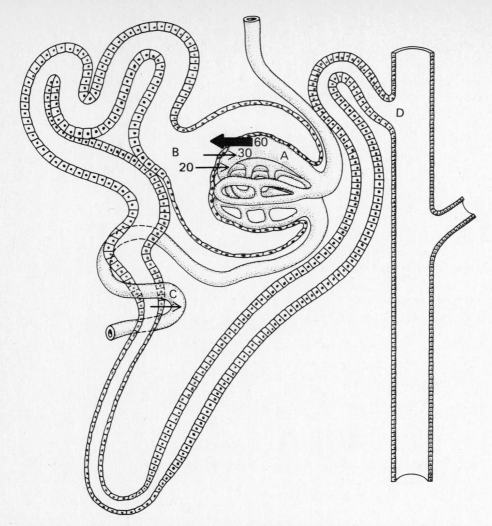

Figure LG 26-3 Diagram of a nephron with renal tubule abbreviated. Numbers refer to Checkpoint B4; letters refer to Checkpoints B3, B7, and B8.

c. Blood pressure (or hydrostatic pressure) in glomerular capillaries is about _____ mm Hg. This value is (*higher? lower?*) than that in other capillaries of the body. This extra pressure is accounted for by the fact that the diameter of the efferent arteriole is (*larger? smaller?*) than that of the afferent arteriole. Picture three garden hoses connected to each other, representing the afferent arteriole, glomerular capillaries, and efferent arteriole. The third one (efferent arteriole) is extremely narrow; it creates such resistance that pressure builds up in the first two. Fluids will be forced out of pores in the middle (glomerular) hose.

■ **B4.** Blood hydrostatic pressure is not the only force determining the amount of filtration occurring in glomeruli.

a. Name two other forces (Figure 26-8, page 835, in your textbook).

_____   _____

b. Normal values for these three forces are given on Figure LG 26-3. Write the formula for calculating effective filtration pressure Peff; draw an arrow beneath each term to show the direction of the force in Figure LG 26-3. Then calculate a normal Peff using the values given.

Peff = (Glomerular blood hydrostatic pressure) − (_____ + _____)

= (_____ mm Hg) − (_____ mm Hg + _____ mm Hg)

= _____ mm Hg

Notice that blood (hydrostatic) pressure pushes (*out of? into?*) glomerular blood and is largely counteracted by the other two forces.

■ **B5.** *A clinical challenge.* Show the effects of alterations of these pressures in pathological situations. Determine Peff of the following patients. Note which values are abnormal and suggest causes.

	Patient A	Patient B
Glomerular blood Hydrostatic pressure	50	73
Osmotic pressure	30	25
Capsular filtrate Hydrostatic pressure	20	20

a. Patient A Peff =

b. Patient B Peff =

■ **B6.** Explain why a person under severe stress might go into a state of oliguria or anuria. Under stress the (*sympathetic? parasympathetic?*) system dominates. It

particulary affects the _____ arteriole, vasoconstricting it. This (*increases? decreases?*) urine output.

■ **B7.** Do this exercise.

a. Glomerular filtration is a(n) (*passive, nonselective? active, selective?*) process. If this were the only step in urine formation, all of the substances in the filtrate would leave the body in urine. Note from Exhibit 26-1 (page 834) that the body

would produce _____ liters of urine each day! And it would contain many valuable substances. Obviously some of these "good" substances must be drawn back into blood and saved.

b. Recall that blood in most capillaries flows into vessels named

_____, but blood in glomerular capillaries flows into

_____ and _____. This unique arrangement permits blood to recapture some of the substances indiscriminately pushed out during filtration. This occurs during the second step of urine formation, called

_____. This process moves substances in the (*same? opposite?*) direction as glomerular filtration as shown by the direction of the arrow at area C of Figure LG 26-3.

■ **B8.** Refer to areas *C* and *D* of Figure LG 26-3 and to the third and fourth columns of Exhibit 26-1 (page 834) in your textbook. Discuss the effects of tubular reabsorption in this learning activity.

a. Which solutes are 100 percent reabsorbed into area *C,* so that virtually none remains in urine (area *D*)?

b. Which substances are mostly, but not entirely, reabsorbed?

About what percent of filtered water is reabsorbed? _____%

c. Which solute is approximately 50 percent resorbed into blood?_____
d. Identify the solute that is filtered from blood, but none of which is reabsorbed.

_____ State the clinical significance of this fact.

e. You can conclude that tubular reabsorption is a (*nonselective? discriminating?*) process which enables the body to save valuable nutrients, ions, and water.

■ **B9.** *A clinical challenge.* Eileen and Marlene are both "spilling sugar" in urine, that

is, they have _____-uria. Discuss possible causes of glycosuria in each case, considering their blood sugar levels.

a. Eileen: glucose of 240 mg/100 ml plasma

b. Marlene: glucose of 80 mg/100 ml plasma

■ **B10.** Do this exercise about reabsorption of electrolytes.

a. Na⁺ is (*actively? passively?*) reabsorbed mostly from the (*proximal? distal?*) convoluted tubule. Its reabsorption in the distal tubule and collecting duct is enhanced by (*high? low?*) levels of aldosterone. Aldosterone production is stimulated by the

_____-angiotensin pathway which is triggered by ____-crease in Na⁺ concentration in ECF. (*For extra review* of this pathway, refer to Chapter 18, Checkpoint E1, page LG 321.)

b. Reabsorption of Na⁺ from the proximal tubule is accompanied by (*active? passive?*) transport of Cl⁻ back into blood, thus achieving electrochemical balance.

c. In the ascending limb of the loop of the nephron, (*Na⁺? Cl⁻?*) is actively reab-

sorbed, while ____ follows passively.

d. The net effect of the processes described in (b) and (c) is (*the same? different?*).

■ **B11.** Identify statements that describe facultative water reabsorption (F) or obligatory water reabsorption (O).

_____ a. Water follows sodium according to osmotic pressure factors.

_____ b. The process is influenced by presence of hormone ADH.

_____ c. About 80 percent of water reabsorption is of this type.

_____ d. Reabsorption occurs specifically across distal and collecting tubules.

■ **B12.** Explain how the juxtaglomerular apparatus regulates glomerular filtration rate (GFR).

a. Recall that approximately _____ ml of blood (or about 650 ml plasma) enter kidneys via renal arteries each minute. Of this, about 125 ml/min of fluids and solutes are removed from blood in the glomeruli and enter the renal tubule. This value

(125 ml/min) is a typical _____ rate (GFR).

b. GFR is maintained in two ways. Both are based on the fact that when GFR is low, (*more? less?*) filtrate is passing through renal tubules. Consequently, filtrate flows more (*slowly? rapidly?*), permitting (*more? less?*) reabsorption of electrolytes such as Cl⁻ back into blood.

c. Less Cl⁻ is then present in tubule cells. Distal tubule cells (called the macula

_____) respond to the low Cl⁻ levels by causing dilation of

(*afferent? efferent?*) arterioles. This ____-creases GFR.

d. Juxtaglomerular cells also respond to the lowered Cl^- level by secreting

_____. Renin is a vaso-(*dilator? constrictor?*). It especially

constricts the ___-ferent arteriole, which increases glomerular blood pressure, again

___-creasing GFR.

e. A term closely related to GFR is filtration _____, which means the percentage of plasma entering glomeruli that does form glomerular filtrate. Calculate a typical filtration fraction using values in Checkpoint B12a: (*1–3%? 15–20%? 50–60%?*). If the endothelial-capsular membrane is especially

porous, as in some kidney conditions, filtration fraction ___-creases.

■ **B13.** The third step in urine production is _____. It involves movement of substances from (*blood to urine? urine to blood?*). In other words, tubular secretion is movement of substances in (*the same? opposite?*) direction as movement occurring in filtration. List four substances that are secreted by the process of tubular secretion.

B14. Explain how tubular secretion helps to control pH. Refer to Figure 26-11 (page 839) in the text and do this exercise.

a. Deborah's respiratory rate is slow. Her blood therefore contains high levels of

_____. It is also somewhat (*acidic? alkaline?*). Explain why.

b. Her kidneys can assist in restoring normal pH (Figure 26-11a). The presence of acid (H^+) stimulates the kidneys to eliminate H^+. In order for this to happen, another cation present in urine must exchange places with H^+. What ion en-

ters kidney tubule cells? _____ It joins HCO_3^- to form

_____.

c. Figure 26-11b shows a second possible mechanism for eliminating acid. H^+

combines with ammonia (_____) to form ammonium

(_____) ion. This cation displaces _____

from NaCl in urine. In this way _____ is conserved and the

H^+ is lost as part of the salt _____.

d. In both of these cases the body eliminated acid (H^+) by exchanging it for another

cation, _____. Consequently, blood pH is raised. Since the body is constantly forming acids during metabolism, this process occurs continually, and so urine is slightly (*acidic? alkaline?*).

■ **B15.** State an example of a situation in which your body needs to produce dilute urine in order to maintain homeostasis.

Now describe how your body can do this by doing this learning activity.

a. To form dilute urine, kidney tubules must reabsorb (*more? fewer?*) solutes than usual. Two factors facilitate this. One is permeability of the ascending limb and part of the distal tubule to Na^+ and Cl^- (*and also to? but not to?*) water. Thus solutes enter interstitium, but water stays in urine.

b. The second requirement for dilute urine is a (*high? low?*) level of ADH. The func-

tion of ADH is to ____-crease permeability of distal and collecting tubules to water. Less ADH forces more water to stay in urine, leading to a (*hyper? hypo?*)-osmotic urine.

■ **B16.** Summarize factors that result in concentrated urine in this exercise. The counter-current mechansim provides a means for producing (*hypotonic? hypertonic?*) urine

during times when the body is _____. This occurs because the (*ascending? descending?*) limb of the loop of Henle is completely impermeable to H_2O, and it is also related to vasa recta permeability. As a result, the interstitial fluid

of the renal _____ is extremely hypertonic, that is, has a (*high?*

low?) osmolality, so pulls H_2O from urine in the _____ tubules. These tubules are made more permeable at this time under the influence of the

hormone _____. As a result, a (*large? small?*) volume of concen-trated urine is produced.

B17. Describe the role of the vasa recta in the concentration of urine.

B18. Contrast *hemodialysis* and *peritonial dialysis* as methods of cleansing body fluids.

C. Homeostasis: other excretory organs (pages 845–846)

C1. Besides the urinary organs, what other body structures perform excretory func-tions? List three and name the substances they eliminate.

C2. As an example of how the activities of the kidneys and other excretory organs are coordinated, describe the adjustments made by the kidneys when the skin increases its output of water (as by sweating on a July day).

D. Ureters, urinary bladder, urethra (pages 846–849)

■ **D1.** Refer to Figure LG 26-1 as you complete the following exercise.

a. Ureters connect _____ to _____. Ureters

are about _____ cm (_____ inches) long.

b. These tubes enter the urinary bladder at two of the three corners of the

_____. The third corner marks the opening into the

_____.

c. The urinary bladder is located in the (*abdomen? pelvis?*). Two sphincters lie just inferior to it. The (*internal? external?*) sphincter is under voluntary control.

d. Urine leaves the bladder through the _____. In females the

length of this tube is about ____ cm; in males it is about ____ cm.

■ **D2.** Ureters, urinary bladder, and urethra are lined with _____ membrane. What is the clinical significance of that fact?

■ **D3.** In the micturition reflex (*sympathetic? parasympathetic?*) nerves stimulate the

_____ muscle of the urinary bladder and cause relaxation of the internal sphincter.

D4. Contrast *incontinence* and *retention*.

E. Urine (pages 849–853)

■ **E1.** Summarize factors that influence urine volume.

a. *Blood pressure.* When blood pressure drops, _____ cells of

the kidney release _____, which leads to formation of

_____. This substance (*increases? decreases?*) blood pressure in two ways: directly, since it serves as a (*vasoconstrictor? vasodilator?*) and, in-

directly, since it stimulates release of _____, which causes

retention of both _____ and _____. Increase in fluid volume will (*increase? decrease?*) blood pressure.

b. *Blood concentration.* The hypothalamic hormone _____ helps to (*conserve? eliminate?*) fluid. When body fluids are hypertonic, ADH is released, causing (*increased? decreased?*) tubular reabsorption of water back into blood.

c. *Medications.* A chemical that mimics secretion of ADH is called a(n)

_____. One that inhibits (or is antagonistic to) ADH is called

a(n) _____. Diuretics cause diuresis which means a (*large? small?*) volume of urine. Removal of extra ECF in this manner is likely to (*increase? decrease?*) blood pressure.

d. *Temperature.* On hot days skin gives up (*large? small?*) volumes of fluid via sweat glands. As blood becomes more concentrated, the posterior pituitary responds by releasing (*more? less?*) ADH, so urine volume (*increases? decreases?*).

E2. Describe the following characteristics of urine. Suggest causes of variations from the normal in each case.

a. Volume: _____ ml/d

b. Color: _____

c. Turbidity: _____

d. Odor: _____

e. pH: _____

f. Specific gravity: _____

■ **E3.** Match the names of tests for renal function with descriptions below.

> BUN. Blood urea nitrogen   SC. Serum creatinine
> RC. Renal clearance

_____ a. Assesses glomerular filtration rate: low value indicates low GFR. Example is creatinine clearance test.

_____ b. Measures levels of urea in blood; level increases in kidney failure.

_____ c. Measures level of a by-product of muscle metabolism that remains in blood; value increases in kidney failure.

E4. The following substances are not normally found in urine. Explain what the presence of each might indicate.

a. Albumin

b. Red blood cells

c. Ketone bodies

d. Bilirubin

E5. Contrast *casts* and *calculi.* Explain what their presence in urine indicates.

F. Aging and development of the urinary system (page 854)

■ **F1.** GFR ____-creases with age. Two other common problems associated with aging

are inability to control voiding (_____) and increased frequency

of _____. Explain why UTIs may occur more often in elderly
males.

Nocturia may also occur more among the aged population. Define *nocturia*.

✗ **F2.** Do this exercise about urinary system development.

a. Kidneys form from _____-derm, beginning at about the

_____ week of gestation.

b. Which develops first? (*Pro? Meso? Meta?*)-nephros. Which extends most superiorly

in location? _____-nephros. Which one ultimately forms the

kidney? _____

c. The urinary bladder develops from the original _____.

G. Disorders, medical terminology (pages 854–857)

■ **G1.** Match each of the terms with the correct description.

C. Cystitis	Po. Polyuria
D. Dysuria	Pt. Ptosis
Gl. Glomerulonephritis	Py. Pyelitis
Go. Gout	U. Uremia

_____ a. Painful urination

_____ b. Inflammation of the urinary bladder

_____ c. Inflammation of the pelvis and
        calyces of kidney

_____ d. Urea in blood

_____ e. Excessive urine

_____ f. Inflammation of the kidney involving
        glomeruli; may follow strep infection

_____ g. Floating kidney; slipping of the kid-
        ney from its normal position

_____ h. High uric acid level of blood; painful
        inflammation of joints caused by
        deposits of the crystallized acid

G2. Discuss and contrast two types of renal failure in this learning activity.

a. (*Acute? Chronic?*) kidney failure is an abrupt cessation (or almost) of kidney function. Write one example of each type of cause of acute renal failure (ARF): *prerenal, renal, postrenal.*

b. (*Acute? Chronic?*) renal failure is progressive and irreversible. Describe changes that occur during the three stages of CRF.

■ **G3.** Urinary tract infections (UTIs) are most often caused by Gram-(*positive?*

negative?) bacteria named _____. List three signs or symptoms of UTIs.

ANSWERS TO SELECTED CHECKPOINTS

A1. (a) Nitrogen-containing products of protein catabolism, such as ammonia and urea; also certain ions and excessive water. (b) Renin, renal erythropoietic factor, D.

A2. (a) Waist; T12, L3; 10–12 cm (4–5 in.). (b) Posterior, retroperitoneal. (c) See key to Figure LG 26-1.

A3. (a) A, glomerular (Bowman's) capsule; B, proximal convoluted tubule; C, descending limb of the loop of the nephron (loop of Henle); D, ascending limb of the loop of the nephron (loop of Henle); E, distal convoluted tubule; F, collecting duct. (b) A, glomerular (Bowman's) capsule; glomerulus, 4. (c) B, C.

A4. (a) CBA. (b) Their size restricts the passage of red blood cells and proteins from blood to urine.

A6. 20 to 25 (1200 ml/min).

A7. 1, Arcuate artery; 2, interlobular artery; 3, afferent arteriole; 4, glomerulus; 5, efferent arteriole; 6A, peritubular capillaries; 6B, vasa recta; 7, interlobular vein; 8, arcuate vein.

A8. Juxtaglomerular apparatus (JGA); af-; distal, blood pressure.

B1. Its volume, electrolyte content, and pH are adjusted; toxic wastes are removed.

B2. Glomerular filtration, tubular reabsorption, tubular secretion.

B3. (a) Pushing, blood, filtrate. (b) All small substances, protein molecules, too large to pass through filtration slits of healthy endothelial-capsular membranes. (c) 70, higher, smaller.

B4. (a) Glomerular blood osmotic pressure and capsular filtrate hydrostatic pressure.

(b) − (capsular hydrostatic pressure + blood osmotic pressure); (60) − (20 + 30); 10; out of.

B5. (a) Peff = 0 mm Hg, (50) − (20 + 30). Anuria due to low blood pressure could be related to hemorrhage or stress. (b) Peff = 28 mm Hg, (73) − (20 + 25). Possibly related to kidney disease such as glomerulonephritis in which protein is allowed to pass out of blood into filtrate, thus lowering blood osmotic pressure.

B6. Sympathetic, afferent, decreases (as in patient A above).

B7. (a) Passive, nonselective; 180. (b) Venules, peritubular capillaries and vasa recta, tubular reabsorption, opposite.

B8. (a) Glucose and proteins. (b) Cl^-, Na^+, HCO_3^-, K^+, uric acid; about 99 percent. (c) Urea. (d) Creatinine; creatinine clearance is often used as a measure of glomerular filtration rate (GFR). 100 percent of the creatinine (a breakdown product of muscle protein) filtered or "cleared" out of blood will show up in urine since 0 percent is reabsorbed. (e) Discriminating.

B9. Glycos. (a) Ingestion of excessive amounts of carbohydrates or hormone imbalance, such as insulin deficiency (diabetes mellitus) or excess growth hormone (GH) may lead to Eileen's hyperglycemia. Such high blood glucose levels exceed capacity of glucose reabsorption mechanisms even in the normal kidney. (b) Although Marlene's blood glucose is in the normal range, kidney malfunction may prevent adequate reabsorption of glucose.

B10. (a) Actively, proximal, high; renin, de. (b) Passive. (c) Cl^-, Na^+. (d) The same.

B11. (a) O. (b) F. (c) O. (d) F.

B12. (a) 1,200, glomerular filtration. (b) Less, slowly, more. (c) Densa, afferent, in. (d) Renin, constrictor, ef. in. (e) Fraction, 15–20% (125/650), in.

B13. Tubular secretion, blood to urine, the same; H^+, K^+, NH_3, creatinine and certain drugs.

B14. (a) CO_2, acidic, increasing levels of CO_2 tend to cause increase of H^+: CO_2 + H_2O → H_2CO_3 → H^+ + HCO_3^-. (Review content on transport of respiratory gas in Chapter 23.) (b) Na^+, $NaHCO_3^-$ (sodium bicarbonate). (c) NH_3, (NH_4^+), Na^+, $NaHCO_3$, NH_4Cl. (d) Na^+, acidic.

B15. During periods of high water intake or decreased sweating. (a) More, but not to. (b) Low, in, hypo.

B16. Hypertonic, dehydrated, ascending, medulla, high, collecting, ADH, small.

D1. (a) Kidneys to urinary bladder, 25–30 (10–12). (b) Trigone, urethra. (c) Pelvis, external. (d) Urethra, 3.8, 20.

D2. Mucous, microbes can spread infection from exterior of body (at urethral orifice) along mucosa to kidneys.

D3. Parasympathetic, detruser.

E1. (a) Juxtaglomerular, renin, angiotensin II, increases, vasoconstrictor, aldosterone, Na^+ and H_2O, increase. (b) ADH, conserve, increased. (c) Antidiuretic, diuretic, large, decrease. (d) Large, more, decreases.

E3. (a) RC. (b) BUN. (c) SC.

F1. De, incontinence, urinary tract infections (UTIs); increased frequency of prostatic hypertrophy leads to urinary retention; excessive urination at night.

F2. (a) Meso, third. (b) Pro, pro, meta. (c) Cloaca (urogenital sinus portion).

G1. (a) D. (b) C. (c) Py. (d) U. (e) Po. (f) Gl. (g) Pt. (h) Go.

G3. Negative; *E. coli;* burning or painful urination, urinary urgency and frequency, back pain, cloudy or blood-tinged urine, urethral discharge, fever, chills, nausea.

MASTERY TEST: Chapter 26

Questions 1–6: Arrange the answers in correct sequence.

_____ _____ _____ 1. From superior to inferior:

    A. Ureter
    B. Bladder
    C. Urethra

_____ _____ _____ _____ 2. From most superficial to deepest:

    A. Renal capsule
    B. Renal medulla
    C. Renal cortex
    D. Renal pelvis

_____ _____ _____ _____ _____ 3. Pathway of glomerular filtrate:

    A. Ascending limb of loop
    B. Descending limb of loop
    C. Collecting tubule
    D. Distal convoluted tubule
    E. Proximal convoluted tubule

_____ _____ _____ _____ _____ 4. Pathway of blood:

    A. Arcuate arteries
    B. Interlobular arteries
    C. Renal arteries
    D. Afferent arteriole
    E. Interlobar arteries

_____ _____ _____ _____ _____ 5. Pathway of blood:

    A. Afferent arteriole
    B. Peritubular capillaries and vasa recta
    C. Glomerular capillaries
    D. Venules and veins
    E. Efferent arteriole

_____ _____ _____ _____ _____ _____ 6. Order of events to restore low blood pressure to normal:

    A. This substance acts as a vasoconstrictor and simulates aldosterone.
    B. Renin is released and converts angiotensinogen to angiotensin I.
    C. Angiotensin I is converted into angiotensin II
    D. Decrease in blood pressure stimulates juxtaglomerular cells.
    E. Under influence of this hormone Na^+ and H_2O reabsorption occurs.
    F. Increased blood volume raises blood pressure.

7. Which of these is a normal constituent of urine?

   A. Albumin
   B. Urea
   C. Glucose
   D. Casts
   E. Acetone

8. Which is a normal function of the bladder?

   A. Oliguria
   B. Nephrosis
   C. Calculi
   D. Micturition

9. Which parts of the nephron are composed of simple squamous epithelium?

   A. Ascending and descending limbs of loop of the nephron
   B. Glomerular capsule (parietal layer) and descending limb of loop of the nephron
   C. Proximal and distal convoluted tubules
   D. Distal convoluted and collecting tubules

10. Choose the one FALSE statement.

    A. The efferent arteriole normally has a larger diameter than the afferent arteriole.
    B. Glomerular capillaries have higher blood pressure than other capillaries of the body.
    C. Blood in glomerular capillaries flows into arterioles, not into venules.
    D. Vasa recta pass blood from the efferent arteriole toward veins.

11. BUN to creatinine ratio in blood is normally about:

    A. 20:1
    B. 1:20
    C. 3:1
    D. 1:3

T F 12. Antidiuretic hormone (ADH) <u>decreases</u> permeability of distal and collecting tubules to water.

T F 13. A ureter is a tube that carries urine from <u>the bladder to the outside of the body.</u>

T F 14. As you run on a hot summer day, you are likely to <u>increase</u> your ADH secretion and therefore to <u>decrease</u> urine production.

T F 15. The juxtaglomerular apparatus consists of cells of the <u>afferent arteriole and distal convoluted tubule.</u>

T F 16. A person taking diuretics is likely to urinate <u>less</u> than a person taking no medications.

T F 17. Blood osmotic pressure is a force that tends to <u>push substances from blood into filtrate.</u>

T F 18. The ascending limb of the loop of the nephron is permeable to <u>Na$^+$ and Cl$^-$, but not to water.</u>

T F 19. The countercurrent multiplier mechanism permits production of <u>small volumes of concentrated urine.</u>

T F 20. In infants 2 years old and under, <u>retention</u> is normal.

_____ 21. The glomerular capsule and its enclosed glomerular capillaries constitute a ____.

_____ 22. A ____ consists of a renal corpuscle and a renal tubule.

_____ 23. Write a value for specific gravity of dilute urine ____

_____ 24. Sympathetic impulses cause greater constriction of ____ arterioles with resultant ____-crease in GFR and ____-crease in urinary output.

_____ 25. A person with chronic renal failure is likely to be in acidosis since kidneys fail to excrete ____, may be anemic since kidneys do not produce ____, and may have symptoms of hypocalcemia since kidneys do not activate ____.

ANSWERS TO MASTERY TEST: ■ Chapter 26

Arrange

1. A B C
2. A C B D
3. E B A D C
4. C E A B D
5. A C E B D
6. D B C A E F

Multiple Choice

7. B   10. A
8. D   11. A
9. B

True-False

12. F. Increases
13. F. A kidney to the bladder
14. T
15. T
16. F. More
17. F. Pull substances from filtrate into blood
18. T
19. T
20. F. Incontinence

Fill-ins

21. Renal corpuscle
22. Nephron
23. 1.008 (or other low value; not high values like 1.040)
24. Afferent, de, de
25. H^+, erythropoietic factor, vitamin D

Fluid, Electrolyte, Acid-Base Balance

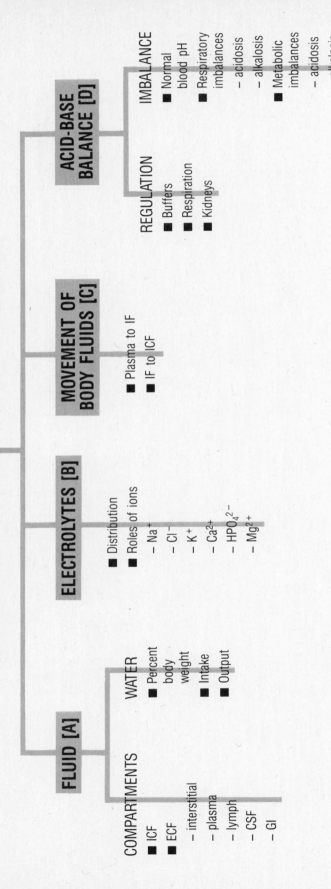

FLUID [A]

COMPARTMENTS
- ICF
- ECF
  – interstitial
  – plasma
  – lymph
  – CSF
  – GI

WATER
- Percent body weight
- Intake
- Output

ELECTROLYTES [B]

- Distribution
- Roles of ions
  – Na^+
  – Cl^-
  – K^+
  – Ca^{2+}
  – HPO_4^{2-}
  – Mg^{2+}

MOVEMENT OF BODY FLUIDS [C]

- Plasma to IF
- IF to ICF

ACID-BASE BALANCE [D]

REGULATION
- Buffers
- Respiration
- Kidneys

IMBALANCE
- Normal blood pH
- Respiratory imbalances
  – acidosis
  – alkalosis
- Metabolic imbalances
  – acidosis
  – alkalosis

Fluid, Electrolyte, and Acid-Base Dynamics

For the maintenance of homeostasis, levels of fluid, ions or electrolytes, acids and bases, must be kept within acceptable limits. Without careful regulation of pH, for example, nerves will become overactive (in alkalosis) or underfunction leading to coma (in acidosis). Alterations in blood levels of electrolytes can be lethal: excess K^+ can cause cardiac arrest, and decrease in blood Ca^{2+} can result in tetany of the diaphragm and respiratory failure. Systems you studied in Chapter 23 (respiratory) and Chapter 26 (urinary), along with buffers in the blood, normally achieve fluid, electrolyte, and acid base balance.

Look first at the Chapter 27 Framework and the essential terms.

TOPIC OUTLINE AND OBJECTIVES

A. Fluid compartments, water

1. Define the processes available for fluid intake and fluid output.
2. Compare the mechanisms regulating fluid intake and fluid output.

B. Electrolytes

3. Contrast the electrolytic concentration of the three major fluid compartments.
4. Explain the functions and regulations of sodium, chloride, potassium, calcium, phosphate, and magnesium.

C. Movements of body fluids

5. Describe the factors involved in the movement of fluid between plasma and interstitial fluid, and between interstitial fluid and intracellular fluid.

6. Explain the relationship between electrolyte imbalance and fluid imbalance.

D. Acid-base balance and imbalance

7. Compare the roles of buffers, respirations, and kidney excretion in maintaining body pH.
8. Define acid-base imbalances and their effects on the body.
9. Explain the appropriate treatments for acidosis and alkalosis.

WORDBYTES

Now study the following parts of words that may help you better understand the terminology in this chapter.

Wordbyte	Meaning	Example		Wordbyte	Meaning	Example
hyper-	above	*hyper*ventilation		-osis	condition of	alkal*osis*
hypo-	below	*hypo*tonic				

Checkpoints

A. Fluid compartments, water (pages 861–862)

■ **A1.** About 66 percent of body fluids are (*inside of? outside of?*) cells. Check your understanding of fluid compartments by circling all fluids that are considered *extracellular fluids* (*ECF*):

Blood plasma   Lymph

Aqueous humor and vitreous humor of the eyes   Pericardial fluid

Endolymph and perilymph of ears   Synovial fluid

Fluid within liver cells   Cerebrospinal fluid (CSF)

Fluid immediately surrounding liver cells

For extra review, refer to Figure LG 1–2, page 13.

■ **A2.** *A clinical challenge.* Is a 45-year-old woman who has a water content of 55 percent likely to be in fluid balance?

■ **A3.** About what percent of body weight consists of water? _____ Circle the person in each pair who is more likely to have a higher proportion of water content.
a. 37-year-old woman/67-year-old woman
b. Female/male
c. Lean person/fat person

■ **A4.** Daily intake and output (I and O) of fluid usually both equal _____ ml

(_____ liters). Of the intake, about _____ ml is ingested in food and drink, and

_____ ml is a product of catabolism of foods (discussed in Chapter 25).

■ **A5.** Now list four systems of the body that eliminate water. Indicate the amounts lost by each system on an average day. One is done for you.
a. Integument (skin): 500 ml/day

b. _____ : _____ ml/day

c. _____ : _____ ml/day

d. _____ : _____ ml/day

On a day when you exercise vigorously, how would you expect these values to change?

■ **A6.** Consider regulation of fluid intake and output in this Checkpoint.

a. Fluid intake is regulated mainly by the sensation of _____,

which is triggered by dryness in the _____ and the pharynx
and also by receptors in the (*medulla? hypothalamus?*).

b. Circle the factors that increase fluid output:

A. Increase in ADH   D. Increase in blood pressure (BP)

B. Increase in aldosterone   E. Fever

C. Diarrhea   F. Hyperventilation

B. Electrolytes   (pages 862–866)

■ **B1.** Circle all of the answers that are classified as electrolytes:

A. Proteins in solutions   D. K^+ and Na^+

B. Anions   E. Compounds containing at least one

C. Glucose   ionic bond

■ **B2.** List three general functions of electrolytes.

_____   _____   _____

B3. (*Electrolytes? Nonelectrolytes?*) exert greater osmotic effect. In one or two sentences, explain why.

■ **B4.** On Figure LG 27-1, write the chemical symbols for electrolytes that are found in greatest concentrations in each of the three compartments. Write one symbol on each line provided. Use the following list of symbols.

Cl^-. Chloride HPO_4^{2-}. Phosphate Mg^{2+}. Magnesium

HCO_3^-. Bicarbonate K^+. Potassium $Protein^-$. Protein

                                                           Na^+. Sodium

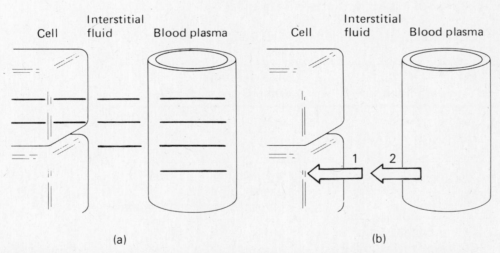

Figure LG 27-1 Diagrams of fluid compartments. (a) Write in electrolytes according to Checkpoint B4. (b) Direction of fluid shift resulting from Na^+ deficit. Color according to Checkpoint C3.

■ **B5.** Analyze the information in Figure LG 27-1 by answering these questions.

a. Circle the two compartments that are most similar in electrolyte concentration.

cell   plasma   interstitial fluid

b. What is the one major difference in electrolyte content of these two similar compartments?

c. Most body protein is located in which compartment? _____

d. Protein anions are more likely to be bound to the cation (Na^+? K^+?) in intracellular fluid.

■ **B6.** Refer to Figure 27-2 (page 864) in your text. Notice that concentrations of electro-

lytes are expressed in units called _____, abbreviated _____.
Write a formula which can be used to determine meq/liter.

meq/liter = (_____) × (_____) / _____

■ **B7.** Calculate meq/liter of each of the plasma electrolytes listed in the table below.
Use information in that table and in Figure 27-2 (page 864) in the text for help.

Electrolyte	mg/liter	Number of Charges	Atomic Weight	Plasma Concentration (meq/liter)
a. Sodium (Na$^+$)	3300	1	23	
b. Chloride (Cl$^-$)	3670		35	
C. Magnesium (Mg^{2+})	30	2	24	

Notice how your answers correlate with values for plasma shown in the graph (Figure 27-2, page 864 in your text); much less Mg^{2+} is present than either Na^+ or Cl^-.

B8. Complete the following table describing important electrolytes.

Electrolyte	Principal Functions	Disorders and Some Symptoms
a. Sodium (Na$^+$)		
b.		Hypochloremia; muscle spasms, alkalosis, depressed respiration, coma
c.	Most abundant cation in ICF; helps maintain fluid volume of cells; functions in nerve transmission	

Electrolyte	Principal Functions	Disorders and Some Symptoms
d. Calcium (Ca^{2+})		
e.	Helps form bones, teeth; important component of DNA, RNA, ATP, and buffer system	
f. Magnesium (Mg^{2+})		

■ **B9.** *A clincal challenge.* Norine is having her serum electrolytes analyzed. Complete this exercise about her results.

a. Her Na$^+$ level is 128. Refer back to Checkpoint B7 and identify a normal value for serum (or plasma) Na$^+$. Norine's value indicates that she is more likely to be in a state of (*hyper? hypo?*)-natremia. Write two symptoms of this electrolyte imbalance.

b. A diagnosis of hypochloremia would indicate that Norine's blood level of

_____ ion is lower than normal. One cause of this is excessive

_____.

c. A normal range for K$^+$ is 3.5–5.0 meq/liter. This range is considerably (*higher? lower?*) than that for Na$^+$. This is reasonable since K$^+$ is the main cation in (*ECF? ICF?*), yet lab analysis of serum electrolytes is examining (*ECF? ICF?*) K$^+$.

d. Norine's potassium (K$^+$) level is 5.6 meq/liter, indicating that she is in a state of

hyper-_____. Hyperkalemia and hyponatremia (as described in *a*) may be caused by (*high? low?*) levels of aldosterone. Explain why.

e. A normal range for Mg^{2+} level of blood is 1.5–2.5 meq/liter. Norine's electrolyte report indicates a value of 1.3 meq/liter for Mg^{2+}. This electrolyte imbalance is

known as _____. This condition, like hyponatremia, may be caused by (*high? low?*) aldosterone levels.

f. Norine's Ca^{2+} level is 3.9 meq/liter. A normal range is 4.3–5.3 meq/liter. This elec-

trolyte imbalance, known as _____, is most closely linked to

_____-creased levels of parathyroid hormone (PTH) and _____-creased secre-

tion of the thyroid hormone, _____. Low blood levels of both Ca^{2+} and Mg^{2+} cause overstimulation of the central nervous system and muscles. List two symptoms of these electrolyte imbalances.

C. Movement of body fluids (pages 866–867)

■ **C1.** Four forces are involved in the movement of water between plasma and interstitial compartments and between interstitial and intracellular compartments. (They include, incidentally, the same types of forces involved in glomerular filtration.) Movement across capillary membranes due to these four forces is according to

_____ law of capillaries.

C2. Review Chapter 21 Checkpoint B9 (page LG 379) on calculation of effective filtration pressure (Peff) controlling movement of fluids between blood and interstitial fluid.

■ **C3.** Show on Figure LG 26-1b the shift of fluid resulting from loss of Na^+, for example by excessive loss of sweat. Color arrows (1 and 2) according to these color code ovals:

○ Shift of fluid possibly leading to hypovolemic shock
○ Shift of fluid leading to overhydration or water intoxication of brain cells, possibly leading to convulsions and coma

D. Acid-base balance and imbalance (pages 867–871)

■ **D1.** Complete this Checkpoint about acid-base regulation.

a. The pH of blood and other ECF should be maintained between _____

and _____.

b. Name the three major mechanisms that work together to maintain acid-base balance.

_____ _____ _____

c. List the four main buffer systems in the body.

_____ _____

_____ _____

■ **D2.** Check your understanding of the carbonic acid–bicarbonate buffer in this Checkpoint. When a person actively exercises, (*much? little?*) CO_2 forms leading to H_2CO_3 production. Consequently, (*much? little?*) H^+ will be added to body tissues. In order to buffer this excess H^+ (to avoid drop in pH and damage to tissues), the (*H_2CO_3? NaHCO_3?*) component of the buffer system will be called upon. The excess H^+ will

then replace the _____ ion of $NaHCO_3$, in a sense "tying up" the potentially harmful H^+.
 For extra review of this buffer system, refer to Chapter 2 Checkpoint C8, and Figure LG 2-3, pages LG 30–31.

■ **D3.** Phosphates are more concentrated in (*ECF? ICF?*). (For help, see Figure 27-2, page 864, in the text.) Name two types of cells in which the phosphate buffer system

is most important. _____ _____

D4. Write the reaction showing how phosphate buffers strong acid.

D5. Explain how the hemoglobin buffer system buffers carbonic acid, so that an acid even weaker than carbonic acid (that is, hemoglobin), is formed.

For extra review. See Chapter 23, Checkpoint E5, page LG 000 and Figure 27-4, (page 868) in the text.

■ **D6.** Which is the most abundant buffer system in the body? _____
Circle the part of the amino acid shown below that buffers acid. Draw a square around the part of the amino acid that buffers base.

$$
\begin{array}{c}
R \\
| \\
NH_2 - C - COOH \\
| \\
H
\end{array}
$$

■ **D7.** Complete this Checkpoint about respiration and kidneys related to pH.

a. Hyperventilation will tend to (*raise? lower?*) blood pH, since as the person exhales

CO_2, less CO_2 is available for formation of _____ acid and free hydrogen ion.

b. A slight decrease in blood pH will tend to (*stimulate? inhibit?*) the respiratory center and so will (*increase? decrease?*) respirations.

c. The kidneys regulate acid-base balance by altering their tubular secretion of

_____ or elimination of _____ ion.

■ **D8.** Complete this exercise about acid-base imbalances.

a. An arterial blood pH of 7.15 indicates _____-osis. In such a case, the bicarbonate-carbonic acid buffer system would buffer the excess H^+, and an increase in (*NaHCO₃? HHCO₃? = H₂CO₃?*) would result.

b. Normal blood pH is maintained by balancing components of this buffer system such that the (*NaHCO₃? H₂CO₃?*) member of the system is in much greater abundance. In fact, the normal ratio of bicarbonate (or sodium bicarbonate) to carbonic

acid is _____ : _____. Write that ratio here:

$$\frac{NaHCO_3}{H_2CO_3} = \text{\underline{\hspace{1cm}}}$$

c. Any condition that increases the ratio (so that this buffer system consists of more

bicarbonate or less carbonic acid) results in _____-osis. A condition that decreases the ratio, for example, by raising (*bicarbonate? carbonic acid?*) in blood, will lead to acidosis.

d. Tell whether each of these ratios indicates alkalosis or acidosis.

A. 30:1 _____

B. 20:0.5 _____

C. 12:1 _____

■ **D9.** An acid-base imbalance caused by abnormal alteration of the respiratory system is classified as (*metabolic? respiratory?*) acidosis or alkalosis. Any other cause of acid-base imbalance, such as urinary or digestive tract disorder, is identified as (*metabolic? respiratory?*) acidosis or alkalosis.

■ **D10.** *A clinical challenge.* Indicate which of the four categories of acid-base imbalances listed in the box is most likely to occur as a result of each condition described below.

| MAcid. Metabolic acidosis | RAcid. Respiratory acidosis |
| MAlk. Metabolic alkalosis | RAlk. Respiratory alkalosis |

_____ a. Decreased blood level of CO_2 as a result of hyperventilation

_____ b. Decreased respiratory rate in a patient taking overdose of morphine

_____ c. Excessive intake of antacids

_____ d. Prolonged vomiting of stomach contents

_____ e. Ketosis in uncontrolled diabetes mellitus

_____ f. Decreased respiratory minute volume in a patient with emphysema or fractured rib

_____ g. Excessive loss of bicarbonate from the body, as in prolonged diarrhea or renal dysfunction

D11. Briefly describe treatments that may be used for:

a. Respiratory acidosis

b. Metabolic acidosis

c. Respiratory alkalosis

d. Metabolic alkalosis

ANSWERS TO SELECTED CHECKPOINTS

A1. Inside of; all answers except fluid within liver cells.

A2. The percentage is appropriate; however, more information is needed since the fluid must be distributed correctly among the three major compartments: cells (ICF) and ECF in plasma and interstitial fluid. For example, excessive distribution of fluid to interstitial space occurs in the fluid imbalance known as edema.

A3. About 60%. (a) 37-year-old woman. (b) Male. (c) Lean person.

A4. 2500 (2.5); 2300, 200.

A5. (Any order) (b) Kidney, 1500. (c) Lungs, 300. (d) GI, 200. Output of fluid increases greatly via sweat glands and slightly via respirations. To compensate, urine output decreases and intake of fluids increases.

A6. (a) Thirst, mouth, hypothalamus. (b) C–F.

B1. All except C (glucose).

B2. Serve as essential minerals, regulate acid-base balance and osmosis between compartments.

B4.

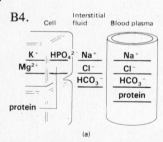

Figure LG 27-1A Diagrams of fluid compartments.

B5. (a) Interstitial fluid and plasma (the two extracellular fluids). (b) Plasma contains more protein. (c) Intracellular. (d) K^+ (since it is more concentrated inside of cells).

B6. Milliequivalents/liter, meq/liter,
$$\frac{\text{mg of ion/liter solution} \times \text{number of valence charges}}{\text{atomic weight}}$$

B7. (a) 143.0 = (3300 × 1)/23. (b) 105 = (3670 × 1)/35. (c) 2.5 = (30 × 2)/24.

B9. (a) Normal serum Na^+ range is 137–150 meq/liter, hypo; examples: headache, confusion, lethargy, eventually coma. (b) Cl^-, vomiting of gastric contents (HCl). (c) Lower, ICF, ECF. (d) Kalemia, low, aldosterone promotes reabsorption of Na^+ and H_2O and also Mg^{2+} into blood in kidneys, and secretion of K^+ into urine. (e) Hypomagnesemia, low. (f) Hypocalcemia, de, in, calcitonin; tremor, tetany and possible convulsions.

C1. Starling's.
C3. 1, Shift of fluid leading to overhydration or water intoxication of brain cells, possibly leading to convulsions and coma; 2, shift of fluid leading to hypovolemic shock.
D1. (a) 7.35 − 7.45. (b) Buffers, respiration, and kidney excretion. (c) Carbonic acid-bicarbonate, phosphate, hemoglobin-oxyhemoglobin, protein.

D2. Much, much, $NaHCO_3^-$, Na^+.
D3. ICF; red blood cells and kidney cells.
D6. Protein;

Buffers acid —(NH₂)— C —[COOH]— Buffers base (with R above and H below)

D7. (a) Raise, carbonic. (b) Stimulate, increase. (c) H^+, HCO_3^-.
D8. (a) Acid, $HHCO_3 = H_2CO_3$. (b) $NaHCO_3$, 20:1, $NaHCO_3/H_2CO_3 = 20/1$. (c) Alkal, carbonic acid. (d) A, alkalosis; B, alkalosis; C, acidosis.
D9. Respiratory, metabolic.
D10. (a) RAlk. (b) RAcid. (c) MAlk. (d) MAlk. (e) MAcid. (f) RAcid. (g) MAcid.

MASTERY TEST: Chapter 27

Questions 1–6: Circle T (true) or F (false). If the statement is false, change the underlined word or phrase so that the statement is correct.

T F 1. Hyperventilation will tend to raise pH.

T F 2. Carbonic acid is a weaker acid than reduced hemoglobin (H·Hb).

T F 3. Under normal circumstances fluid intake each day is greater than fluid output.

T F 4. During edema there is increased movement of fluid out of plasma and into interstitial fluid.

T F 5. Parathyroid hormone causes a high blood level of calcium.

T F 6. When aldosterone is in high concentrations, sodium is conserved (in blood) and potassium is excreted (in urine).

T F 7. In general, electrolytes cause greater osmotic effects than nonelectrolytes.

T F 8. The body of an infant contains a higher percentage of water than the body of an adult.

T F 9. Starch, glucose, HCO_3^-, Na^+, and K^+ are all electrolytes.

T F 10. Loss of blood via hemorrhage is likely to decrease blood hydrostatic pressure and also decrease Peff.

T F 11. Protein is the most abundant buffer system in the body.

T F 12. Sodium plays a much greater role than magnesium in osmotic balance of ECF since more meq's of sodium are in ECF.

T F 13. The carbonic acid portion of the bicarbonate-carbonic acid system buffers base.

T F 14. A bicarbonate-carbonic acid ratio of 10:1 indicates acidosis.

T F 15. The pH of arterial blood should be higher than that of venous blood and should be about pH 7.45.

T F 16. Hydrostatic pressure of blood and hydrostatic pressure of interstitial fluid tend to move substances in the same direction: out of blood and into interstitial fluid.

17. On an average day the greatest volume of fluid output is via:

    A. Feces   C. Lungs
    B. Sweat   D. Urine

18. Which term refers to a lower than normal blood level of sodium?

    A. Hyperkalemia   D. Hyponatremia
    B. Hypokalemia   E. Hypercalcemia
    C. Hypernatremia

19. Which two electrolytes are in greatest concentration in ICF?

    A. Na^+ and Cl^- D. K^+ and HPO_4^{2-}
    B. Na^+ and K^+ E. Na^+ and HPO_4^{2-}
    C. K^+ and Cl^- F. Ca^{2+} and Mg^{2+}

20. All of the following factors will tend to increase fluid output EXCEPT:

    A. Fever
    B. Vomiting and diarrhea
    C. Hyperventilation
    D. Increased glomerular filtration rate
    E. Decreased blood pressure

21. Interstitial fluid, blood plasma, and lymph are all ____-cellular fluids.

22. About ____ percent of body fluids form part of ____-cellular fluid, and ____ percent are in ECF.

23. ____ are chemicals that have at least one ionic bond and dissociate into positive and negative ions.

24. To compensate for acidosis, kidneys will increase secretion of ____, and in alkalosis kidneys will eliminate more ____.

25. Normally blood serum contains about ____ meq/liter of Na^+ and ____ meq/liter of K^+.

ANSWERS TO MASTERY TEST: ■ Chapter 27

True-False

1. T
2. F. Stronger
3. F. Equals
4. T
5. T
6. T
7. T
8. T
9. F. HCO_3^-, Na^+, K^+ (Starch and glucose are nonelectrolytes.)
10. T
11. T
12. T
13. T
14. T
15. T
16. F. Move substances in opposite directions: BHP pushes substances out of blood and IFHP pulls substances from IF into blood.

Multiple Choice

17. D   19. D
18. D   20. E

Fill-ins

21. Extra
22. 67, intra-, 33
23. Electrolytes
24. H^+, HCO_3^-
25. 137–150, 3.5–5.0

Continuity

The Reproductive System

MALE REPRODUCTION [A]

FEMALE REPRODUCTION

SEXUAL INTERCOURSE, BIRTH CONTROL [E]

AGING, DEVELOPMENT [F]

DISORDERS [G]

TESTES
- In scrotum
- Seminiferous tubules
- Interstitial endocrinocytes
- Sustenacular cells

DUCTS
- Epididymis
- Ductus (vas) deferens
- Ejaculatory duct
- Urethra

ACCESSORY GLANDS
- Seminal vesicles
- Prostate
- Bulbourethral glands

PENIS
- Corpus cavernosum & spongiosum
- Glans
- Prepuce

INTERNAL ORGANS [B]
- Ovaries
- Uterine tubes
- Uterus
  - endometrium
  - myometrium
  - Pap smear

EXTERNAL GENITALIA [D]
- Vagina
- Vulva
- Perineum
- Mammary glands

ENDOCRINE RELATIONS [C]

Hormonal Control
- Hypothalamus
- Anterior pituitary
- Ovary

Menstrual Cycle
- Menstrual phase (days 1–5)
- Preovulatory phase (days 6–13)
- Ovulation (day 14)
- Postovulatory phase (days 15–28)

Menarche, Menopause

Sexual activity
- erection
- lubrication
- orgasm
- Birth control

The Reproductive Systems

Unit V introduces the system that provides for continuity of the human species, not simply homeostasis of the individual. The reproductive system produces sperm and ova that may unite and develop to form offspring. In order for sperm and ova to reach a site for potential union, they must each pass through duct systems and be nourished and protected by secretions. External genitalia include organs responsive to stimuli and capable of sexual activity. Hormones play significant roles in the development and maintenance of the reproductive system. The variations of hormones in ovarian and uterine (menstrual) cycles are considered in detail, along with methods of birth control. As we have done for all other systems, we discuss development of the reproductive system and normal aging changes. Start your study with the Chapter 28 Framework and become familiar with the terms presented there.

TOPIC OUTLINE AND OBJECTIVES

A. Male reproduction

1. Define reproduction and classify the organs of reproduction by function.
2. Explain the structure, histology, and functions of the ducts, accessory sex glands, and penis.

B. Female reproductive system: ovaries, uterine tubes, uterus

3. Describe the location, histology, and functions of the ovaries, uterine (Fallopian) tubes, uterus, vagina, vulva, and mammary glands.

C. Female reproductive system: endocrine relations

4. Compare the principal events of the menstrual and ovarian cycles.

D. Female reproductive system: vagina, external genitalia, mammary glands

E. Sexual intercourse and birth control

5. Explain the roles of male and female in sexual intercourse.

6. Contrast the various kinds of birth control (BC) and their effectiveness.

F. Aging and development of the reproductive systems

7. Describe the effects of aging on the reproductive systems.
8. Describe the development of the reproductive systems.

G. Disorders and medical terminology

9. Explain the symptoms and causes of sexually transmitted diseases (STDs) such as gonorrhea, syphilis, genital herpes, chlamydia, trichomoniasis, and genital warts.
10. Describe the symptoms and causes of male disorders (testicular cancer, prostate dysfunctions, impotence, and infertility) and female disorders (amenorrhea, dysmenorrhea, premenstrual syndrome (PMS) toxic shock syndrome (TSS), ovarian cysts, endometriosis, infertility, breast tumors, cervical cancer, pelvic inflammatory disease (PID), and vulvovaginal candidiasis).
11. Define medical terminology associated with the reproductive systems.

WORDBYTES

Now study the following parts of words that may help you better understand terminology in this chapter.

Wordbyte	Meaning	Example	Wordbyte	Meaning	Example
-arche	beginning	men*arche*	-metrium	uterus	endo*metrium*
cervix	neck	*cervix* of uterus	mito	thread	*mito*sis
corp-	body	*corp*us cavernosum	oo-, ov-	egg	*oo*genesis, *ov*um
crypt-	hidden	*crypt*orchidism	orchi-	testis	*orchi*dectomy
-genesis	formation	spermato*genesis*	rete	network	*rete* testis
hyster(o)-	uterus	*hyster*ectomy	salping-	tube, trumpet	*salping*itis
meio-	smaller	*meio*sis	troph-	nutrition	*troph*oblast
men-	month	*men*opause, *men*ses			

CHECKPOINTS

A. Male reproduction (pages 879–891)

A1. Define each of these groups of reproductive organs.

a. Gonad

b. Ducts

c. Accessory glands

d. Supporting structures

■ **A2.** As you go through the chapter, note which structures of the male reproductive system are included in each of the above categories.

■ **A3.** Describe the scrotum and testes in this exercise.

a. Arrange coverings around testes from most superficial to deepest

_____ _____ _____ D. Dartos: smooth muscle that wrinkles skin
TA. Tunica albuginea: dense fibrous tissue
TV. Tunica vaginalis: derived from peritoneum

b. The temperature in the scrotum is about 3°C (*higher? lower?*) than the temperature in the abdominal cavity. Of what significance is this?

c. During fetal life, testes develop in the (*scrotum? posterior of the abdomen?*).

Undescended testes is a condition known as _____. Write one possible effect of this condition.

d. Muscles that elevate testes closer to the warmth of the pelvic cavity are known as (*cremaster? dartos?*) muscles.

■ **A4.** Discuss spermatogenesis in this exercise.

a. Spermatozoa are one type of gamete; name the other type:

_____. Gametes are (*haploid? diploid?*); human gametes

contain ____ chromosomes.

b. Fusion of gametes produces a cell called a _____. This cell, and all cells of the organism derived from it, contain the (*haploid? diploid?*) number of chromosomes, written as (*n? 2n?*). These chromosomes, received from both

sperm and ovum, are said to exist in _____ pairs.

c. Gametogenesis assures production of haploid gametes from otherwise diploid

humans. In males, this process is called _____-genesis. It

occurs in the _____, and requires about 2–3 (*days? months? years?*) for full maturation of spermatozoa.

■ **A5.** Refer to Figure LG 28-1 and do the following Checkpoint. Use the following list of cells for answers:

Early spermatid	Secondary spermatocyte
Interstitial endocrinocyte (cell of Leydig)	Spermatogonium
Late spermatid	Spermatozoon (sperm)
Primary spermatocyte	Sustentacular (Sertoli) cells

a. Label cells *A* and *B* and *1–6*.
b. Mark cells *1–6* according to whether they are diploid (2n) or haploid (n). Use lines on figure.

c. Meiosis is occurring in cells numbered _____ and _____. Color these meiotic cells according to color ovals on the figure. Which cell undergoes reduction division

(meiosis 1?) _____ Which undergoes equatorial division (meiosis II)? _____
For extra review of the process of meiosis, review Chapter 3 Checkpoint F4, pages LG 55–56.
d. From this figure, you can conclude that as cells develop and mature, they move (*toward? away from?*) the lumen of the tubule.

e. Which cells are in the process of spermiogenesis? _____ and _____.

f. Which cells secrete testosterone? _____
g. Which cells produce nutrients for sperm, phagocytose degenerating sperm, produce

inhibin, and form the blood–testes barrier? _____

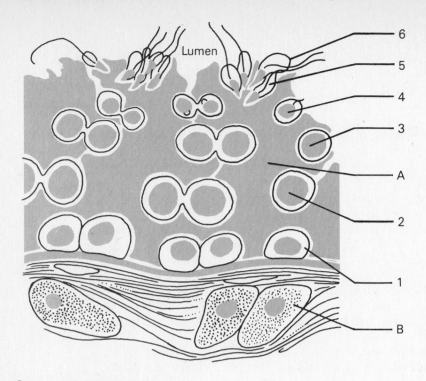

○ Primary spermatocyte
○ Secondary spermatocyte

Figure LG 28-1 Diagram of a portion of a cross section of a seminiferous tubule showing the stages of spermatogenesis. Color and label as directed in Checkpoint A5.

A6. Explain how cytoplasmic bridges between cells formed during spermatogenesis may facilitate survival of Y-bearing sperm, and therefore permit generation of male offspring.

■ **A7.** Describe spermatozoa (sperm) in this Checkpoint.

a. Number produced in a normal male: _____/day

b. Life expectancy of sperm once inside female reproductive tract: _____ hours.

c. Portion of sperm containing nucleus: _____

d. Part of sperm containing mitochondria: _____

e. Function of acrosome: _____

f. Function of tail: _____

■ **A8.** Match the hormones listed in the box with descriptions below.

> FSH   LH
> GnRH   Testosterone
> Inhibin

a. Released by the hypothalamus; stimulates release of two anterior pituitary hormones:

   _____

b. Stimulates seminiferous tubules to start sperm production; also stimulates sustancular cells

   to nourish sperm: _____

c. Stimulates release of testosterone:

   _____

d. Principal male hormone:

   _____

e. Inhibits release of GnRF:

   _____

f. Testicular hormone that inhibits FSH

   production: _____

A9. List five functions of testosterone.

■ **A10.** Refer to Figure LG 28-2. Structures in the male reproductive system are numbered in order along the pathway that sperm take from their site of origin to the point where they exit from the body. Match the structures in the figure (using numbers 1 to 9) with descriptions below.

_____ a. Ejaculatory duct

_____ b. Epididymis

_____ c. Prostatic urethra

_____ d. Spongy (cavernous) urethra

_____ e. Membranous urethra

_____ f. Ductus (vas) deferens (portion within the abdomen)

_____ g. Ductus (vas) deferens (portion within the scrotum)

_____ h. Urethral orifice

_____ i. Testis

■ **A11.** Which structures numbered 1–9 in Figure LG 28-2 are paired? Circle their numbers:

1   2   3   4   5   6   7   8   9

Which structures are single? Write their numbers: _____

■ **A12.** Answer these questions about the ductus (vas) deferens.

a. The ductus deferens is about _____ cm (_____ in.) long.

b. It is located (*entirely? partially?*) in the abdomen. It enters the abdomen via the

   _____. Protrusion of abdominal contents through this

   weakened area is known as _____.

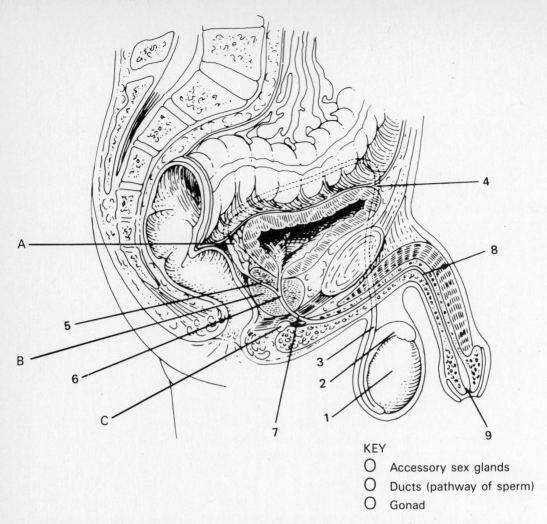

Figure LG 28-2 Male organs of reproduction seen in sagittal section. Structures are numbered in order along the pathway taken by sperm. Color and label as directed in Checkpoints A10, A11, A15, and A20.

KEY
O   Accessory sex glands
O   Ducts (pathway of sperm)
O   Gonad

c. What structures besides the vas compose the seminal cord?

d. A vasectomy is performed at a point along the section of the vas that is within the

_____. Must the peritoneal cavity be entered during this procedure? (*Yes? No?*) Will the procedure affect testosterone level of the male? (*Yes? No?*) Why?

■ **A13.** Which portion of the male urethra is longest? (*Prostatic? Membranous? Spongy,* or *cavernous?*)

■ **A14.** Match the accessory sex glands with their descriptions.

B. Bulbourethral glands	P. Prostate gland	S. Seminal vesicles

_____ a. Paired pouches posterior to the urinary bladder

_____ b. Structures that empty secretions (along with contents of vas deferens) into ejaculatory duct

_____ c. Single doughnut-shaped gland located inferior to the bladder; surrounds and empties secretions into urethra

_____ d. Contribute about 60 percent of seminal fluid

_____ e. Pair of pea-sized glands located in the urogenital diaphragm

■ **A15.** Color parts of the male reproductive system on Figure LG 28-2 according to color code ovals. Label the *seminal vesicle, prostate gland,* and *bulbourethral gland* on the figure.

■ **A16.** *A clinical challenge.* Explain the clinical significance of the shape and location of the prostate gland.

■ **A17.** Complete the following exercise about semen.

a. The average amount per ejaculation is _____ ml.

b. The average range of number of sperm is _____/ml. When

the count falls below _____/ml, the male is likely to be sterile. Only one sperm fertilizes the ovum. Explain why such a high sperm count is required for fertility.

c. The pH of semen is slightly (*acid? alkaline?*). State the advantage of this fact.

d. What is the function of *seminalplasmin?*

A18. List six characteristics of semen studied in a semen analysis.

◀ **A19.** Refer to Figure 28-10 (page 892) in the text and do this exercise.

a. The _____, or foreskin, is a covering over the

_____ of the penis. Surgical removal of the foreskin is known

as _____ .

b. The names *corpus* _____ and *corpus*

_____ indicate that these bodies of tissue in the penis contain

spaces that distend in the presence of excess _____. As a result, blood is prohibited from leaving the penis, since (*arteries? veins?*) are compressed. This temporary state is known as (*erect? flaccid?*) state.

c. The urethra passes through the corpus _____. The urethra

functions in the transport of _____ and

_____. During ejaculation, what prevents sperm from entering the urinary bladder and urine from entering the urethra?

A20. Label these parts of the penis on Figure LG 28-2: *prepuce, glans, corpus cavernosum, corpus spongiosum,* and *bulb of the penis.*

B. Female reproductive system: ovaries, uterine tubes, uterus (pages 892–900)

■ **B1.** Refer to Figure LG 28-3 and match the structures numbered 1–5 with their descriptions. Note that structures are numbered in order along the pathway taken by ova from the site of formation to the point of exit from the body.

_____ a. Uterus (body)   _____ d. Uterine (Fallopian) tube

_____ b. Uterus (cervix)   _____ e. Vagina

_____ c. Ovary

■ **B2.** Indicate which of those five structures in Figure LG 28-3 are paired.
Circle the numbers: 1   2   3   4   5.

Which structures are singular? Write their numbers: _____

■ **B3.** Describe the ovaries in this Checkpoint.
a. The two ovaries most resemble:
   A. Almonds (unshelled)   C. Pears
   B. Peas   D. Doughnuts

b. Ovaries descend to the brim of the pelvis during the ____ month of fetal life.
   A. Third   C. Ninth
   B. Sixth

c. Position of the ovaries is maintained by _____. Name two of

these: _____ _____
d. Arrange in sequence according to chronological appearance, from first to develop

to last: ____ ____ ____
   CL.   Corpus luteum
   OF.   Ovarian follicle
   VOF. Vesicular ovarian (Graafian) follicle

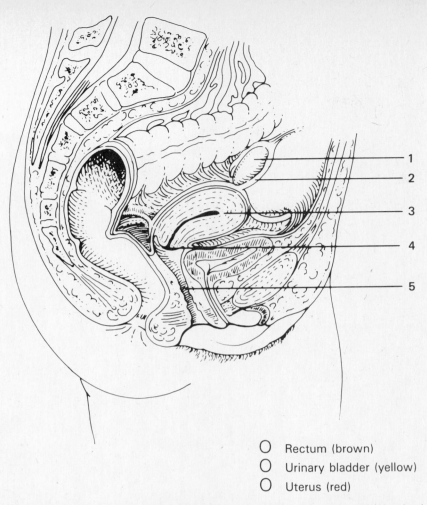

1 _____
2 _____
3 _____
4 _____
5 _____

○ Rectum (brown)
○ Urinary bladder (yellow)
○ Uterus (red)

Figure LG 28-3 Female organs of reproduction. Structures are numbered in order in the pathway taken by ova. Color and label as directed in Checkpoints B1, B2, B6, and B8.

■ **B4.** Color the four cells on Figure LG 28-4 according to the color code ovals.

■ **B5.** Complete this exercise about the uterine tubes.

a. Uterine tubes are also known as _____ tubes. They are about

_____ cm (_____ inches) long.

b. The funnel-shaped open end of each tube is known as the

_____. Its fingerlike projections, called

_____, are close to (but not in direct contact with) the

_____.

c. What structural features of the tube enhance passage of ova into and through the tubes?

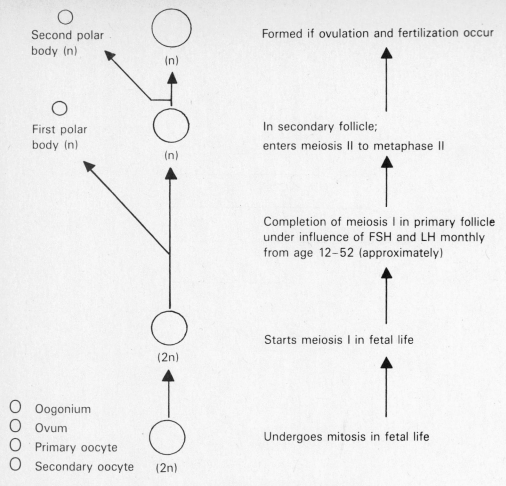

Second polar body (n)

First polar body (n)

(n)

(n)

(2n)

○ Oogonium
○ Ovum
○ Primary oocyte
○ Secondary oocyte

(2n)

Formed if ovulation and fertilization occur

In secondary follicle; enters meiosis II to metaphase II

Completion of meiosis I in primary follicle under influence of FSH and LH monthly from age 12–52 (approximately)

Starts meiosis I in fetal life

Undergoes mitosis in fetal life

Figure LG 28-4 Oogenesis. Color as directed in Checkpoint B4.

d. List two functions of uterine tubes.

e. Define *ectopic pregnancy* and list possible causes of this condition.

B6. Color the pelvic organs on Figure LG 28-3. Use color code ovals on the figure.

■ **B7.** Answer these questions about uterine position.
   a. The organ that lies anterior and inferior to the uterus is the

   _____ . The _____ lies posterior to it.
   b. The rectouterine pouch lies (*anterior? posterior?*) to the uterus, whereas the

   _____-uterine pouch lies between the urinary bladder and the uterus.
   c. The fundus of the uterus is normally tipped (*anteriorly? posteriorly?*). If it is malpositioned posteriorly, it would be (*anteflexed? retroflexed?*).

d. The _____ ligaments attach the uterus to the sacrum. The

_____ ligaments pass through the inguinal canal and anchor

into external genitalia (labia majora). The _____ ligaments are
broad, thin sheets of peritoneum extending laterally from the uterus.
e. The most important ligaments in prevention of drooping (prolapse) of the uterus

are the _____ ligaments which are attached to the base of the
uterus.

■ **B8.** Refer to Figure LG 28-3 and do this exercise about layers of the wall of the
uterus.

a. Most of the uterus consists of _____-metrium. This layer is
(*smooth muscle? epithelium?*). It should appear red in the figure (as directed
above).
b. Now color the peritoneal covering over the uterus green. This is a (*mucous?
serous?*) membrane.

c. The innermost layer of the uterus is the _____-metrium.
Which portion of it is shed during menstruation? Stratum (*basalis? functionalis?*).
Which arteries supply the stratum functionalis? (*Spiral? Straight?*)

■ **B9.** *A clinical challenge.* Name the procedures described.
a. Dilation of the cervix of the uterus with scraping of the uterine lining:

_____.

b. Relatively painless procedure in which a small number of cells are removed from
the cervical area and examined microscopically for possible changes indicating

malignancy: _____.

c. Surgical removal of the uterus: _____.

C. Female reproductive system: endocrine relations **(pages 900–905)**

C1. Contrast *menstrual cycle* and *ovarian cycle*.

■ **C2.** Select hormones produced by each of the endocrine glands listed below.

E. Estrogen	LH. Luteinizing hormone
FSH. Follicle-stimulating hormone	P. Progesterone
GnRH. Gonadotropin-releasing hormone	R. Relaxin
I. Inhibin	

Hypothalamus: _____   Ovary: _____

Anterior pituitary: _____

■ **C3.** Now select the hormones listed in C2 that fit descriptions below.

_____ a. Stimulates release of FSH

_____ b. Stimulates release of LH

_____ c. Inhibits release of FSH

_____ d. Inhibits release of LH

_____ e. Stimulates development and maintenance of uterus, breasts, and secondary sex characteristics; enhances protein anabolism

_____ f. Prepares endometrium for implantation by fertilized ovum

_____ g. Dilates cervix for labor and delivery

_____ h. *B*-Estradiol, estrone, and estriol are the primary forms of this hormone

■ **C4.** Refer to Figure LG 28-5, and check your understanding of events in female cycles in this Checkpoint.

a. The average duration of the menstrual cycle is ____ days. The menstrual phase

usually lasts about ____ days. On Figure LG 28-5d, color red the endometrium in the menstrual phase, and note the changes in its thickness. Also write *menses* on the line in that phase on Figure LG 28-5d.

b. Following the menstrual phase is the _____ phase. During

both of these phases, _____ begin to develop in the ovary. At birth of a female infant, about (*20? 400? 5,000? 200,000?*) primary follicles are

present in her ovaries. Each month about _____ of these follicles become

secondary follicles, while only _____ usually matures each month.

c. Follicle development occurs under the influence of the hypothalamic hormone

_____, which regulates the anterior pituitary hormones _____ and later _____.

d. Write the two main functions of follicles: _____ and

_____.

e. How does estrogen affect the endometrium?

For this reason, the preovulatory phase is also known as the

_____ phase. Since follicles reach their peaks during this

phase, it is also known as the _____ phase. Write those labels on Figure LG 28-5d and 28-5b.

f. Toward the end of this phase estrogen level is very high. What effect does a high level of a target hormone exert?

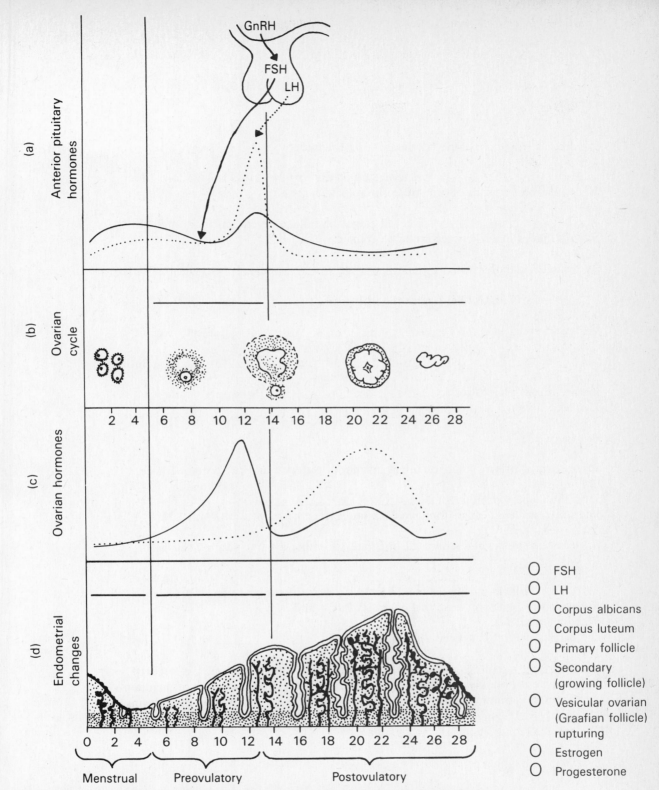

Figure LG 28-5 Correlations of female cycles. (a) Anterior pituitary hormones. (b) Ovarian changes: follicles and corpus luteum. (c) Ovarian hormones. (d) Uterine (endometrial) changes. Color and label as directed in Checkpoint C4.

g. The principle of this negative feedback effect is utilized in oral contraceptives. By starting to take "pills" (consisting of target hormones estrogen and progesterone) early in the cycle (day 5), and continuing until day 25, a woman will maintain a very low level of the tropic hormone _____. Without FSH, follicles and ova will not develop, and so the woman will not _____.

h. The second tropic hormone released during the cycle is

_____. LH level surges just (before? after?) ovulation. LH stimulates release of the ovum from the follicle. This event is known as

_____. Color all stages of follicle development through ovulation on Figure LG 28-5b according to color code ovals.

i. Following ovulation is the _____ phase. It lasts until day

_____ of a 28-day cycle. Under the influence of tropic hormone _____, follicle

cell are changed into the corpus _____. These cells secrete

two hormones: _____ and _____. Both

prepare the endometrium for _____. What preparatory changes occur?

j. What effect do rising levels of target hormones estrogen and progesterone have

upon the tropic hormone LH? _____
k. One function of LH is to form and maintain the corpus luteum. So as LH

____-creases, the corpus luteum disintegrates, forming a white scar on the ovary

known as the corpus _____. Color these two final stages of the original follicle on Figure LG 28-5b. Also write alternative names for the postovulatory phase on lines in Figure LG 28-5b and d.

l. The corpus luteum had been secreting _____ and

_____. With the demise of the corpus luteum, the levels of these hormones rapidly (increase? decrease?). Since these hormones were maintaining the endometrium, endometrial tissue now deteriorates and will be shed during

the next _____ phase.

m. If fertilization should occur, estrogens and progesterone are needed to maintain the

endometrial lining. The _____ around the developing embryo

secretes a hormone named _____. It functions much like LH

in that it maintains the corpus _____, even though LH has
been inhibited (step j above). The corpus luteum continues to secrete estrogen and
progesterone for several months until the placenta itself can secrete sufficient
amounts. Incidentally, hCG is present in the urine of pregnant women only, and so

is routinely used to detect _____.

n. Review hormonal changes by coloring blood levels of hormones in Figure LG 28-5a
and c, according to color code ovals. Then note uterine changes by coloring red the
remainder of the endometrium in Figure LG 28-5d.

■ **C5.** To check your understanding of the events of these cycles, list letters of the major
events in chronological order beginning on day 1 of the cycle. To do this exercise it
may help to list on a separate piece of paper days 1 to 28 and write each event next to
the approximate day on which it occurs.

___ ___ ___ ___ ___ ___ ___ ___ ___ ___ ___

A. Ovulation occurs.
B. Estrogen and progesterone levels drop.
C. FSH begins to stimulate follicles.
D. LH surge is stimulated by high estrogen levels
E. High levels of estrogens and progesterone inhibit LH.
F. Estrogens are secreted for the first time during the cycle.
G. Corpus luteum dies and becomes corpus albicans.
H. Endometrium begins to deteriorate leading to next month's menses.
I. Follicle is converted into corpus luteum.
J. Rising level of estrogen inhibits FSH secretion.
K. Corpus luteum secretes estrogens and progesterone.

■ **C6.** Describe three signs of ovulation in this Checkpoint.

a. Ovulation is likely to occur during the 24-hour period after the basal temperature

____-creases by about 0.4–0.6°F.

b. Under the influence of (*estrogen? progesterone?*) cervical mucus increases in quan-
tity and becomes (*more stretchy? thicker?*) at midcycle. Then just about the time of
ovulation, the mucus decreases and becomes (*more stretchy? thicker?*), under the
influence of (*estrogen? progesterone?*).

c. Some women experience a pain around ovulation known as

_____.

C7. Contrast *menarche, menopause,* and *climacteric.*

D. Female reproductive system: vagina, external genitalia, mammary glands (pages 905–911)

■ **D1.** Refer to Figure LG 28-2. In the normal position, at what angle does the uterus

join the vagina? _____

D2. Describe these vaginal structures.

a. Fornix

b. Hymen

■ **D3.** The pH of vaginal mucosa is (*acid? alkaline?*). What is the clinical significance of this fact?

■ **D4.** Refer to Figure LG 28-6 and do the following exercise.
   a. Label these structures: *anus, clitoris, labia major, labia minor, mons pubis, urethra,* and *vagina.*
   b. Draw the borders of the perineum, using landmarks labeled on the diagram. Draw a line between the right and left ischial tuberosities. Label the triangles that are formed. In which triangle are all of the female external genitalia located?

   _____

   c. Draw a dotted line where an *episiotomy* incision would be performed for childbirth.

■ **D5.** Describe mammary glands in this Checkpoint.

Each breast consists of 15–20 _____ composed of lobules.

Milk-secreting cells, known as _____, empty into a duct system that terminates at the (*areola? nipple?*).

■ **D6.** Match the hormones with their roles in breast development and lactation.

E. Estrogens	P. Progesterone
OT. Oxytocin	PRL. Prolactin

_____ a. Causes duct development in the breast beginning at puberty

_____ b. Stimulates alveolar (gland) development of the breast beginning during adolescence

_____ c. Stimulates secretion of milk in alveoli of lactating mother

_____ d. Stimulates ejection of milk from alveoli to ducts

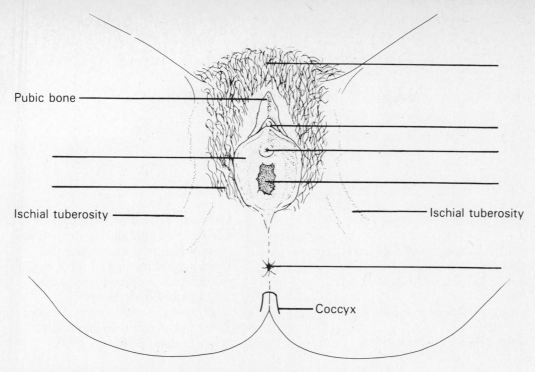

Pubic bone ——————

Ischial tuberosity ——————   —— Ischial tuberosity

—— Coccyx

Figure LG 28-6   Vulva. Complete as directed in Checkpoint D4.

■ **D7.** Do this exercise about breast cancer detection.

a. Identify the most promising method for increasing the survival rate for breast cancer.

b. _____ is a screening technique that shows small calcium deposits by low-dose radiation. Ultrasound procedures utilize (*radiation? sound*

waves?*) to distinguish malignant tumors from _____ tumors or cysts.

E. Sexual intercourse and birth control   (pages 911–914)

E1. Define *sexual intercourse* or *coitus*.

■ **E2.** Answer these questions about the process.

a. (*Sympathetic? Parasympathetic?*) impulses from the (*brain stem? sacral cord?*) cause dilation of arteries of the penis, leading to the (*flaccid? erect?*) state.

b. (*Sympathetic? Parasympathetic?*) impulses cause sperm to be propelled into the urethra. This process is called (*emission? ejaculation?*). Expulsion from the urethra

to the exterior of the body is called _____.

■ **E3.** Most of the lubricating fluid that facilitates sexual intercourse is produced by the (*male? female?*). What structure produces most of this fluid?

■ **E4.** Match the methods of birth control listed in the box with description below.

Ch. Chemical methods	O. Oral contraceptive
Co. Condom	R. Rhythm
D. Diaphragm	T. Tubal ligation
IUD. IUD	V. Vasectomy

_____ a. Spermicidal foams, jellies, and creams

_____ b. "The pill," combination of progesterone and estrogens, causing decrease in FSH and LH so ovulation does not occur

_____ c. Removal of a portion of the ductus (vas) deferens

_____ d. Tying off of the uterine tubes

_____ e. A natural method of birth control involving abstinence during the period when ovulation is most likely

_____ f. A mechanical method of birth control in which a dome-shaped structure is placed over the cervix

_____ g. Small object, an intrauterine device, placed in the uterus by physician

_____ h. A rubber sheath over the penis

E5. List several risks of oral contraceptives.

F. Aging and development of the reproductive systems (pages 914–917)

F1. (*Many? Relatively few?*) age-related changes involve the reproductive system. Describe these changes in:

a. Females

b. Males

F2. Which type of cancer is more common among women over 65 years? (*Cervical? Uterine?*)

■ **F3.** Name the reproductive structures (if any) that develop from the following embryonic ducts.

a. Mesonephric duct in females: _____

b. Mesonephric duct in males: _____

c. Paramesonephric (Müller's) duct in females: _____

d. Paramesonephric (Müller's) duct in males: _____

■ **F4.** Write *E* (endoderm) or *M* (mesoderm) next to the structures below to indicate their embryonic origins.

_____ a. Ovaries and testes

_____ b. Prostate and bulbourethral glands

_____ c. Greater (Bartholin's) and lesser vestibular glands

■ **F5.** Identify male homologues for each of the following.

a. Labia majora: _____

b. Clitoris: _____

c. Lesser vestibular (Skene's) glands: _____

d. Greater vestibular (Bartholin's) glands: _____

G. Disorders, medical terminology (pages 917–923)

G1. Define *sexually transmitted disease* and *veneral disease*.

G2. Describe the main characteristics of these sexually transmitted diseases by completing the table.

Disease	Causative Agent	Main Symptoms
a. Gonorrhea		
b.	*Treponema pallidum*	
c.	Type II herpes simplex virus	
d.	*Trichomonas vaginalis*	

G3. Discuss effects of syphilis on children born to mothers with untreated syphilis.

■ **G4.** Check your understanding of the disorders listed in the box by matching them with the correct description.

A. Amenorrhea	G. Gonorrhea
D. Dysmenorrhea	S. Syphilis
E. Endometriosis	T. Trichomoniasis
F. Fibroadenoma	

_____ a. Painful menstruation, partly due to contractions of uterine muscle

_____ b. Possible cause of blindness in newborns if bacteria transmitted to eyes during birth

_____ c. Spreading of uterine lining into abdominopelvic cavity via uterine tubes

_____ d. Caused by bacterium *Treponema pallidum*; may involve many systems in tertiary stage

_____ e. Absence of menstrual periods

_____ f. Benign tumor with firm, rubbery consistency; easily moved within breast

_____ g. Caused by flagellated protozoan

G5. Fill in the term (*impotence* or *infertility*) that answers the description. Then list several causes of each condition.

a. Inability to fertilize an ovum: _____.

b. Inability to attain and maintain an erection: _____.

G6. List six symptoms of *premenstrual syndrome* (PMS).

_____   _____

_____   _____

_____   _____

G7. Name the microorganism associated with *toxic shock syndrome* (*TSS*).

_____ Relate this condition to use of tampons.

G8. What is PID?

It is most commonly caused by the sexually transmitted disease (STD) known as _____.

■ **G9.** Oophorectomy is removal of a(n) _____, whereas

_____ refers to removal of a uterine (Fallopian) tube.

ANSWERS TO SELECTED CHECKPOINTS

A2. (a) Testes. (b) Seminiferous tubules, epididy- mis, vas deferens, ejaculatory duct, urethra. (c) Prostate, seminal vesicles, bulbourethral glands. (d) Scrotum, penis.

A3. (a) D TV TA. (b) Lower; lower temperature is required for normal sperm production. (c) Posterior of the abdomen; cryptor- chidism; sterility. (d) Cremaster.

A4. (a) Ova, haploid, 23. (b) Zygote, diploid, 2n, homologous. (c) Spermato, seminiferous tubules of testes, months.

A5. (a-b) See Figure LG 28-1A. (c) 2, 3; 2; 3. (d) Toward. (e) 4,5. (f) B. (g) A.

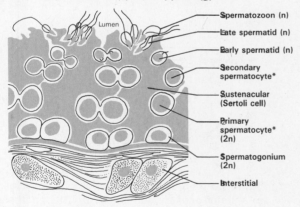

○ Primary spermatocyte
○ Secondary spermatocyte
*Cells in meiosis

Figure LG 28-1A Diagram of a portion of a cross section of a seminiferous tubule showing the stages of spermatogenesis.

A7. (a) 300 million. (b) 48. (c) Head. (d) Mid- piece. (e) Releases enzymes that help sperm to penetrate ovum. (f) Propels sperm.

A8. (a) GnRH. (b) FSH. (c) LH. (d) Testoster- one. (e) Testosterone. (f) Inhibin.

A10. (a) 5. (b) 2. (c) 6. (d) 8. (e) 7. (f) 4. (g) 3. (h) 9. (i) 1.

A11. Paired: 1–5; singular: 6–9.

A12. (a) 45, 18. (b) Partially, inguinal canal and rings, inguinal hernia. (c) Testicular artery, veins, lymphatics, nerves, and cremaster muscle. (d) Scrotum, no, no; hormones exit through testicular veins, which are not cut during vasectomy.

A13. Spongy, or cavernous.

A14. (a) S. (b) S. (c) P. (d) S. (e) B.

A15. Accessory sex glands, A (seminal vesicles), B (prostate), C (bulbourethral glands); ducts, 2–9; gonad, 1.

A16. Since it surrounds the urethra (like a doughnut), an enlarged prostate can cause painful and difficult urination (dysuria) and bladder infection (cystitis) due to incom- plete voiding.

A17. (a) 2.5–5 ml. (b) 50–150 million; 20 million; a large number of sperm is required to release sufficient enzyme to digest the barri- er around the ovum. (c) Alkaline; protects sperm against acidity of male urethra and female vagina. (d) Contains antibiotic that destroys bacteria.

A19. (a) Prepuce, glans, circumcision. (b) Cavernosum, spongiosum, blood, veins, erect. (c) Spongiosum, urine, sperm; sphincter action at base of bladder.

B1. (a) 3. (b) 4. (c) 1. (d) 2. (e) 5.

B2. 1, 2: paired; 3–5, singular.

B3. (a) A. (b) A. (c) Ligaments; mesovarian, ovarian ligament, and suspensory ligament. (d) OF VOF CL.

B4.

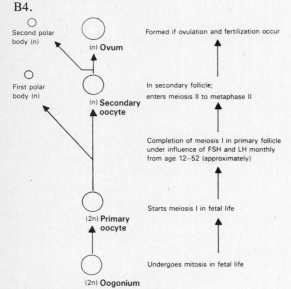

Figure LG 28-4A   Oogenesis.

B5. (a) Fallopian, 10, 4. (b) Infundibulum, fimbriae, ovary. (c) Cilia lining tubes and contractions of smooth muscle in wall of tubes. (d) Passage of ovum to uterus and site of fertilization. (e) Pregnancy following implantation at any location other than in the body of the uterus, for example, tubal pregnancy.

B7. (a) Bladder, rectum. (b) Posterior, vesico-. (c) Anteriorly, retroflexed. (d) Uterosacral, round, broad. (e) Cardinal.

B8. (a) Myo, smooth muscle. (b) Serous. (c) Endo, functionalis, spiral.

B9. (a) D & C (dilation and curettage). (b) Papanicolaou smear. (c) Hysterectomy.

C2. Hypothalamus: GnRH; anterior pituitary: FSH, LH; Ovary: E, I, P, R.

C3. (a) GnRH. (b) GnRH, E. (c) I, E (indirectly since inhibits GnRH). (d) I, P (indirectly since inhibits GnRH). (e) E. (f) P. (g) R. (h) E.

C4. (a) 28, 5; see Figure LG 28-5A. (b) Preovulatory, follicles; 200,000, 20, 1. (c) GnRH; FSH, LH. (d) Secretion of estrogens and development of the ovum. (e) Cause cells of the endometrium to grow;

(proliferate); proliferative, follicular; see Figure LG 28-5A. (f) Negative feedback effect inhibits GnRH and FSH. (g) FSH, ovulate. (h) LH, before, ovulation; see Figure LG 28-5A. (i) Postovulatory, 28, LH; luteum; progesterone, estrogens; implantation; thickening, increased retention of fluid, secretory activity and vascularization. (j) Inhibition of LH. (k) De, albicans; see Figure LG 28-5A. (l) Progesterone, estrogens; decrease, menstrual. (m) Placenta, human chorionic gonadotropin (hCG), luteum, pregnancy. (n) See Figure LG 28A.

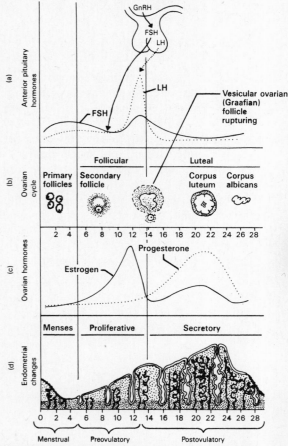

Figure LG 28-5A   Correlations of female cycles. (a) Anterior pituitary hormones. (b) Ovarian changes: follicles and corpus luteum. (c) Ovarian hormones. (d) Uterine (endometrial) changes.

C5. C F J D A I K E G B H.

C6. (a) In. (b) Estrogen, more stretchy, thicker, progesterone. (c) Mittelschmerz.

D1. Close to 90°.

D3. Acid, decreases risk of infection but may injure sperm.

D4. (a–c) See Figure LG 28-6A. (b) Urogenital triangle.

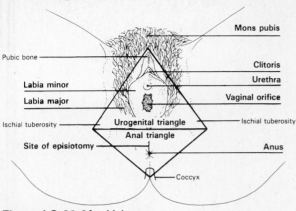

Figure LG 28-6A Vulva.

D5. Lobes, alveoli, nipple.

D6. (a) E. (b) P. (c) PRL. (d) OT.

D7. (a) Early detection, as by self-breast examination (SBE), just after the menstrual phase. (b) Mammography, sound waves, benign.

E2. (a) Parasympathetic, sacral cord, erect. (b) Sympathetic, emission, ejaculation.

E3. Female, vaginal lining.

E4. (a) Ch. (b) O. (c) V. (d) T. (e) R. (f) D. (g) IUD. (h) Co.

F3. (a) None. (b) Ducts of testes and epididymides, ejaculatory ducts and seminal vesicles. (c) Uterus and vagina. (d) None.

F4. (a) M. (b) E. (c) E.

F5. (a) Scrotum. (b) Penis. (c) Prostate. (d) Bulbourethral glands.

G4. (a) D. (b) G. (c) E. (d) S. (e) A. (f) F. (g) T.

G9. Ovary, salpingectomy.

MASTERY TEST: Chapter 28

Questions 1–10: Arrange the answers in correct sequence.

_____ _____ _____ 1. From most external to deepest:

   A. Endometrium
   B. Serosa
   C. Myometrium

_____ _____ _____ 2. Portions of the urethra, from proximal (closest to bladder) to distal (closest to outside of the body):

   A. Prostatic
   B. Membranous
   C. Spongy

_____ _____ _____ 3. Thickness of the endometrium, from thinnest to thickest:

   A. On day 5 of cycle
   B. On day 13 of cycle
   C. On day 23 of cycle

_____ _____ _____ 4. From anterior to posterior:

   A. Uterus and vagina
   B. Bladder and urethra
   C. Rectum and anus

_____ _____ _____ _____ 5. Order in which hormones begin to increase level, from day 1 of menstrual cycle:

   A. FSH
   B. LH
   C. Progesterone
   D. Estrogen

_____ _____ _____ _____ 6. Order of events in the female monthly cycle, beginning with day 1:

   A. Ovulation
   B. Formation of follicle
   C. Menstruation
   D. Formation of corpus luteum

_____ _____ _____ _____ 7. From anterior to posterior:

   A. Anus
   B. Vaginal orifice
   C. Clitoris
   D. Urethral orifice

_____ _____ _____ _____ _____ 8. Pathway of sperm:

   A. Ejaculatory duct
   B. Testis
   C. Uretha
   D. Ductus (vas) deferens
   E. Epididymis

_____ _____ _____ _____ _____ 9. Pathway of milk in breasts:

   A. Ampullae
   B. Alveoli
   C. Secondary tubules
   D. Mammary ducts
   E. Lactiferous ducts

_____ _____ _____ _____ _____ _____ 10. Pathway of sperm entering female reproductive system:

   A. Uterine cavity
   B. Cervical canal
   C. External os
   D. Internal os
   E. Uterine tube
   F. Vagina

Questions 11–12: Circle the letter preceding the one best answer to each question.

11. Eighty to 90 percent of seminal fluid (semen) is secreted by the combined secretions of:

   A. Prostate and epididymis
   B. Seminal vesicles and prostate
   C. Seminal vesicles and seminiferous tubules
   D. Seminiferous tubules and epididymis
   E. Bulbourethral glands and prostate

12. In the normal male, there are two of each of the following structures EXCEPT:

   A. Testis   D. Ductus deferens
   B. Seminal vesicle   E. Epididymis
   C. Prostate   F. Ejaculatory duct

Questions 13–15: Circle letters preceding ALL correct answers to each question.

13. Which of the following are produced by the testes?

   A. Spermatozoa   D. GnRH
   B. Testosterone   E. FSH
   C. Inhibin   F. LH

14. Which of the following are produced by the ovaries and then *leave* the ovaries?

   A. Follicle   D. Corpus albicans
   B. Ovum   E. Estrogen
   C. Corpus luteum F. Progesterone

15. Which of the following are functions of LH?

   A. Begin the development of the follicle.
   B. Stimulate change of follicle cells into corpus luteum cells.
   C. Stimulate release of ovum (ovulation).
   D. Stimulate corpus luteum cells to secrete estrogens and progesterone.
   E. Stimulate release of GnRH.

T F 16. Spermatogonia and oogonia both continue to divide throughout a person's lifetime to produce new primary spermatocytes or primary oocytes.

T F 17. The menstrual cycle refers to a series of changes in the uterus, whereas the ovarian cycle is a series of changes in the ovary.

T F 18. Meiosis is a part of both oogenesis and spermatogenesis.

T F 19. Each oogonium beginning oogenesis produces four mature ova.

T F 20. A high level of a target hormone (such as estrogen) will inhibit release of its related releasing factor and target hormone (GnRH and FSH).

Questions 21–25: fill-ins. Complete each sentence with the word or phrase that best fits.

_____ 21. ____ is a term that means undescended testes.

_____ 22. Most oral contraceptives consist of the hormones ____ and ____ .

_____ 23. ____ cells in testes phagocytose degenerating spermatogenic cells and secrete androgen-binding proteins that concentrate testosterone in seminiferous tubules.

_____ 24. Menstrual flow occurs as a result of a sudden ____-crease in the hormones ____ and ____ .

_____ 25. The three main functions of estrogens are promotion of growth and maintenance of ____, ____ balance, and stimulation of protein ____-bolism.

ANSWERS TO MASTERY TEST: ■ Chapter 28

Arrange

1. B C A
2. A B C
3. A B C
4. B A C
5. A D B C
6. C B A D
7. C D B A
8. B E D A C
9. B C D A E
10. F C B D A E

Multiple Choice

11. B
12. C

Multiple Answers

13. A, B, C
14. B, E, F
15. B, C, D

True-False

16. F. Spermatogonia; primary spermatocytes (Oogonia do not divide after birth of female baby.)
17. T
18. T
19. F. One
20. T

Fill-ins

21. Cryptorchidism
22. Progesterone and estrogens
23. Sustantacular
24. De, progesterone, and estrogens
25. Reproductive organs, fluid and electrolytes, ana

Development & Inheritance

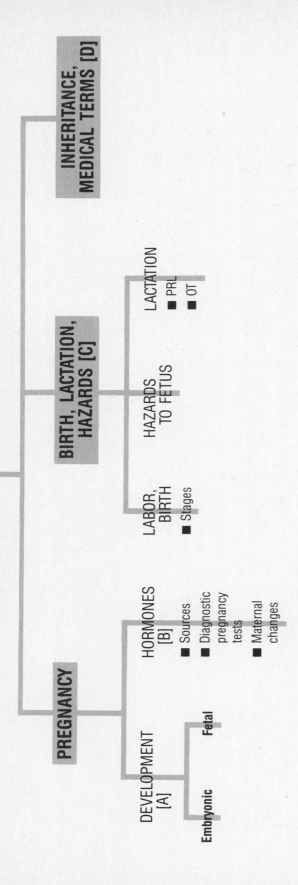

PREGNANCY

DEVELOPMENT
[A]

Embryonic

Fetal

HORMONES
[B]

■ Sources
■ Diagnostic
 pregnancy
 tests
■ Maternal
 changes

BIRTH, LACTATION,
HAZARDS [C]

LABOR,
BIRTH

■ Stages

HAZARDS
TO FETUS

LACTATION

■ PRL
■ OT

INHERITANCE,
MEDICAL TERMS [D]

Development and Inheritance

Twenty-eight chapters ago you began a tour of the human body, starting with the structural framework and varieties of buildings (chemicals and cells) and moving on to the global communication systems (nervous and endocrine), transportation routes (cardiovascular), and refuse removal systems (respiratory, digestive, and urinary). This last chapter synthesizes all the parts of the tour: A new individual, with all those body systems, is conceived and develops during a nine-month pregnancy. Adjustments are made at birth for survival in an independent (outside of mother) world. But parental ties are still evident in inherited features, signs of intrauterine care (mother's avoidance of alcohol or cigarettes), and provision of nourishment, as in lactation.

Look at Framework 29 first. And congratulations on having completed your first tour through the wonders of the human body.

TOPIC OUTLINE AND OBJECTIVES

A. Development during pregnancy

1. Explain the activities associated with fertilization, morula formation, blastocyst development, and implantation.
2. Describe how in vitro fertilization (IVF) is performed.
3. Discuss the principal body changes associated with embryonic and fetal growth.

B. Hormones of pregnancy, gestation, prenatal diagnostic techniques

4. Compare the sources and functions of the hormones secreted during pregnancy.
5. Describe how pregnancy is diagnosed and describe some of the anatomical and physiological changes associated with gestation.

C. Parturition and labor, adjustments at birth, potential hazards during development, lactation

6. Explain the respiratory and cardiovascular adjustments that occur in an infant at birth.

7. Discuss potential hazards to an embryo and fetus associated with chemicals and drugs, irradiation, alcohol, and cigarette smoking.
8. Discuss the physiology and control of lactation.

D. Inheritance and medical technology

9. Define inheritance and describe the inheritance of several traits.
10. Define medical terminology associated with development and inheritance.

Now study the following parts of words that may help you better understand terminology in this chapter.

Wordbyte	Meaning	Example		Wordbyte	Meaning	Example
morula	mulberry	*mor*ula		natus	birth	*post*natal

CHECKPOINTS

A. Development during pregnancy (pages 928–939)

■ **A1.** Answer these questions about the process of fertilization.

a. How many sperm are ordinarily introduced into the vagina during sexual inter-

course? _____ About how many reach the area where the

ovum is located? _____

b. List four mechanisms that may be involved in sperm transport within the female reproductive organs?

_____   _____

_____   _____

c. Name the usual site of fertilization. _____

d. How many sperm fertilize an ovum? What prevents further sperm from entering the ovum?

■ **A2.** Describe roles of the following structures in fertilization.

a. A male pronucleus present in a (*secondary spermatocyte? spermatozoon?*) combines with a female pronucleus from a (*secondary oocyte? ovum?*). Each pronucleus contains the (*n? 2n?*) number of chromosomes.

b. The product of fusion of the two pronuclei is known as a

_____ nucleus, containing _____ chromosomes. Completion of meiosis II by the secondary oocyte results in the final product of fertilization, consisting of segmentation nucleus, cytoplasm, and zona pellucida. This structure is

known as the _____

■ **A3.** Contrast types of multiple births by choosing the answer that best fits each description.

DTwin. Dizygotic twins   MTwin. Monozygotic twins

_____ a. Also known as fraternal twins   _____ b. Two infants derived from a single fertilized ovum that divides once

A4. Summarize the main events during the week following fertilization. Describe each of the following stages. (For help, refer to Figures 29-3 and 29-4, page 931, in the text.)

a. Cleavage

c. Morula

b. Blastomere

d. Blastocyst

A5. Draw a diagram of a blastocyst. Label: *blastocoele, inner cell mass,* and *trophoblast.* Color red the cells that will become the embryo. Color blue the cells that will become part of the placenta. (Refer to Figure 29-4c, page 931, for help.)

■ **A6.** Attachment of the blastocyst to the endometrium is a process known as

_____; this occurs about 7–8 days after ovulation, or about day

_____ of a 28-day cycle. Note that the developing baby is implanted in the uterus even before mother "misses" the first day of her next menstrual cycle. Cells of the (*inner cell mass? trophoblast?*) secrete enzymes that enable the blastocyst to burrow into the uterine lining. Following implantation, the blastocyst can absorb nutrients

from _____.

■ **A7.** *Hyperemesis gravidarum* is a term meaning _____. List a possible cause.

A8. Describe the following aspects of in vitro fertilization (IVF).

a. In what year was the first live birth as a result of this procedure? (*1962? 1970? 1978? 1986?*)

b. Briefly describe this procedure.

c. Contrast these two forms of IVF: *embryo transfer* and *GIFT.*

■ **A9.** A human embryo refers to the developing individual during the (*first two? last seven?*) months of development in utero. Define *fetus.*

■ **A10.** Complete this exercise about the embryonic period. Refer to Figure 29-6 (pages 933–934) in your text.

a. Identify what structures will ultimately form from these two parts of the implanted

blastocyst. Trophoblast: _____ Inner cell mass:

_____ Between these two portions of the blastocyst is the

_____ cavity.

b. The inner cell mass first forms the embryonic _____ consist-
ing of two layers. The upper cells (closer to the implantation site toward the right

on Figure 29-6b, page 934, of the text) are called _____-derm
cells. The lower cells (towards the left in the figure) are called

_____-derm cells. Eventually, a layer of

_____-derm will form between these. Collectively the three

layers are called _____ layers.

c. The process of formation of the primary germ layers is known as

_____ and it occurs by the end of the

_____ week following fertilization.

d. _____-derm cells line the amniotic cavity. The delicate fetal

membrane formed from these cells is called the _____. It is

sometimes called the "_____" surrounding the fetus and it is
broken at birth. (Ectoderm cells also form fetal structures: see Checkpoint A12.)

e. The lower cells of the embryonic disc, called _____ cells, line

the primitive _____ and for a while extend out to form the

_____ sac. This sac (*is? is not?*) as significant in humans as it
is in birds.

f. Some mesoderm cells move around to line the original trophectoderm cells
(Figure 29-4, page 931, of the text). Together these cells form the

_____ layer of the placenta. Which membrane is more
superficial? (*Amnion? chorion?*) The space between these layers is called the

_____.

A11. Each of the primary germ layers is involved with formation of both fetal organs
and fetal membranes. Name the major structures derived from each primary germ layer: endoderm, mesoderm, and ectoderm. (Consult Exhibit 29-1 (page 935) in your
text.)

Endoderm	Mesoderm	Ectoderm

■ **A12.** *For extra review.* Do this matching exercise on primary germ layer derivatives.

> Ecto. Ectoderm   Endo. Endoderm   Meso. Mesoderm

_____ a. Epithelial lining of all of digestive, respiratory, and genitourinary tracts except near openings to the exterior of the body

_____ b. Epidermis of skin, epithelial lining of entrances to the body (such as mouth, nose, and anus), hair, nails

_____ c. All of the skeletal system (bone, cartilage, joint cavities)

_____ d. All muscle (skeletal, smooth, and cardiac)

_____ e. Blood and all blood and lymphatic vessels

_____ f. Entire nervous system, including posterior pituitary

_____ g. Thyroid, parathyroid, thymus, and pancreas

■ **A13.** Write the name of each fetal membrane next to its description.

a. Originally formed from ectoderm, this membrane encloses fluid that acts as a

shock absorber for the developing baby: _____.

b. Derived from mesoderm and trophectoderm; it becomes the principal part of

placenta: _____.

c. Endoderm-lined membrane serving as exclusive nutrient supply for embryos of

some species: _____.

d. A small membrane that forms umbilical blood vessels: _____.

■ **A14.** Further describe the placenta in this exercise.

a. The placenta has the shape of a _____ embedded in the wall

of the uterus. Its fetal portion is the _____ fetal membrane.

Fingerlike projections, known as _____, grow into the uterine lining. Consequently maternal and fetal blood are (*allowed to mix? brought into close proximity but do not mix?*). Fetal blood in umbilical vessels is bathed by

maternal blood in _____ spaces.

b. The maternal aspect of the placenta is called the (*chorion? decidua basalis?*). This

is a portion of the _____-metrium of the uterus.

c. What is the ultimate fate of the placenta?

■ **A15.** *A clinical challenge.* Following a birth, vessels of the umbilical cord are carefully

examined. A total of _____ vessels should be found, surrounded by a mucous jelly-like connective tissue. The vessels should include (*1? 2?*) umbilical artery(-ies) and (*1? 2?*) umbilical vein(s). Lack of a vessel may indicate a congenital malformation.

A16. List three uses of human placentas following birth.

■ **A17.** Study Exhibit 29-2 (page 939) in your textbook. Write the number of the month when each of the following events occurs.

_____ a. Heart starts to beat.

_____ b. Heartbeat can be detected.

_____ c. Backbone and vertebral canal form.

_____ d. Eyes are almost fully developed; eyelids are still fused.

_____ e. Eyelids separate; eyelashes form.

_____ f. Limb buds develop.

_____ g. Ossification begins.

_____ h. Limb buds become distinct as arms and legs; digits are well formed.

_____ i. Nails develop.

_____ j. Fine (lanugo) hair covers body.

_____ k. Lanugo hair is shed.

_____ l. Fetus is capable of survival.

_____ m. Subcutaneous fat is deposited.

_____ n. Testes descend into scrotum.

B. Hormones of pregnancy, gestation, prenatal diagnostic techniques
(pages 939–945)

■ **B1.** Complete this exercise about the hormones of pregnancy.

a. During pregnancy, the level of progesterone and estrogens must remain (*high? low?*) in order to support the endometrial lining. During the first two or three months or

so, these hormones are produced principally by the _____ lo-

cated in the _____, under the influence of the tropic hormone

_____. (It might be helpful to review the section on endocrine relations in Chapter 28.)

b. hCG reaches its peak between the _____ and

_____ month. Since it is present in blood, it will be filtered

into _____ where it can readily be detected as an indication of pregnancy.

c. In addition, the chorion produces the hormone hCS; these initials stand for

_____ chorionic _____. List two functions of this hormone.

d. _____ is a hormone that relaxes pelvic joints and helps dilate the cervix near the time of birth. This hormone is produced by the

_____ and _____.

■ **B2.** Contrast *presumptive evidence* with *probable evidence* of pregnancy.

■ **B3.** The normal human gestation period is _____ days. This period is defined as the time from (*the beginning of the last menstrual period? conception?*) until birth.

■ **B4.** Complete this exercise about maternal adaptations during pregnancy. Describe normal changes that occur.

a. Pulse ____-creases.

b. Blood volume ____-creases by about ____ percent.

c. Tidal volume ____-creases.

d. Gastrointestinal (GI) motility ____-creases, which may cause (*diarrhea? constipation?*).

e. Pressure upon the inferior vena cava ____-creases which may lead to

_____ .

B5. Discuss effects of exercise on pregnancy and lactation.

■ **B6.** Complete this exercise about amniocentesis and CVS.

a. Amniocentesis involves withdrawal of _____ fluid, usually at about (*2–4? 8–10? 14–16?*) weeks after conception. Amniotic fluid is constantly recycled (by drinking and excretion) through the fetus; it (*does? does not?*) contain fetal cells and products of fetal metabolism. For what purposes is amniotic fluid collected and examined?

b. CVS (meaning _____) is a procedure most often performed at about (*2–4? 8–10? 14–16*) weeks. It (*does? does not?*) involve penetration of the

uterine cavity. Cells of the _____ layer of the placenta embedded in the uterus are studied. These (*are? are not?*) derived from the same sperm and egg that is forming the fetus. Therefore this procedure, like amniocentesis, can

be used to diagnose _____ .

B7. Describe uses of the following techniques in pregnancy. State one reason why each may be utilized.

a. Ultrasonography

b. Alphafetoprotein

C. Parturition and labor, adjustments at birth, potential hazards during development, lactation (pages 945–949)

C1. Contrast these pairs of terms.

a. Labor/parturition

b. False labor/true labor

■ **C2.** Do the following hormones stimulate (S) or inhibit (I) uterine contractions during the birth process?

_____ a. Progesterone   _____ c. Oxytocin

_____ b. Estrogens

C3. What is the "show" produced at the time of birth?

■ **C4.** Identify descriptions of the three phases of labor (first, second, or third).

_____ a. Stage of expulsion: from complete cervical dilation through delivery of the baby

_____ b. Time after the delivery of the baby until the placenta ("after-birth") is expelled; the placental stage

_____ c. Time from onset of labor to complete dilation of the cervix; the stage of dilation

C5. Define these terms:

a. Dystocia

b. Electronic fetal monitoring (EFM)

■ **C6.** A premature infant ("preemie") is considered to be one who weighs less than

_____ gm (_____ lb) at birth. These infants are at high risk for the RDS

(or _____). *For extra review* of RDS, refer to Chapter 23, Checkpoint H3, page LG 431.

C7. What adjustments must the newborn make at birth as it attempts to cope with its new environment? Describe adjustments of the following:

a. Respiratory system (What serves as a stimulus for the respiratory center of the medulla?)

The respiratory rate of a newborn is usually about (*one-third? three times?*) that of an adult.

b. Heart and blood vessels. (*For extra review* go over changes in fetal circulation, Chapter 21, pages LG 388–389.)

c. Blood cell production

C8. Explain why perhaps 50 percent of newborns may experience a temporary state of jaundice soon after birth.

■ **C9.** Complete the exercise about potential hazards to the embryo and fetus. Match the terms at right with the related descriptions.

C. Cigarette smoking	FAS. Fetal alcohol syndrome	T. Teratogen

_____ a. An agent or influence that causes defects in the developing embryo

_____ b. May cause defects such as small head, facial irregularities, defective heart, retardation

_____ c. Linked to a variety of abnormalities such as lower infant birth weight, higher mortality rate, GI disturbances, and SIDS

■ **C10.** Complete this exercise about lactation.

a. The major hormone promoting lactation is _____ which is

secreted by the _____. During pregnancy the level of this hormone increases somewhat, but is inhibited during pregnancy while levels of estrogens and progesterone are (*high? low?*).

b. Following birth and the loss of the placenta, a major source of estrogens and progesterone is gone, so prolactin (*increases? decreases?*) dramatically.

c. The sucking action of the newborn faciitates lactation in two ways. What are they?

d. What hormone stimulates milk letdown? _____

C11. What is colostrum?

C12. Explain how lactation may inhibit ovulation and therefore decrease likelihood of pregnancy in a breast-feeding mother.

C13. List five advantages offered by breast-feeding.

D. Inheritance and medical terminology (pages 950–954)

D1. Define these terms.

a. Inheritance

b. Genetics

c. Homologous chromosomes

D2. Contrast:

a. Genotype/phenotype

b. Homozygous/heterozygous

■ **D3.** Answer these questions about genes controlling PKU. (See Figure 29-16, page 950 in the text.)

a. The dominant gene for PKU is represented by (*P? p?*). The dominant gene is for the (*normal? PKU?*) condition.

b. The letters PP are an example of a genetic makeup or (*genotype? phenotype?*). A person with such a genotype is said to be (*homozygous dominant? homozygous recessive? heterozygous?*). The phenotype of that individual would be (*normal? normal, but serve as a "carrier" of the PKU gene? PKU?*).

c. The genotype for a heterozygous individual is _____. The phenotype is

_____.

d. Use a Punnett square to determine possible genotypes of offspring when both parents are Pp.

■ **D4.** In most traits the normal gene is (*dominant? recessive?*). Name several exceptions to this rule.

D5. Contrast *autosome* and *sex chromosome*.

■ **D6.** Answer these questions about sex inheritance.

a. All (*sperm? ova?*) contain the X chromosome. About half of the sperm produced

by a male contain the _____ chromosome, and half contain the _____ .

b. Every cell (except those forming ova) in a normal female contains (*XX? XY?*) genes on sex chromosomes. All male cells (except those forming sperm) are said to be (*XX? XY?*).

c. Who determines the sex of the child? (*Mother? Father?*) Show this by using a Punnett square.

■ **D7.** Complete this exercise about X-linked inheritance.

a. Sex chromosomes contain other genes besides those determining sex of an individu-

al. Such traits are called _____ traits. Y chromosomes are shorter than X chromosomes and lack some genes. One of these is the gene con-

trolling ability to _____ colors. Thus this ability is controlled

entirely by the _____ gene.

b. Write the genotype for females of each type: color blind, _____ ; carrier,

_____ ; normal, _____ .

c. Now write possible genotypes for males: color blind, _____ ; normal, _____ .

d. Determine the results of a cross between a color-blind male and a normal female.

e. Now determine the results of a cross between a normal male and a carrier female.

f. Color blindness and other X-linked, recessive traits are much more common in (*males? females?*).

D8. Name another X-linked trait. _____

ANSWERS TO SELECTED CHECKPOINTS

A1. (a) Several hundreds of millions; probably several hundred to thousands. (b) Peristaltic contractions and cilia of uterine (Fallopian) tubes; flagella; acrosin that stimulates sperm motility; prostaglandins in semen. (c) Uterine tube. (d) One; electrical charges and enzymes produced by the fertilized ovum.

A2. (a) Spermatozoon, secondary oocyte, n. (b) Segmentation, 46, zygote.

A3. (a) DTwin. (b) MTwin.

A6. Implantation, 21–23, trophoblast, mother's uterine blood and glands.

A7. Morning sickness in early months of pregnancy; high levels of hCG.

A9. First two; developing individual during the last seven months of human gestation.

A10. (a) Trophoblast: part of chorion of placenta; inner cell mass: embryo and parts of placenta; amniotic. (b) Disc, ecto, endo, meso, primary germ. (c) Gastrulation, third. (d) Ecto, amnion, bag of waters. (e) Endoderm, gut, yolk, is not. (f) Chorion, chorion, extraembryonic coelom.

A12. (a) Endo. (b) Ecto. (c) Meso. (d) Meso. (e) Meso. (f) Ecto. (g) Endo.

A13. (a) Amnion. (b) Chorion. (c) Yolk sac. (d) Allantois.

A14. (a) Flattened cake (or thick pancake), chorion, chorionic villi, brought into close proximity but do not mix, intervillous. (b) Decidua basalis, endo. (c) It is discarded from mother at birth (as the "afterbirth"); separation of the decidua from the remainder of endometrium accounts for bleeding during and following birth.

A15. 3, 2, 1.

A17. (a) 1. (b) 3. (c) 1. (d) 3. (d) 3. (e) 6. (f) 1. (g) 2. (h) 2. (i) 3. (j) 5. (k) 9. (l) 7. (m) 8. (n) 8.

B1. (a) High, corpus luteum, ovary, hCG. (b) First, third, urine. (c) Human (chorionic) somatomammotropin, stimulates development of breast tissue and alters metabolism during pregnancy. (d) Relaxin, placenta, ovaries.

B3. 280, the beginning of the last menstrual period.

B4. (a) In. (b) In, 30–50. (c) In. (d) De, constipation. (e) In, varicose veins or hemorrhoids.

B6. (a) Amniotic, 14–16, does; to determine fetal maturity and diagnose genetic or chromosomal defects, such as Down's syndrome. (b) Chorionic villi sampling, 8–10, does not, chorion, are, genetic and chromosomal defects.

C2. (a) ↓. (b) ↑. (c) ↑.

C4. (a) Second. (b) Third. (c) First.

C6. 2500 (5.5), respiratory distress syndrome.

C9. (a) T. (b) T, FAS. (c) T, C.

C10. (a) Prolactin (PRL), anterior pituitary, high. (b) Increases. (c) Maintains prolactin level by inhibiting PIF and stimulating release of PRL; stimulates release of oxytocin. (d) Oxytocin.

D3. (a) P, normal. (b) Genotype, homozygous dominant, normal. (c) Pp, normal but a carrier of PKU gene. (d) 25 percent PP, 50 percent Pp, and 25 percent pp.

D4. Dominant, see recessive traits in Exhibit 29-3 (page 951) of the text.

D6. (a) Ova, X, Y. (b) XX, XY. (c) Father, as shown in the Punnett square (Figure 29-18, page 951 in your text) since only the male can contribute the Y chromosome.

D7. (a) X-linked, differentiate, X. (b) X^cX^c; X^CX^c; X^CX^C. (c) X^cY; X^CY. (d) All females are carriers and all males are normal. (e) Of females, 50 percent are normal and 50 percent are carriers; of males, 50 percent are normal and 50 percent are color blind. (f) Males.

MASTERY TEST: Chapter 29

Questions 1–2: Arrange the answers in correct sequence.

_____ _____ _____ _____ 1. From most superficial to deepest (closest to embryo):

    A. Amnion
    B. Amniotic cavity
    C. Chorion
    D. Decidua

_____ _____ _____ _____ _____ 2. Stages in development:

    A. Morula
    B. Blastocyst
    C. Zygote
    D. Fetus
    E. Embryo

Questions 3–10: Circle the letter preceding the one best answer to each question.

3. All of the following are hormones made by the placenta EXCEPT:

    A. hCS   D. Relaxin
    B. hCG   E. Estrogen
    C. GnRH   F. Progesterone

4. All of the following statements are true of males EXCEPT:

    A. Their genotype is XY.
    B. They determine sex of their offspring.
    C. They are more likely than females to be color blind.
    D. They usually produce more lubricating fluid during sexual intercourse than females do.

5. All of the following structures are developed from mesoderm EXCEPT:

    A. Aorta   D. Humerus
    B. Biceps muscle E. Sciatic nerve
    C. Heart   F. Neutrophil

6. Which part of the structures surrounding the fetus is developed from maternal cells (not from fetal)?

    A. Allantois   D. Decidua
    B. Yolk sac   E. Amnion
    C. Chorion

7. All of the following events occur during the first month of embryonic development EXCEPT:

    A. Endoderm, mesoderm, and ectoderm are formed.
    B. Amnion and chorion are formed.
    C. Limb buds develop.
    D. Ossification begins.
    E. The heart begins to beat.

8. Implantation of a developing individual (blastocyst stage) usually occurs about ____ after fertilization.

    A. Three weeks   D. Seven hours
    B. One week   E. Seven minutes
    C. One day

9. The high level of estrogens present during pregnancy is responsible for all of the following EXCEPT:

   A. Stimulates release of prolactin
   B. Prevents ovulation
   C. Maintains endometrium
   D. Prevents menstruation
   E. Stimulates uterine contractions

10. Which of the following is a term that refers to discharge from the birth canal during the days following birth?

   A. Autosome   D. Lochia
   B. Cautery   E. PKU
   C. Culdoscopy

Questions 11–20: Circle T (true) or F (false). If the statement is false, change the underlined word or phrase so that the statement is corrrect.

T F 11. Oxytocin and progesterone both stimulate uterine contractions.

T F 12. Of the primary germ layers, only the ectoderm forms part of a fetal membrane.

T F 13. Maternal blood mixes with fetal blood in the placenta.

T F 14. Amniocentesis is a procedure in which amniotic fluid is withdrawn for examination at about 8–10 weeks during the gestation period.

T F 15. The normal gestation period is about 266 days following the beginning of the last menstrual period (LMP).

T F 16. Stage 2 of labor refers to the period during which the "after-birth" is expelled.

T F 17. Maternal pulse rate, blood volume, and tidal volume all normally increase during pregnancy.

T F 18. The decidua is part of the endoderm of the fetus.

T F 19. Inner cell mass cells of the blastocyst form the chorion of the placenta.

T F 20. A phenotype is a chemical or other agent that causes physical defects in a developing embryo.

Questions 21–25: fill-ins. Complete each sentence with the word or phrase that best fits.

_____ 21. The newborn infant generally has a respiratory rate of ____, pulse of ____, and a WBC count that may reach ____, all values that are ____ than those of an adult.

_____ 22. As a prerequisite for fertilization, sperm must remain in the female reproductive tract for at least 4–6 hours so that ____ can occur.

_____ 23. hCG is a hormone made by the ____; its level peaks at about the end of the ____ month of pregnancy, when it can be used to detect ____.

_____ 24. Determine the probable genotypes of children of a couple in which the man has hemophilia and the woman is normal. ____

_____ 25. This is the ____ mastery test question in this book.

558

ANSWERS TO MASTERY TEST: ■ Chapter 29

Arrange

1. D C A B
2. C A B E D

Multiple Choice

3. C   7. D
4. D   8. B
5. E   9. A
6. D   10. D

True-False

11. F. Ocytocin but not progesterone stimulates
12. F. All three layers form parts of fetal membranes: ectoderm (trophectoderm) forms part of chorion; mesoderm forms parts of the chorion and the allantois; endoderm lines the yolk sac.
13. F. Does not mix but comes close to fetal blood
14. F. Amniocentesis, 14–16
15. F. 280
16. F. Fetus
17. T
18. F. Endometrium of the uterus
20. F. Trophectoderm
21. F. Teratogen

True-False

21. 45, 120–160, up to 45,000, higher
22. Capacitation (dissolving of the covering over the ovum by secretion of acrosomal enzymes of sperm)
23. Chorion of the placenta, second, pregnancy
24. All male children normal; all female children carriers
25. Last or final. Congratulations! Hope you learned a great deal.